D1479926

NATIONAL
IDENTITY

The Role of Science and Technology

EDITED BY

Carol E. Harrison
and Ann Johnson

OSIRIS | **24**

A Research Journal Devoted to the
History of Science and Its Cultural Influences

Osiris

Series editor, 2002–2012

KATHRYN OLESKO, *Georgetown University*

Volumes 17 to 27 in this series are designed to dissolve boundaries between history and the history of science. They cast science in the framework of larger issues prominent in the historical discipline but infrequently treated in the history of science, such as the development of civil society, urbanization, and the evolution of international affairs. They aim to open up new categories of analysis, to stimulate fresh areas of investigation, and to explore novel ways of synthesizing major historical problems that demand consideration of the role science has played in them. They are written not only for historians of science, but also for historians and other scholars who wish to integrate issues concerning science into courses on broader themes, as well as for readers interested in viewing science from a general historical perspective. Special attention is paid to the international dimensions of each volume's topic.

Cover Illustration:

Mahatma Ghandi at the microscope observing leprosy germs at Sevagram Ashram (1942). Copyright Dinodia. Reproduced with the permission of Dinodia.

OSIRIS   2009   SECOND SERIES   VOLUME 24

NATIONAL IDENTITY: THE ROLE OF SCIENCE AND TECHNOLOGY

Introduction:
Science and National Identity

By Carol E. Harrison and Ann Johnson[*]

In 1947, before India's formal independence, Jawaharlal Nehru formed his first cabinet and, in addition to assuming the prime ministership, reserved for himself the ministry of science and technology. This was a suitable distribution of power, Nehru believed, because what he called the "scientific approach" was "fundamental" to the new nation. "You cannot change man by law," he proclaimed upon laying the foundation stone of the National Institute of Sciences in 1948, but "[y]ou can create an atmosphere where his actions are governed by a scientific approach."[1] As both prime minister and minister for science and technology, Nehru would make law suited to the "scientific temper" of the citizens of a new and independent India.[2] As he inaugurated construction of the institute, Nehru laid claim to an Indian share of the universal value of science.

Nehru's gesture drew on a long tradition of founding national identity on a particular claim to an allegedly universal ideal of scientific competence. Scientific ventures featured among the foundational projects of the sovereign French nation after the Revolution of 1789, for instance. Revolutionaries sought to efface the traces of the Old Regime from the map by replotting the newly national territory, erasing traditional and irrational boundaries and marking off new and uniform departments. They rejected the particularism of France's hodgepodge of weights and measures and replaced it with rational and universal units, most notably the meter, a unit of length drawn from the size of the earth itself that would be "for all people, for all time."[3] First and foremost, however, the meter was for the French and by the French, and the National Assembly rejected plans that called for international cooperation in establishing a meter based on the beat of a pendulum in favor of a survey of a meridian

[*] Department of History, University of South Carolina, Gambrell Hall, Columbia, SC 29208; ceharris @mailbox.sc.edu and ajohnson@mailbox.sc.edu.

We would like to thank Kathryn Olesko and two anonymous reviewers for their comments on earlier drafts. Special thanks to Thomas Brown and Holly Groover for careful and constructive readings of this chapter. Participants at the 2007 workshop on Science, Technology, and National Identity in Columbia, South Carolina, especially Michael Adas and John Krige, offered thoughtful and spirited discussion. The University of South Carolina, through the College of Arts and Sciences, the Walker Institute for International and Area Studies, and the Departments of History and Philosophy, provided critical financial and logistical support for that workshop.

[1] Jawaharlal Nehru, *Selected Works of Jawaharlal Nehru*, ed. S. Gopal, 2nd ser. (New Delhi, 1987), 6:347.

[2] Jawaharlal Nehru, *The Discovery of India* (New York, 1946), 532.

[3] Jean Antoine Nicolas Caritat, marquis de Condorcet, quoted by Ken Alder, *The Measure of All Things: The Seven-Year Odyssey and Hidden Error That Transformed the World* (New York, 2002), 1.

© 2009 by The History of Science Society. All rights reserved. 0369-7827/08/2009-0001$10.00

passing through France.[4] The surveying expedition was one of a number of projects involved in what Albert Mathiez called the "scientific mobilization of the savants"[5] on behalf of the new nation-state.

Charles Coulston Gillispie has recently argued that during the French Revolution "the density of the intersections [between science and state] increased to a degree that is characteristic of modern polity in general"[6]—that is, a characteristically "modern" relationship between science and state emerged in new national states in the wake of the revolutionary proclamation of the sovereign nation.[7] Nehru's strategy of combining political leadership with the management of science in the emerging Indian national state thus had venerable origins, dating back to the emergence of modern nation-states in the Enlightenment West.

Nehru and his French predecessors all believed that a scientific approach was crucial to the kind of revolution they were trying to create; science would teach citizens to be sovereign and would draw them into a close relationship with their new state. Frenchmen of the late eighteenth century built on the Enlightenment conviction that the institution of the state had to be brought into line with the basic rationality of the human mind; science was the means to make the French state conform most closely to human reason. Nehru, in contrast, understood himself to be following a model established by modernization in an array of European and American states.[8] To be a nation-state, India needed science along with a middle class, an industrial economy, a national education system, representative government, a press—all elements taken from the example of Western modernization. Both nations saw science as catalyzing other necessary developments. Moreover, in turbulent early national times, both India and France could make science happen by creating national bodies such as institutes and public commissions. Other, more complex social developments—such as dechristianization of the peasantry or the formation of a middle class—proved beyond the immediate reach of the early national state. Leaders of both states, however, believed that scientific institutions would create a scientific public, which in turn

---

[4] Ibid., 92–94.

[5] Albert Mathiez, quoted by Jean Dhombres and Nicole Dhombres, *Naissance d'un pouvoir: Sciences et savants en France (1793–1824)* (Paris, 1989), 47. See also Patrice Bret, *L'Etat, l'armée, la science: L'invention de la recherche publique en France (1793–1830)* (Rennes, France, 2002), chap. 2.

[6] Charles Coulston Gillispie, *Science and Polity in France: The Revolutionary and Napoleonic Years* (Princeton, N.J., 2004), 1.

[7] On the characteristically modern relationship between science and the state, see Hans Blumenberg, *The Legitimacy of the Modern Age*, trans. Robert M. Wallace (Cambridge, Mass., 1983); Chandra Mukerji, *A Fragile Power: Scientists and the State* (Princeton, N.J., 1989), especially chap. 2; Bruno Latour, *We Have Never Been Modern*, trans. Catherine Porter (Cambridge, Mass., 1993); M. Norton Wise, introduction to *Values of Precision*, ed. M. Norton Wise (Princeton, N.J., 1994); Margaret Jacobs, *Scientific Culture and the Making of the Industrial West* (New York, 1997); Gabrielle Hecht, *The Radiance of France: Nuclear Power and National Identity* (Cambridge, Mass., 1998); David Edgerton, "Science in the United Kingdom: A Study in the Nationalization of Science," in *Companion to Science in the Twentieth Century*, ed. Dominique Pestre and John Krige (London, 2003), 759–75; Greg Eghigian, Andreas Killen, and Christine Leuenberger, eds., *The Self as Project: Politics and the Human Sciences in the Twentieth Century, Osiris* 22 (2007); and Paul Forman, "The Primacy of Science in Modernity, of Technology in Postmodernity, and of Ideology in the History of Technology," *History and Technology* 23 (2007): 1–152.

[8] The model of modernization formally put forth by W. W. Rostow often guided the codevelopment of states and science during the 1950s and 1960s. See W. W. Rostow, *Stages of Economic Growth: A Non-Communist Manifesto* (New York, 1960); for the interaction of statecraft with social science crucial to modernization, see Michael E. Latham, *Modernization as Ideology: American Social Science and Nation Building in the Kennedy Era* (Chapel Hill, N.C., 2000).

would effect necessary future transformations such as the full realization of human rational potential and the creation of a middle class.

The authors of the essays collected here similarly make the case for science in the creation of national identity. With examples drawn from a variety of Eastern and Western nation-states, they argue that scientific ability became a marker of national character. The volume explores the ways in which modern science and the nation-state grew up together, each fully aware of the other's capabilities and prepared to use them. The Enlightenment was the moment when a mathematized modern science and the bureaucratized national state joined forces.[9] The first states to take advantage of the power this nexus produced became globally dominant and were widely imitated. In this collection, we trace the Western origins of the relationship between science and the nation-state as well as the dissemination and adaptation of that relationship in the postcolonial world.

In these chapters, nations celebrate their modernity by displaying their scientific prowess. Since the groundbreaking publication of Eric Hobsbawm and Terence Ranger's *The Invention of Tradition*, research on national identity has focused on nations' putative relationships with the past. *The Invention of Tradition* argued that national legitimacy was anchored in a continuous relationship with the past and that the continuity of that relationship was often fabricated wholesale. Nationalism, Hobsbawm argues, was "so unprecedented that . . . historic continuity had to be invented." Scholars following this lead have focused on uncovering that process of invention.[10] Although the essays in this book do not obviate these previous claims, they do present a parallel discourse of national identity rooted in modernity and oriented toward the future rather than toward the past.[11] These essays locate national identity in the conjunction of modern professional science, state sponsorship, and an engaged citizenry. They echo Ernest Renan, who in "What is a Nation?" asserted that nations look forward as well as backward. In addition to the "glorious heritage" of their past, they need "a shared program to implement" in a common future. The national imaginary often defines that future in terms of scientific and technological achievement.[12] As nations cultivate the camouflage of ancient tradition, they simultaneously celebrate their scientific character, which they present as unprecedented. The unique achievement of the nation, they maintain, lies in its citizens' scientific competence, their position on the cutting edge of discovery. We argue, then, that national identity is rooted in a tradition of invention as well as in the invention of tradition.

When nationalists proclaimed this tradition of invention, they construed science broadly; their notions of science included, among other subjects, medicine, engineering, technology, and public health. Science in these contexts meant a rational, universal understanding as well as an approach to problem solving that would leverage the physical world and control nature for human benefit. Thus, when Nehru or the French revolutionaries appealed to "science," they referred to a common intellectual and political enterprise, even though their conceptions of science obviously lacked common

---

[9] Wise, introduction (cit. n. 7), 5–6.

[10] Eric Hobsbawm, "Introduction: Inventing Traditions," in *The Invention of Tradition,* 2nd ed., ed. Eric Hobsbawm and Terence Ranger (1983; repr., New York, 2000), 7.

[11] Konstantinos Chatzis, "Introduction: The National Identities of Engineers," *Hist. & Tech.* 23 (2007): 194.

[12] Ernest Renan, "What Is a Nation?" in *Becoming National*, ed. Geoff Eley and Ronald Grigor Suny, trans. Martin Thom (New York, 1996), 41–55.

content. "Science" acted as an empty term, allowing the state to apply that label to different kinds of projects. Because of the elasticity of "science" in its multiple interactions with modern polities, it is important to pay close attention to the specific content and meaning of the science that particular nation-states and citizenries value. Thus, by "science" we refer to a broad scientific enterprise that takes on specific, local meanings attuned to national settings and circumstances.[13]

Research on nationalism has largely ignored the nexus between science and national identity. In recent decades, scholars have lavished attention on the historical project of nation building, calling attention to its complexity and to the multifaceted relationships between nation-states and national citizenship. Scholars have been remarkably reticent about the place of science and technology in the construction of national identities, however. The articles presented here fill that gap. They address the coincidence between the emergence of modern, professional science and the nation-state both in the West and in postcolonial Asia, and they argue that chronological overlap was not happenstance but coproduction, an affiliation between science and the nation that went beyond funding streams and economic development to speak to national identity. The papers in this volume focus on different scientific enterprises—ranging from natural history collecting to nanotechnology—but they all consider the ways in which a national public functioned as one of the intended audiences for scientific production. By inserting science into the formation of national identities, these authors address a significant lacuna in an area of growing scholarly attention.

Nationalism and national identity have emerged as key categories of historical analysis in response to global political upheavals, leading scholars to interrogate the nation as history's obvious subject. In the late twentieth century, historians began dissecting the process of nation building, and they rejected the nation's claim to be a primordial form of social organization. The foundation of the process of nation formation, they discovered, was not the timeless unity of culture, ethnicity, language, and religion that allegedly constituted a nation. Rather, historians and social scientists maintained that the nation, no less than the nation-state, required political will and calculation. History itself as a discipline originated in the nineteenth century as part of the nation-building project, its purpose to provide the nation with its past.[14] Nations required efforts of construction, invention, and imagination rather than simply identification; they were never already there waiting to enter history.

In the past forty years, scholarly disagreements have focused on the nature of the effort involved in producing a nation. In some research, nations feature as the product of elite inspiration and state planning; the patriotic masses arrive on the scene last, not first.[15] Other scholars emphasize popular loyalties and demands, often generated by the discontents of the uneven distribution of the fruits of modern capitalism.[16] In some accounts, nations emerged out of the political processes of liberal revolutions

---

[13] Dominique Pestre similarly details the multiple meanings that science takes on for different social actors in "Science, Political Power, and the State," in Pestre and Krige, *Companion to Science* (cit. n. 7), 61.

[14] Stefan Berger, Mark Donovan, and Kevin Passmore, eds., *Writing National Histories: Western Europe since 1800* (New York, 1999).

[15] Eric Hobsbawm, *Nations and Nationalism since 1780: Programme, Myth, Reality,* 2nd ed. (New York, 1992); John Breuilly, *Nationalism and the State* (Manchester, UK, 1982); Miroslav Hroch, *Social Preconditions of National Revival in Europe: A Comparative Analysis of the Social Composition of Patriotic Groups among the Smaller European Nations,* trans. Ben Fowkes (New York, 1985).

[16] Ernest Gellner, *Nations and Nationalism* (Ithaca, N.Y, 1983).

and debates about citizenship; while in others, nations were the result of industrial-ization and the imposition of bourgeois hegemony. For other scholars, looking for na-tionhood in deliberate action either from the top or from the bottom misses the point. Thus for Rogers Brubaker, post-Soviet nations are unplanned events, accidental and unexpected products of bureaucratic practices.[17] Similarly for Benedict Anderson, the nation is the result of print capitalism and an identity available for use either by elites or by disenfranchised groups.[18] For some researchers, nation is a purely mod-ern form that replaces premodern forms of community, while for others, the nation derives its affective power from its affinities with earlier ethnic or religious commu-nities from which it derives its symbols and myths.[19] Counter to the claims of nation-alists themselves, scholars emphasize that nations are anything but necessary and that nationalities are not essential human attributes, even though they are designed to appear as such.[20]

The intense late-twentieth-century professional activity surrounding nations and nationalism shows no sign of abating. New peer-reviewed journals, Web sites, dis-cussion lists, and conferences generate an ever-increasing quantity of international and interdisciplinary scholarship.[21] Greater volume, however, does not appear to bring this research any closer to questions about the relationship between scientific enterprise and the national project. Although nationalism scholarship has been catho-lic in its research methodologies and subjects, it has overlooked the extent to which an allegedly national capacity to excel in the production of science forms one of the affective bonds between states and citizens.

Recent work on nation formation has focused on how representations of the nation such as flags, vernacular languages, rituals, and ceremonies mediate the relationship between the nation-state and its people. This emphasis on the symbolic has not en-couraged historians to include science and technology in the story of nation building. Nineteenth-century Germans, to take a well-researched example, forged their na-tional identity from the celebration of rural homelands, Bach's genius, nature preser-vation, physical culture, female domesticity, and Christmas trees—symbols that gave "Germany" historical depth and a usable past.[22] This literature has been remarkably

[17] Rogers Brubaker, *Nationalism Reframed: Nationhood and the National Question in the New Eu-rope* (New York, 1996).

[18] Benedict Anderson, *Imagined Communities: Reflections on the Origins and Spread of National-ism* (1983; repr., New York, 1991).

[19] Anthony Smith, *The Ethnic Origins of Nations* (New York, 1986); John Hutchinson, *Nations as Zones of Conflict* (Thousand Oaks, Calif., 2005); Adrian Hastings, *The Construction of Nationhood: Ethnicity, Religion, and Nationalism* (New York, 1997); Anthony Marx, *Faith in Nation: Exclusionary Origins of Nationalism* (New York, 2003). Cf. Liah Greenfeld, *Nationalism: Five Roads to Modernity* (Cambridge, Mass., 1992).

[20] Gellner, *Nations and Nationalism* (cit. n. 16), 6.

[21] For journals, see, e.g., *Nationalities Papers* (founded 1994), *Nations and Nationalism* (1995), *Nationalism and Ethnic Politics* (1995), *National Identities* (1999), and *Studies in Ethnicity and Na-tionalism* (2000). The Nationalism Project Web page (http://www.nationalismproject.org) acts as a clearinghouse for relevant scholarship, listing conferences and providing links to the many associa-tions dealing with nationalism. A search of the Nationalism Project site for "science" produces virtu-ally no hits that refer to anything other than social science.

[22] For a useful overview, see Nancy R. Reagin, "Recent Work on German National Identity: Re-gional? Imperial? Gendered? Imaginary?" *Central European History* 37 (2004): 273–89. In a litera-ture too extensive to cite completely, see, especially, Celia Applegate, *Bach in Berlin: Nation and Cul-ture in Mendelssohn's Revival of the St. Matthew Passion* (Ithaca, N.Y., 2005); Applegate, *A Nation of Provincials: The German Idea of Heimat* (Berkeley, Calif., 1990); Joseph B. Perry, "The Private Life of the Nation: Christmas and the Invention of Modern Germany" (PhD diss., Univ. of Illinois, 2001);

attentive to deconstructing the nation's claim to obviate other differences, and it focuses on the ways in which identification with the nation articulates with other forms of identity such as race, gender, religion, ethnicity, and class. The citizen's ties to the nation find symbolic form not only on explicitly political occasions such as national holidays but also in the texture of daily life, in the citizen-consumer's choice of products, the schoolchild's homework, and the nightly news.[23]

An emphasis on the symbolic and the ritual is an important element of the scholarly project of deconstructing the primordial nature of the nation. Revealing national traditions—kilts, pumpkin pie, champagne—as invented is the central mode of a great deal of historical research on national identity.[24] In what has become a common historiographical operation, scholars take allegedly ancient symbols of national identity and expose their conscious fabrication and novelty. Symbols of great affect are revealed as the products of consumer capitalism. What remains of the fabric of national identity is essentially kitsch—empty symbols filled with the vapid sentimentality of nationalism.

The project of exposing invented traditions and reducing national symbols to kitsch discourages scholars from focusing on the kinds of icons of national identity that might lead to a discussion of science and technology. Mobilizing science in the construction of national identity proclaims novelty rather than concealing it and is essentially future oriented; a scientific icon of national achievement such as the space program is not subject to the same kind of deconstruction as pumpkin pie. In his account of cold war astronomy, W. Patrick McCray, for instance, undermines the heroic narrative of American professional scientific genius by emphasizing the role of amateur scientists and curious citizens, from ham radio enthusiasts to housewives. Taking apart the space program's claim to have built America as a technoscientific superpower does not, however, reduce rockets to consumer kitsch.[25] Scientific and technological icons have a material reality that resists their complete reduction to empty sentiment. The scientific, or at least engineering, principles behind artifacts such as the Autobahn and satellites continue to have a technological history even if their cultural meaning can be read as an instance of an invented tradition.

The infrastructure through which states build themselves is a good example of the polyvalent nature of the technological artifact.[26] The Erie Canal, for instance, is

---

Thomas M. Lekan, *Imagining the Nation in Nature: Landscape Preservation and German Identity, 1885–1945* (Cambridge, Mass., 2004); Alon Confino, *The Nation as a Local Metaphor: Württemberg, Imperial Germany, and National Memory, 1871–1918* (Chapel Hill, N.C., 1997); Jean Quataert, *Staging Philanthropy: Patriotic Women and the National Imagination in Dynastic Germany, 1813–1916* (Ann Arbor, Mich., 2001).

[23] On national holidays, see, e.g., Confino, *Nation as a Local Metaphor* (cit. n. 22). On consumer choice, see, e.g., Leora Auslander, *Taste and Power: Furnishing Modern France* (Berkeley, Calif., 1996); and Erica Carter, *How German Is She? Postwar West German Reconstruction and the Consuming Woman* (Ann Arbor, Mich., 1997). On education, see, e.g., Stephen L. Harp, *Learning to be Loyal: Primary Schooling as Nation Building in Alsace and Lorraine, 1850–1940* (De Kalb, Ill., 1998).

[24] Hugh Trevor-Roper, "The Invention of Tradition: The Highland Tradition of Scotland," in Hobsbawm and Ranger, *Invention of Tradition* (cit. n. 10); Cynthia Ott, "Squashed Myths: The Cultural History of the Pumpkin in North America" (PhD diss., Univ. of Pennsylvania, 2002); Kolleen M. Guy, *When Champagne Became French: Wine and the Making of a National Identity* (Baltimore, 2003).

[25] W. Patrick McCray, *Keep Watching the Skies!: The Story of Operation Moonwatch and the Dawn of the Space Age* (Princeton, N.J., 2008).

[26] Peter Kroes, Anthonie Meijers, and other philosophers at the Delft University of Technology have explored the dual nature of technological artifacts, a concept through which they explore the ontological and epistemological relationships between artifacts' physical and functional definitions. See Peter

simultaneously an engineering project, a political undertaking, and a national icon. Describing the coproduction of these elements is both extraordinarily difficult and, on the whole, not what most historians are trying to do.[27] Carol Sheriff's *Artificial River* discusses the Erie Canal's place in antebellum culture, exploring its symbolic meaning as an example of the young nation's ability to compress time and distance. The canal was a beacon of progress and ingenuity that also brought America's vices to rural New York, causing upstaters to question progress itself. Sheriff is primarily interested in the cultural work of nation building and is less attentive to engineering.[28] Jonathan Larson's *Internal Improvement*, by contrast, investigates the politics of infrastructure building and the canal's economic function, again with little attention to engineering.[29] Both historians are breaking with the "just-so story" of how the canal was built, a body of literature that focuses on the minutiae of canal engineering.[30] The Erie Canal could lend itself to a study of its coproduction as an engineering project, political undertaking, and national icon, but in both Sheriff and Larson's studies, the engineering project drops out.

In fact, the gap between the technological artifact and Sheriff and Larson's political and cultural treatments points to one of the tensions in the history of technology. On the one hand, cultural and political historians overlook the literally dirty business of making technologies. On the other, historians write the "internalist" history of technology as a just-so story about how things come into being, a simple narrative of invention or engineering in which the cultural and the symbolic play little role. Both accounts have serious flaws, and the interaction of the state, science and technology, and citizens or the public falls into an interstice.

Recent research on maps and mapping further illustrates this mismatch between cultural and political histories of nationalism and the history of science and technology and suggests ways in which scientific production and national identity can illuminate one another without slighting the significance of either. Benedict Anderson, drawing on Thongchai Winichakul's research, famously proposed the national map as an object of study. Anderson and Thongchai's argument, however, presents the map as logo rather than as a problem for scientific knowledge. According to Thongchai, "In terms of most communication theories and common sense, a map is a scientific abstraction of reality. A map merely represents something which already exists objectively 'there.'" The relationship between maps and their nations, however, upsets this common sense, because national maps "anticipated spatial reality, not vice versa. In other words, a map was a model for, rather than a model of, what it purported to represent."[31] The "map-as-logo" was "pure sign, no longer compass to the world." The map naturalized the bordered state, haven of the nation that resided within it.[32] Anderson and Thongchai's formulation accounts for the powerful

---

Kroes and Anthonie Meijers, "The Dual Nature of Technological Artifacts: Presentation of a New Research Programme," *Techné* 6 (2002): 4–8. We propose that there are more than two aspects that warrant attention and thus refer to the polyvalent nature of technological artifacts.

[27] Hecht, *Radiance of France* (cit. n. 7), is a notable exception.

[28] Carol Sheriff, *The Artificial River: The Erie Canal and the Paradox of Progress, 1817–1862* (New York, 1996).

[29] Jonathan Lauritz Larson, *Internal Improvement: National Public Works and the Promise of Popular Government in the Early United States* (Chapel Hill, N.C., 2001).

[30] Ronald Shaw, *Erie Water West: A History, 1792–1862* (Lexington, Ky., 1990).

[31] Thongchai Winichakul, quoted in Anderson, *Imagined Communities* (cit. n. 18), 173.

[32] Ibid., 175.

symbolic resonance of the outline of national territory, but it effaces the role of science and technology in producing this image.

In contrast, Matthew Edney's account of mapmaking includes the science of cartography in the production of the national image.[33] The shape of what became India was the result of engagement between the British Raj, British science, and Indian workers. Thongchai and Edney treat maps in fundamentally different ways. For Thongchai, maps are essentially tropes; the process by which landscape features are inscribed in two dimensions matters very little. In their most obviously logo form, they even border on kitsch—the nation with all its affective resonance captured in simple graphic form. Edney's account focuses on the production of cartographic images that come to represent the nation; indeed, Edney's history amounts to labor history, although the labor involved is scientific.

The Erie Canal and the making of the Indian map, like Nehru's institute and the revolutionary French meter, suggest that science has been present as a matter of national pride at the origins of many nation-states. Since the Enlightenment, nation-states have enlisted science as one of the bonds connecting citizens to the state. Science and engineering become "proof" that the nation can improve the living conditions of citizens, "proof" that the nation and its citizens are modern, and "proof" that they are economically and militarily competitive in a global world.[34] As a result, the capacity to produce science becomes an indicator of national superiority. Enlightenment western Europe distinguished itself from the irrational East just as in a later era the scientifically advanced North sees itself in opposition to a technologically deficient South. Taken together, these ways of fashioning science have created a powerful alliance between science and the nation-state, a project in which scientists and engineers readily participate. This alliance is passed to the public to show the power of the state to beget development and improvement—often cast to show the superiority of one nation's science and technology over those of other nations.

Often, however, scholars who examine the relationship between the state and the scientific community abstract scientists from the broader body of citizens; interactions between governments and scientists thus appear to take place in a vacuum. These studies often emphasize the role of the state in the construction of scientific institutions, investigating state funding of science and the concomitant scientific lobbying for public resources. In studies of totalitarian states, this research tends to highlight the state's "abuse" or "perversion" of science, as in the cases of Nazi racial science or Lysenkoism.[35] Other studies focus on the state as a patron of scientific research, which may or may not involve a discussion of state manipulation of science

[33] Matthew Edney, *Mapping an Empire: The Geographical Construction of British India, 1765–1843* (Chicago, 1997), 1.

[34] James Scott's notion of "High Modernist Ideology" as a tool of the state has been suggestive in underlining the role of the scientific capacity of the modern state as a technique of control, often a disastrous one. James Scott, *Seeing Like a State: How Certain Schemes to Improve the Human Condition Have Failed* (New Haven, Conn., 1998), chap. 3.

[35] Loren Graham, *Science and the Soviet Social Order* (Cambridge, Mass., 2002); Valery N. Soyfer, *Lysenko and the Tragedy of Soviet Science* (New Brunswick, N.J., 1994); Robert N. Proctor, *Racial Hygiene: Medicine under the Nazis* (Cambridge, Mass., 1988); and John Cornwell, *Hitler's Scientists: Science, War, and the Devil's Pact* (New York, 2003).

for ideological purposes. Invariably, these approaches demonstrate the importance of economic and military motives in state patronage of science.[36]

To explore the role of science in nation-state building and identity formation, historians have to follow science and scientists out of the lab. Science cannot inflect national identity unless it finds its way on to public agendas, either through the lobbying of scientists, the projects of the state, or the demands of citizen-consumers. The authors of these essays maintain that significant discussions of science occur not only in laboratories or in political corridors of power but also in the "intellectual commons," a space where "multisided conversations about scientific topics take place . . . [in] which various publics can . . . seek to engage scientific professionals in conversation."[37] The "public" that matters to the authors of the essays in this collection is the nation.

Recent research on colonial science and science education offers models of how scholars might fit science and scientists into the public. Although the literature on colonial science has extensively documented scientists' special role in the operations of the imperial state, recent research has uncovered the participation of colonial subjects in the process of making knowledge—a strategy that has the effect of integrating "science" into a broader community, in this case colonial society.[38] Science was "inextricably woven into the whole fabric of colonialism" because the authority of science and of the imperial state were mutually reinforcing.[39] New research emphasizes hybridity and co-option in the production of colonial science. As the civilizing mission deployed science and scientists to specify the nature of colonial otherness in its social, biological, and topographical forms, it created an intellectual commons that translated indigenous knowledge into the languages of state officials and Western scientists. This hybrid knowledge of the colonial world was not only a tool of imperial authority but also a potential basis for postcolonial national identity.

---

[36] Wise, introduction (cit. n. 7); David Cahan, *An Institute for an Empire: The Physikalisch-Technische Reichsanstalt, 1871–1918* (New York, 2004); Jeffery A. Johnson, *The Kaiser's Chemists: Science and Modernization in Imperial Germany* (Chapel Hill, N.C., 1990); Itty Abraham, *The Making of the Indian Atomic Bomb* (London, 1998); Stuart W. Leslie, *The Cold War and American Science* (New York, 1993); John Krige, *American Hegemony and the Postwar Reconstruction of Science in Europe* (Cambridge, Mass., 2006); Krige and Kai-Henrik Barth, eds., *Global Power Knowledge: Science and Technology in International Affairs, Osiris* 21 (2006); Morris Low, *Building a Modern Japan: Science, Technology, and Medicine in the Meiji Era and Beyond* (London, 2005).

[37] Katherine Pandora and Karen Rader, "Science in the Everyday World: Why Perspectives from the History of Science Matter," *Isis* 99 (2008): 360.

[38] C. A. Bayly, *Empire and Information: Intelligence Gathering and Social Communication in India, 1780–1870* (New York, 1996); Susan Scott Parish, *American Curiosity: Cultures of Natural History in the Colonial British Atlantic World* (Chapel Hill, N.C., 2006); Londa Schiebinger, *Plants and Empire: Colonial Bioprospecting in the Atlantic World* (Cambridge, Mass., 2007); and Kapil Raj, *Relocating Modern Science: Circulation and the Construction of Knowledge in South Asia and Europe, 1650–1900* (New York, 2007).

[39] Deepak Kumar, *Science and the Raj: A Study of British India*, 2nd ed. (New York, 1995), 15. In a large literature, see, e.g., Roy McLeod, ed., *Nature and Empire: Science and the Colonial Enterprise, Osiris* 15 (2000); Jorge Cañizares-Esguerra, *Nature, Empire, and Nation: Explorations of the History of Science in the Iberian World* (Stanford, Calif., 2006); Harold Cook, *Matters of Exchange: Commerce, Medicine, and Science in the Dutch Golden Age* (New Haven, Conn., 2007); James E. McClellan, *Colonialism and Science: Saint Domingue in the Old Regime* (Baltimore, 1992); Michael Osborne, *Nature, the Exotic, and the Science of French Colonialism* (Bloomington, Ind., 1994); Richard Drayton, *Nature's Government: Science, Imperial Britain, and the "Improvement" of the World* (New Haven, Conn., 2000); Lewis Pyenson, *Civilizing Mission: Exact Sciences and French Overseas Expansion, 1830–1940* (Baltimore, 1993).

Research on educational systems similarly locates science within an intellectual commons—in this case a space defined in explicitly national terms. Educating citizens is one of the projects that modern states undertake, and the education of scientists, in particular, has been the domain of the nation-state. Education "serves as a crucible for reproducing cultural, political and moral values"; that is, students learn not only how to be good scientists but also how to be good subjects or citizens of the states that sponsor their training.[40] Like the essays in this collection, research on education examines the alignment between scientific and more broadly cultural values; educational institutions transmit both universal scientific knowledge and particular, local priorities.[41]

The essays in this volume, like research on colonial science and scientific education, locate scientists within a public sphere. Science participates in forming national identity not because scientists talk to the public but because they speak within the public, as participants in a rational-critical discussion of the affairs of state.[42] Certainly, scientists' position within the body of citizens is privileged, in part because the value the state places on the utility of scientific knowledge gives them access to an effective soapbox. Recognizing the benefits of this privileged relationship with the state, scientists seek out the state's approval for their work, and this mutual benefit pushes scientists, whether they intend it or not, into the public sphere.

Our focus on the interactions between science and the formation of national identity instantiated in the nation-state leads us to emphasize the modernity of the national project. The public sphere, the bureaucratized state, and modern science began to cohere in the eighteenth century. Although scholars with an interest in the significance of ethnicity and the persistence of premodern forms of community have challenged the nation's modern chronology, we argue that the importance of science in defining the nation suggests the modernity of both.[43] Thus, the question of national identity as presented in this volume is essentially an issue for the eighteenth, nineteenth, and twentieth centuries. The bulk of the essays in the collection focus on this period, exploring the relationship between the national and the scientific enterprise. Articles on protonational and postnational scientific projects bookend the collection, raising the possibility that science tracks both the rise and the fall of the national paradigm.

The volume opens with a study of protonational identity formation, Chandra Muk-

---

[40] David Kaiser, "Introduction: Moving Pedagogy from the Periphery to the Center," in *Pedagogy and the Practice of Science: Historical and Contemporary Perspectives*, ed. David Kaiser (Cambridge, Mass., 2005), 2. In a field that Kaiser characterizes as underdeveloped, see also William H. Brock, *Science for All: Studies in the History of Victorian Science and Education* (Brookfield, Vt., 1996); Antoine Picon, *L'invention de l'ingénieur moderne: L'Ecole des ponts et chaussées, 1747–1851* (Paris, 1992); Ken Alder, *Engineering the Revolution* (Princeton, N.J., 1997); Terry Shinn, *Savoir scientifique et pouvoir social: L'Ecole Polytechnique, 1794–1914* (Paris, 1980); Kathryn M. Olesko, "Science Pedagogy as a Category of Historical Analysis: Past, Present, and Future," *Science and Education* 15 (2006): 863–80.

[41] David Kaiser, "The Postwar Suburbanization of American Physics," *American Quarterly* 56 (2004): 851–88; Kathryn M. Olesko, *Physics as a Calling: Discipline and Practice in the Koenigsberg Seminar for Physics* (Ithaca, N.Y., 1991); Andrew Warwick, *Masters of Theory: Cambridge and the Rise of Mathematical Physics* (Chicago, 2003).

[42] Jürgen Habermas, *Structural Transformation of the Public Sphere: An Inquiry into a Category of Bourgeois Society*, trans. Thomas Burger, with the assistance of Frederick Lawrence (Cambridge, Mass., 1989); Bernadette Bensaude-Vincent, "A Genealogy of the Increasing Gap between Science and the Public," *Public Understanding of Science* 10 (2001): 99–113.

[43] In addition to Anderson, *Imagined Communities* (cit. n. 18), see also Patrick J. Geary, *The Myth of Nations: The Medieval Origins of Europe* (Princeton, N.J., 2002).

erji's essay on the seventeenth-century construction of the Canal du Midi in the south of France. The notion of France as a "new Rome" engaged in engineering projects whose genius exceeded even that of the ancients fell short of being "national" in many ways—in particular, it excluded the peasant builders of the canal whose tacit engineering knowledge was vital to the success of the project. Although many of the pieces of the characteristic relationship between modern science and the state were in place in Louis XIV's engineering projects, they did not generate modern forms of national identity. In particular, contemporaries conceived of the canal as the product of the labor of royal subjects directed by individual engineering genius, not as an expression of the will and ability of French citizens.

The next section turns to the emergence of modern nations and modern science in the West. Carol Harrison's contribution demonstrates the distance between Mukerji's hydraulic engineers of the seventeenth century and a fully formed national identity. Where the "new Romans" were autonomous aristocrats, the citizens, both sailors and scientists, who traveled on revolutionary France's expeditions to the south Pacific acted as representatives of the newly sovereign nation. Essays by Ann Johnson and Katherine Pandora use the case of the early United States to explore how a new nation-state generated identity among its citizens in the absence of an ancient lineage. Johnson examines the formation of an American engineering community to argue that scientists' confrontation with the variety and unfamiliarity of American nature led them to see American science—and American identity—as exceptional. Pandora similarly argues that American approaches to science contributed to convincing Americans of their exceptionality. The vernacular science of American children's literature produced scientific citizens; the boy and girl scientists of these stories represented America's capacity to democratize science and science's capacity to democratize America. Finally, Michael Gordin compares two late-emerging nations, Ireland and Russia. His paper examines the scientist as national icon, exploring why Dmitrii Mendeleev came to represent Russian national character while his contemporary, the equally distinguished Thomas Andrews, never acquired national significance. In each of these papers, new nations deploy scientific competence as an attribute of emerging national character; in France and the United States, the association between the nation and scientific ingenuity established itself early on, while in imperial Russia and Ireland, the scientific character of the nation never achieved the same rhetorical potency.

The second major section of the book addresses science in mature Western states but in national contexts that challenge simple conceptions of national identity. Asif Siddiqi explores the postwar Soviet Union's failed attempt to incorporate German rocket scientists into a secret nationalist project. Siddiqi points to the tension of trying to take up German science while discarding the scientists and shows the necessity of that move in the context of articulating a cold war Soviet identity. Edward Jones-Imhotep also discusses postwar big science but on the other side of the cold war. Ionospheric research underscored place-dependent ideas of Canadian national identity. Canada was neither a linguistic nor an ethnic entity, and it was overshadowed by the superpower to the south, but its "northernness" became the defining element of its national identity in part because of the science that it allowed Canadians alone to carry out. Bruno J. Strasser argues that the definition of certain areas of international cooperation as "technical" rather than "political" allowed Switzerland to be both Western and neutral during the cold war. Swiss leadership in postwar scientific institutions such as CERN allowed them to accept participants from East and West, and this form of neutrality—the national space defined as an international meeting ground—became central to Swiss identity. The essays in this section remind us that the mutually reinforcing processes of scientific development and the formation of national identity remain active in the West well past the classic nineteenth-century "age of nation building."

Emerging postcolonial Asian countries similarly posited that science was indispensable and that the state and its citizens needed to collaborate in the project of creating the scientific nation. Contributions on Indonesia, India, and Republican China illustrate the intensity of this desire to deploy science in the project of identity formation as well as that of state building. The section opens with the late colonial period in Pratik Chakrabarti's essay on the development of medical research in India. Public health enjoyed prestige for its claim to utility, particularly its ability to protect the population from epidemic disease; it was therefore important to Indian nationalists that medical research should be Indian rather than British. The location of research laboratories became a point of contention in the early twentieth century, with nationalists insisting that facilities be moved away from colonial hill stations to locations that actually suffered from epidemics. Ross Bassett's paper on Indians trained at MIT also looks at science and technology as a tool of resistance to imperialism. By opting for American engineering education, young Indians chose an alternative version of modernity, one that, to their eyes, countered British imperialism. The experiences of these engineers in both the colonial and early national periods did not reflect a division between Gandhian and Nehruvian visions of India, the former hostile to technology and the latter built on the mastery of Western science. Grace Y. Shen explores the consciousness of scientific inadequacy that featured in Chinese shame following the fall of the empire. Western-trained Chinese geologists argued that knowledge of the national territory was an indispensable step to reclaiming a sense of pride that in the new China would be national rather than ethnic or dynastic. Suzanne Moon considers technology transfer with her paper on the Krakatau steel company. Krakatau was supposed to spearhead the industrialization of Indonesia, and it symbolized the unity of the nation by referring to a precolonial, archipelagic empire. Locating the plant in a remote and impoverished region that had previously been the center of the Majapahit empire established the Sukarno government's commitment to social justice and the equitable distribution of wealth. Despite ideological differences between Sukarno and Suharto's New Order, the steel mill functioned for both as a representation of national identity. The papers on Asia make clear the extent to which nation building utilizes common building blocks, including steel plants, scientific societies, and infrastructure. Whether in Indonesia or in France, the nationalist's goal is to construct a nation that resembles and, therefore, can compete with other nations. Nations may have their own distinctive characteristics, but they resemble nothing so much as other nations.

Finally, the collection closes with Alfred Nordmann's essay on European Union emerging technology initiatives, a contribution that addresses the role of science in a potentially postnational world. Nordmann's essay highlights the irony of the persistence of a rhetoric of national scientific competitiveness in a transnational setting; his essay raises the possibility that science and technology may now be in the forefront of the creation of a postnational, global society. The European Union (EU) constructed its European Knowledge Society as a way to distinguish European research in the area of converging technologies from the American Nano-Bio-Info-Cogno (NBIC) program. Although the two projects accomplished similar goals from a scientific perspective, the EU wanted to be sure it did so in an identifiably European way; the initiative directed the familiar rhetoric of national scientific competitiveness toward the creation of a transnational identity.

Nordmann's paper gestures toward the uncertain status of the relationship of science and national identity in the contemporary world. On the one hand, recent scholarship on the nation suggests that its moment has passed and that its history was, ultimately, brief. In this literature, modern science and technology appear to be forces provoking the dissolution of nations. "Postnational" scholars build on Benedict Anderson's argument that nineteenth-century media—notably the novel and the newspaper—created the sense of imagined community among people unlikely ever to meet face to face. The contemporary heirs to modernity's print culture—the globalized media of the moving image and the Internet—fragment the nation and render it meaningless.[44] On the other hand, the nationalist voice persists in science policy: bureaucrats around the world continue to produce documents with titles such as *Establishing Strong Korea in Science and Technology Human Resources* and *Rising above the Gathering Storm: Energizing and Employing America for a Brighter Economic Future.* It is clear that science and nation often continue in tandem as they move toward the future; whether in that future the nation will prove dispensable remains to be seen.

The essays in the collection strike a balance between well-known examples and challenging and less familiar cases in both national and scientific terms. Many of the pieces focus on countries and periods for which there are well-developed literatures in both the history of science and national identity, for example, Enlightenment France, the Soviet Union, and India. A second group treats countries for which a robust scholarship concerning national identity formation exists but only a thin treatment of science and technology, such as the early American republic and Republican China. The third category contains cases that challenge ideas about the nation, such as Canada, Indonesia, and Switzerland, and about whose science little has been written.

The essays similarly range from treatment of paradigmatic cases of big science to little-known scientific developments. The collection includes classic cases of big science, such as Strasser's paper on CERN and Siddiqi's on rocketry. Harrison's essay proposes an eighteenth-century analogue for big science: voyages of exploration that mobilized massive scientific workforces and budgets. Nuclear physics, natural history, and missiles are well-trodden ground, although their connections to Swiss, French, and Soviet identity are less explored. In contrast, Johnson's paper on American engineering explores a largely unknown dimension of American science—materials research produced by the Corps of Engineers—while Pandora develops the neglected subject of vernacular science. Focusing on the nation opens the door to a wide range of fields across the physical and life sciences, engineering, and medicine because of the voracity of the nation-building project. Virtually anything can be used to build the nation.

---

[44] Arjun Appadurai, *Modernity at Large: Cultural Dimensions of Globalization* (Minneapolis, Minn., 1996). Tom Nairn, *Faces of Nationalism: Janus Revisited* (New York, 1997); and Michael Billig, *Banal Nationalism* (Thousand Oaks, Calif., 1995), chap. 6, counter the postnationalist view, insisting, in Nairn's words, that nationalism "is unlikely to have laid empires low in order to succumb to 'Macdonaldisation,' Islam or breakfast television" (72).

Responding to Ernest Geller's claim that the "true subject of modern philosophy is industrialization," Tom Nairn asserted that "the true subject of modern philosophy is nationalism, not industrialization; the nation, not the steam engine or the computer."[45] Like Gellner, Nairn links nation and industrialization as the two key elements of modernity. Nonetheless, assigning the nation priority by placing it in opposition to the steam engine and the computer—defining technologies of the nineteenth and twentieth centuries, respectively—is a telling rhetorical move. The authors of this volume propose alternatives to Nairn's depiction of the computer and the nation as either-or choices. The computer participates in the building of the nation, we maintain, and various nations around the world hold up their computer engineers as exemplars of national achievement.[46] If the nation is, as Ernest Renan famously claimed, "a daily plebiscite," then steam engines and computers and other icons of scientific progress and achievement are significant reasons why citizens regularly vote in its favor; these artifacts represent citizens' belief in the benefits of a common national future.[47] Scholarship on the nation needs to be as attentive to nations' anticipated futures as to their imagined pasts. The nation that seeks to demonstrate its long unbroken connection to an illustrious past also endeavors to present itself as oriented toward a scientific future.

[45] Nairn, *Faces of Nationalism* (cit. n. 44), 17.

[46] Jon Agar, *The Government Machine: A Revolutionary History of the Computer* (Cambridge, Mass., 2003); Atsushi Akera, *Calculating a Natural World: Scientists, Engineers, and Computers during the Rise of U.S. Cold War Research* (Cambridge, Mass., 2006).

[47] Renan, "What Is a Nation?" (cit. n. 12), 53.

The New Rome:
Infrastructure and National Identity on the Canal du Midi

*By Chandra Mukerji**

ABSTRACT

Before France was a nation, France was promoted as a New Rome and the French as descendents of ancient Gaul. Under Louis XIV, the idea of becoming French was not generally appealing because of opposition to monarchical power. But a campaign to make France a New Rome inadvertently provided a basis for French national identity. The Canal du Midi, one of the infrastructural projects for the New Rome, came to define France and stand for French native genius. By the nineteenth century, Pierre-Paul Riquet, the canal's entrepreneur, had become a national hero; the New Rome and the French nation had become one.

The development of national identity is often treated in Benedict Anderson's terms as mainly a matter of communication and political imagination.[1] People who do not know each other come to conceive of themselves as part of a shared community. They do not have to have very much in common as long as they can create narratives of commonality that define their unity. The problem with forging national identities, from this perspective, is that people must find compelling narratives that encourage them to abandon or modify previously held views of themselves. How this happens is not so transparent.

This paper tries to address the issue empirically by focusing on one case: the development of French identity from the 1660s through the 1680s around a program of territorial politics designed to make France an empire in the image of ancient Rome. The late seventeenth century was well before France had a national identity,[2] so the move toward an empire in this period was not designed to serve the French nation.

* Department of Communication, University of California at San Diego, 9500 Gilman Drive, La Jolla, CA 92093-0503; cmukerji@weber.ucsd.edu.

I would like to thank Simon Werret for encouraging me to think about the importance of Rome in science from this period into the eighteenth century. His observations helped me see the protonationalist implications of seeing France as Rome.

[1] Benedict Anderson, *Imagined Communities: Reflections on the Origins and Spread of Nationalism* (London, 1983). See also Vincente L. Raphael, *White Love and Other Events in Filipino History* (Durham, N.C., 2000).

[2] For ways of thinking about national identity and collective memory in France and elsewhere, see Eugen Weber, *Peasants into Frenchmen* (Stanford, Calif., 1976); Barry Schwartz, *Abraham Lincoln and the Forge of National Memory* (Chicago, 2000); Lyn Spillman, *Nation and Commemoration: Creating National Identities in the United States and Australia* (New York, 1997); Nikolas Rose, *Powers of Freedom: Reframing Political Thought* (Cambridge, UK, 1999), 51–60.

© 2009 by The History of Science Society. All rights reserved. 0369-7827/09/2009-0002$10.00

Still, the program to make France a New Rome brought together diverse people from different parts of the kingdom in state engineering projects, the successes of which defined the participants as a group. Examining how these New Romans worked and how their achievements were used to define French character and culture allows one to consider how state building and political identity were connected before the two processes were permanently and deeply entangled in the idea of a nation.

No one expected the French state under Louis XIV to reflect or serve the people of the kingdom; the political administration was designed to exercise monarchical will. As Jean Bodin made clear,[3] this was the period of personal rule, not of politics based on impersonal categories such as the nation or collective processes such as national identity formation. Still, Louis XIV's will to become emperor of a New Rome inadvertently provided a basis for French national identity formation[4] by defining the French people as descendents of Gaul and taking their engineering abilities as evidence of this heritage.

The pursuit of the New Rome through engineering—roads, ports, ship-building facilities, and canals—proved a particularly effective tool for empowering the administration because it was difficult to counter through normal patrimonial politics.[5] Infrastructures were surprisingly transformative because they exercised power in dislocating, material ways. They were mute media that did not exploit patron-client ties or engage in political debates. Instead, they silently and impersonally changed conditions of life for the communities around them, making the countryside different from what it had been before—and an artifact of government. They also testified to the intelligence of those who built them, not the king or God but rather ordinary men and women, indicating a potential power of people to change the world through thoughtful collective action.

Focusing on the seventeenth-century engineering of the New Rome and its consequences for national identity highlights the role of science and technology in the formation of modern patterns of political life. This study helps indicate the close connections between political territoriality, engineering, and identity—the material bases of an imagined community—in the period before national identities became routine. This analytic approach focuses on the territoriality of nations as well as states and on the importance of communities of practice (engineering) for the representation of political community (nation).

The minister of the treasury and navy under Louis XIV, Jean-Baptiste Colbert, intended territorial engineering for the New Rome to empower the state, not define a nation. Infrastructural work was a way to gain political advantage for the administration vis-à-vis the nobility by indemnifying noble land for state projects and giving the state places to administer. But Colbert's efforts at territorial engineering engendered a shift in political culture, too. France became a collective accomplishment, not just a royal holding.[6] Interestingly, it was not the natural landscape central to romantic

---

[3] Jean Bodin, *Six Books of the Commonwealth* (New York, 1967); and the legal commentary on his work in Roland Mousnier, *The Institutions of France under the Absolute Monarchy, 1598–1789: Society and the State,* trans. Brian Pearce (Chicago, 1979), 645–720.

[4] Peter Burke, *The Fabrication of Louis XIV* (New Haven, Conn., 1992).

[5] Alain Degage, "Le Port de Sète: Proue Méditerreanéenne du Canal de Riquet," in Jean-Denis Bergasse, *Canal du Midi* (Cessenon, France, 1985), 4:265–306, especially, 272.

[6] Chandra Mukerji, *Territorial Ambitions and the Gardens of Versailles* (Cambridge, UK, 1997), 1–38.

forms of national identity in the nineteenth century[7] but rather the built environment in the seventeenth century that associated people in France with the cultural heritage of ancient Gaul.

The Canal du Midi (1662–81) was the most dramatic example of infrastructure engineering in pursuit of the New Rome under Louis XIV, and one later taken as an object of national pride. The waterway cut across Languedoc, connecting the Mediterranean Sea to the Atlantic Ocean through the Garonne River, just above the Pyrenees. It was seemingly impossible by period engineering standards, so it easily came to stand for native genius. In fact, in the eighteenth and nineteenth centuries, its entrepreneur, Pierre-Paul Riquet, came to exemplify French character as an engineer of natural genius (a true man of Gaul).[8]

Louis XIV authorized the Canal du Midi and Colbert nominally oversaw it, but as a technical enterprise, the waterway was fundamentally beyond their political control. Engineering power was not like patrimonial power. Territorial improvements were not so much strategic exercises as logistical ones, in which expert knowledge more than networks of allies or enemies affected the outcome.[9] The Canal du Midi exemplified a new kind of impersonal, material power that stood (sometimes visibly) in contrast to the king's personal rule and the patrimonial politics exercised at court. It was a creation of engineers and thousands of unnamed peasants and artisans—low-status laborers who knew how to work the land. Louis XIV's New Rome could not easily be administered or its powers contained, so its political consequences moved well beyond the king's dreams of empire, fostering a protonational identity around the brilliance of the project.

The Canal du Midi was a project of enormous scale and ambition. It was daunting in size, stretching more than 240 kilometers through Languedoc and crossing the continental divide to join the two seas. The waterway required 100 locks, many of them double or triple locks of oval form, laid out in steps. It also needed a mountain water supply to keep filling the canal at its high point, as water continually drained from the divide down the Atlantic and Mediterranean watersheds toward the two seas. Yet in spite of its difficulty, the Canal du Midi was finished in roughly twenty years. It was made possible by a collaboration of French men and women—mainly northern engineers and southern laborers with Roman skills in their local cultures and hands.[10]

The resulting waterway had an uncanny, inhuman presence in the landscape that mutely testified to the new efficacy of state administration and the powers over nature that could be wielded in the name of the state. Disconcertingly, the Canal du Midi carried boats through mountains and arid regions where water did not normally flow, and it exercised power over local life in the name of a king who lived far away. It

[7] Compare Simon Schama, *Landscape and Memory* (New York, 1995), with Schama, *Embarrassment of Riches* (London, 1988), chap. 1.

[8] Jean-Denis Bergasse, "Le 'culte' de Riquet en Languedoc au XIXe siècle," in Bergasse, *Canal du Midi* (cit. n. 5), 1:217–31.

[9] For an extended description and analysis of patrimonial politics in this time and place, see Sharon Kettering, *Patrons, Brokers, and Clients in Seventeenth-Century France* (New York, 1986).

[10] Michel Adgé, "Chronologie des Principaux Événements de la Construction du Canal (1662–1694)," in Bergasse, *Canal du Midi* (cit. n. 5), 4:176–77; "La première navigation sur le canal du Langudedoc, fait par ordre du Roy pour la jonction des deux Mers, depuis Toulouse jusques au port de Cette" (Toulouse, France, c.1691), reprinted in Arnaud Ramière de Fontanier, "Du Grand Canal au Grand Bassin: Fêtes sur le canal du Midi aux XVIIe et XVIIIe siècles" in *Le Canal du Midi et les Voies Navigables dans le Midi de la France*. Actes du congrès des fédérations historiques languedociennes (Carcassonne, France, 1998), 243–44.

stood for the state, but unlike the king's emissaries, it could not be shouted away or shot to death.[11] Practically, the waterway was designed to allow French ships to avoid Gibraltar, undercutting the power of the Hapsburg empire and Spain; symbolically, it passed through Languedoc roughly where Julius Cesar had crossed Gaul, marking on the ground the historical events that first made France Roman. In terms of power and prestige, it was a perfect enterprise for the New Rome.[12]

The administration did not initiate the Canal du Midi; an entrepreneur, Pierre-Paul Riquet, proposed it to Colbert. In this period, when treasury funds were limited, it was normal for financiers to apply for civil engineering contracts. Riquet appealed to the minister by describing himself as a good Catholic with church patronage and by presenting his project as worthy of a great king and useful to the navy and merchant marine.[13]

Pierre-Paul Riquet was an unlikely candidate to head the project, both as an architect of the New Rome and as an icon of French national identity; he was a rich tax farmer who had money enough to invest in state infrastructure but no engineering skills and little social credibility. What motivated Riquet to assume the work was the desire to increase his fortune and advance his sons. Initially, he did not realize (because of his lack of technical background) what a large enterprise he had taken on.[14] But Pierre-Paul Riquet had one great asset: he recognized the skills of the ordinary people of Languedoc. He might not know how to build a canal, but he could see that knowledge and ability in others. He found among the peasants and artisans in his tax region the New Romans he needed for his canal.[15] Colbert was suspicious of Riquet's reliance on peasant advice;[16] still, the entrepreneur had faith in what they could do. And he came to stand in the historical record—both inappropriately and somewhat justly—for the collective ability of ordinary artisans and laborers: a genius that would define the French.

The prospect of cutting a canal across Languedoc through noble estates stimulated immediate local political opposition. Elites did not want their fields or or-

---

[11] Louis de Froidour, *Les Pyrenées centrales au XVIIe siècle: Lettres par M. de Froidour . . . à M. de Haericourt . . . et à M. de Medon . . . publiées avec des notes par Paul de Casteran* (Auch, France, 1899), 30–31.

[12] See the map by Nicolas de Fer, "Partie de France, et d'Italie par rapport a la Route de Cesar pendant la premiere campagne dans les Gaules," Bancrost Library, University of California, Berkeley.

[13] Riquet à Colbert à Bonrepos, 15 Nov. 1662, 20-2, Archives du Canal du Midi, Toulouse, France (hereafter cited as ACM); "L'Arrêt d'adjudication des ouvrages à faire pour le canal de communication des Mers en Languedoc est promulgué. Ce même jour, le Roi 'fait bail et délivrance à M. de Riquet des ouvrages contenues au Devis,'" préalablement défini sous l'autorité du Chevalier de Clerville, 14 Oct. 1666, 03-10, ACM. L. T. C. Rolt, *From Sea to Sea: The Canal du Midi* (London, 1973), 24–26; P. Burlats-Brun and J.-D. Bergasse, "L'Oligarchie gabelière, soutien financier de Riquet," in Bergasse, *Canal du Midi* (cit. n. 5), 3:123–41; J. Dent, *Crisis in Finance: Crown, Financiers, and Society in Seventeenth-Century France* (New York, 1973).

[14] "L'Arrêt d'adjudication des ouvrages à faire pour le canal de communication des Mers en Languedoc est promulgué" (cit. n. 13), especially, 20–21.

[15] L. Malavialle, "Une Excursion dans la Montagne Noire," *Bulletin de la Société Languedocienne de Géographie* 14 (1891): 7:287, 8:135; Bertrand Gabolde, "Revel: Des eaux du Sor à la rigole de la plaine," in Bergasse, *Canal du Midi* (cit. n. 5), 4:241–44.

[16] A letter from Colbert to Riquet on 12 Dec. 1664 (94-14, ACM) illustrates Riquet's lack of credibility with the minister. Colbert says he is pleased with Riquet's zeal in developing the project, but he must wait for the Chevalier de Clerville, France's highest-ranking engineer, who served Colbert in improving French infrastructure, and in his company, not by himself, come up to Paris to talk about the project. For peasant knowledge and the water supply, see Rolt, *From Sea to Sea* (cit. n. 13), 45–46.

chards indemnified and feared further losses of privilege. Worse, Louis XIV insisted that the provincial governing body, the Etats du Languedoc, help pay for the enterprise. But in spite of opposition to Riquet's canal, making France a New Rome was a program with political appeal even in Languedoc. Colbert successfully hijacked for the state the bases of regional identity. The province had been the center of the former Narbonnaise, a vital area of ancient Gaul. Languedoc was full of classical remains that fascinated educated men of rank. One popular genre of books in the south of France described local cities, comparing maps of the towns in Roman times and the classical monuments that remained in the modern streets.[17] In this historical moment, then, educated elites who opposed the king's canal and monarchical power in their region were still prone to think of themselves as descendents of Gaul and were primed to identify themselves in terms of engineering monuments such as the Canal du Midi.

THE NEW ROME

The promotion of France as the New Rome was a broad campaign that began early in the reign of Louis XIV. It was first used as symbolism for his marriage. The bride and groom arrived in Paris dressed in costumes like those of the ancients, and they were greeted by a city adorned with material symbols of Roman power—including a new triumphal arch serving as a door to the city. The king arrived as a conquering hero with his Hapsburg bride as a token of his accomplishments. The marriage in principle aligned the Hapsburg empire with France and seemed to hold the promise of giving the young king the power to forge or take over an empire. Louis XIV was happy to embrace this program of both politics and propaganda, so the imagery and classical imitation began.[18]

The propaganda theme was reiterated—again materially—at Versailles. The chateau was designed to represent the palace of the sun, using imagery from Ovid;[19] the royal gardens around the palace had their triumphal arches, neoclassical statuary, and friezes that depicted Roman helmets, arrows, and spears. Rows of statues brought from the French Academy in Rome were neoclassical figures representing the elements, the seasons, the continents, and other classificatory themes associated with classical texts. Some of the *bosquets* were set out with islands shaped like the footprints of fortresses derived from the Italian tradition of military architecture based on Vitruvius. And in the center of the garden below the terrace and before the Grand Canal stood the fountain of Apollo, the sun god, representing Louis XIV, rising with his chariot to bring light into the world.[20]

Not only did the chateau and garden at Versailles evoke Rome, so did the ritual life

---

[17] Jean Poldo d'Albenas, *Discours Historial de l'Antiqve et Illvstre Cite de Nismes, En la Gaule Narbonnaise, auecs les portraitz des plus antiques & insignes bastimens dudit lieu, reduitz à leur vraye mesure & proportion, ensemble de l'antique & moderne ville* (Lyons, France, 1560); Henri-Louis Baudrier, *Bibliotheque lyonnaise: Recherché sur les impremeurs, libraries, relieurs et fondeurs de letters de Lyon au XVIe-XVIIe siècles*, 13 vols (Paris, 1964–65), 9:93–102, 268–69.

[18] See, e.g., Nocret's 1670 painting *L'Assemblée des Dieux*, showing the king's whole family as classical gods.

[19] J. P. Neraudau, *L'Olympe du Roi Soleil* (Paris, 1986), 234–37.

[20] Mukerji, *Territorial Ambitions* (cit. n. 6), 39–97.

at court.[21] The *divertissements* at Versailles normally had classical themes. Nobles were costumed as gods and heroes taking roles in plays written by Molière and based on classical themes. They danced in ballets to music by Jean-Baptiste Lully, demonstrating a form of agility associated with military ability, while also enacting classical themes. In these cultural events, nobles echoed the political evocations of Rome that Colbert was mirroring elsewhere in the kingdom.[22]

Nature was needed for the propaganda efforts to make France more Roman because the classical heritage was understood as centrally organized around engineering. The gardens at Versailles were decorated not simply with neoclassical sculpture but also with a massive array of fountains that alluded to the abundant water supply for Rome. In fact, the fountains of the *petit parc* were so numerous that they taxed the local water supply. Their political and aesthetic value lay in their conspicuous waste of water, so an illusion of endless abundance was reproduced by gardeners who turned the fountains on and off in succession during promenades.[23]

This manipulated symbolic play at court could not and did not in itself create a national identity. The participants in these events were courtiers, members of a small, exclusive club. Even the bourgeoisie of Paris who came to witness the divertissements at Versailles were required to stay out of the garden and view the events only from the terrace near the palace. Peasants could be closer, sitting in the trees, because these demonstrations of superiority needed witnesses, but the peasants were not to tread on the ground where their betters performed. Space in the garden nominally defined the kingdom as a whole but practically kept groups apart in these ritual moments, underscoring social differences rather than providing a common ground for national identity.[24] France might have been offered up symbolically as the New Rome in the gardens of Versailles, but that Rome was far from the engineered landscape of the kingdom on which a shared New Rome began to grow.

Material efficacy was important to the formation of a political identity around Gaul, not only because the classical past already had a material presence in the kingdom but also because political life in the period was deeply performative. Power needed to be demonstrated and witnessed and was routinely enacted in court society. Members of the aristocracy in this period did not just name their superiority by rank; they also ritually ratified their high status in spectacles of power such as jousts and divertissements that placed them in narratives as gods and heroes.

Becoming Roman again also required performance, but ritual enactment was not enough. Making France Roman put engineering in the center of the action. Links to Rome had to be demonstrated with grand projects of infrastructural development, not simply claimed in propaganda. So, Colbert engaged in multiple projects of territorial improvement in the classical mold, helping to make the New Rome a credible object of collective imagination and defining a community of New Romans around collective action on the land. The Canal du Midi was among the most dramatic of these projects, and one realized in arguably the most dissident province of the kingdom.

---

[21] Ibid., 257–58, 297–99.

[22] Hélène Vérin, "Technology in the Park: Engineers and Gardeners in Seventeenth-Century France," in *The Architecture of Western Gardens*, ed. Monique Mosser and George Teyssot (Cambridge, Mass., 1991), 135–46.

[23] L. A. Barbet, *Les Grandes Eaux de Versailles: Installations méchaniques et éstangs artificiels, descriptions des fontaines et de leurs origines* (Paris, 1907), 31–48.

[24] Mukerji, *Territorial Ambitions* (cit. n. 6), 203–19, especially, 207.

Thus, its story provides a good basis for understanding how French national identity was conjured up through infrastructural engineering that began a slow political process of detaching political identity from princes and locating it in places that embodied a heritage defining a people.

ENGINEERING THE NEW ROME AND THE CANAL DU MIDI

Engineering the Canal du Midi required and brought into existence a collaboration that defined for future generations what the French could do. The project would not have existed without the state. The mountain peasants from the Pyrenees who came as seasonal laborers to the valleys of Languedoc and the educated engineers from northern France whom Colbert asked to assess and supervise the work had no reason to interact until the minister and Riquet brought them together. In this sense, the Canal du Midi was a political product, an artifact of the state not only because it was an important piece of infrastructure but also because it was the product of a social process that only a state could initiate.

Those who worked on the Canal du Midi bridged enormous social divides. The military engineers on the waterway were generally from the north, formally educated, good Catholics who spoke French. The highest ranking of them were even gentlemen with courtly manners and patrimonial connections to Versailles. In contrast, the people of Languedoc who worked on the canal were mainly illiterate peasants or artisans who spoke the langue d'oc, or Occitan, not French. Many were women; even more were probably Huguenots. They also lived in a region economically and psychologically organized around the Mediterranean world, not around the Atlantic economy that preoccupied those in the north. How people from such different social worlds communicated or accomplished so much together is difficult to know, but nonetheless they made this "impossible" canal possible after all.

The Canal du Midi was designed from the beginning through collaboration. Riquet worked with a local peasant, Pierre Campmas, to design the water supply, understanding full well that without such help the canal would not work.[25] Once Riquet submitted the plan to Colbert, the minister appointed a commission to study it, led by France's highest-ranking military engineer, the Chevalier de Clerville. The rest of the commission consisted of another broad mix of people: local notables, a set of experts in surveying and construction, and lower-status men of experience. In the end, the enterprise depended on an unusual community of practitioners—noble soldiers, mountain peasants, seasoned military engineers, artisans, and financiers—who learned to dream in common about the restoration of Gaul.

STRUCTURAL ENGINEERING IN THE CLASSICAL TRADITION

Ironically, a great deal of the engineering used for the Canal du Midi was in fact derived from classical principles but not recognized as such. The military engineers clearly knew their Vitruvius,[26] but they also worked with artisans in the building trades who carried traditions of classical construction in their hands. Together, the

---

[25] Chandra Mukerji, "Entrepreneurialism, Land Management, and Cartography during the Age of Louis XIV," in *Merchants and Marvels*, ed. Paula Findlen and Pamela Smith (New York, 2002), 248–76.

[26] Vitruvius, *The Ten Books on Architecture*, trans. M. H. Morgan (New York, 1960).

two groups helped build the novel structures needed for the waterway, such as the dam for the water supply, a port along the Garonne River, oval lock staircases for carrying ships up hills, and overflows for the waterway. The builders not only drew on classical techniques of structural engineering but also helped demonstrate the value of this heritage for engineering structures that none of them knew in advance how to design.

Informal traditions of practice in the building trades had helped keep classical methods alive in the region. The landscape contained remains of classical engineering that served as primers for artisans. The region of Languedoc had an abundance of classical ruins. Many structures had been abandoned after the fall of Rome; some had been torn down during the Middle Ages to build new churches and towns. But even this destruction did not erase the classical past. Pulling apart ancient buildings taught lessons about classical building methods; it was a dissection process as well as a form of destruction.[27] And to the extent that locals used stones from ruins to reduce the need for new quarrying and stonecutting, they were also encouraged to reproduce classical construction methods. The stones and bricks extracted from antique buildings were easier to reuse with comparable building methods than to recut for different types of structures. The result was a robust tradition of construction based on Roman methods that kept classical techniques in circulation.[28]

As Elting Morison has argued,[29] Roman construction could be reproduced long after the end of the ancient empire in part because it was efficient and made sense to builders. For example, Roman buildings were generally constructed with walls made of two exterior shells of stone or brick masonry with cement, filled with a mixture of cement and rubble. When the cement dried, the result was a strong structure that did not depend on precision stonecutting and good contact between surfaces.[30] The mortar was made with lime or calcium oxide mixed with sand and water, the calcium oxide coming from sandstone or marble heated in hot furnaces until it turned soft. This lime could be produced from chips from quarrying, making the method economical as well as efficient.

Walls were also held together with special tricks. The corners on buildings were made with rectangular stones or bricks alternating between the two walls. They created a sequence of interspersed "fingers" that helped secure the two faces of the building. On arches, some of the stones forming the interior of the vault were also made longer, creating a pattern of radiating spokes inside the structure that helped connect the walls above to the arches themselves. Long keystones on the top of arches also worked to secure the superstructure to the arches.[31]

The early collaboration between military engineers and artisans produced a capacity for structural engineering that surpassed the traditions of both groups, yielding a social intelligence that was innovative as well as an assemblage of Roman methods.

---

[27] Pierre Aupert, Raymond Monturet, and Christine Dieulafait, *Saint-Bertrand-de-Comminges: Les thermes de forum* (Pessac, 2001), especially, 203–305.

[28] Cf. Pamela O. Long, *Openness, Secrecy, Authorship: Technical Arts and the Culture of Knowledge from Antiquity to the Renaissance* (Baltimore, 2001), 72–101.

[29] Elting Morison, *From Know-How to Nowhere* (New York, 1974).

[30] Jacques Heyman, *The Stone Skeleton: Structural Engineering of Masonry Architecture* (New York, 1995), 12–26.

[31] Gustavo Giovannoni, "Building and Engineering," in *The Legacy of Rome*, ed. Cyril Bailey (1923; repr., Oxford, 1940), 429–74.

The result was not only a transformed landscape but also an engineering ability that was a product of the state that would become a defining quality of the French nation.

HYDRAULIC ENGINEERING AND INDIGENOUS PRACTICES

Although essential to the enterprise, the structural engineering for the Canal du Midi was not enough to ensure a working canal. Building the waterway required knowledge of hydraulics that neither military nor civil engineers had and that even academically trained experts did not possess. To move water through the landscape, the canal's designers required a surprising source of local knowledge: peasants. These low-status laborers routinely worked the land, often used waterworks for agriculture and other purposes, and carried their indigenous traditions of classically based hydraulics.

Hydraulics existed in the 1660s, but most of the literature addressed problems of mining and the design of pumps. Canals were used for irrigation and drainage projects as well as for navigation, but the literature on canals was scarce. They were peasant projects, not worthy of analytic attention. Military engineers wrote on hydraulics after the Dutch flooded low-lying areas around their cities with seawater during sieges, drowning their attackers.[32] But this literature appeared in the eighteenth century and did not provide models of how to proceed in Languedoc.

In the 1660s, there also existed no reliable calculations for determining how much water was needed for a navigation canal with so many locks or for estimating water losses due to splashing, evaporation, or seepage into sandy soils. It was not clear how many reservoirs or holding tanks one would need to hold to ensure an adequate supply during the long, dry summers of southwestern France.[33] Most of all, it was not clear how to run water through the mountains, either for the water supply at the continental divide or for the canal proper in the mountainous area approaching the Mediterranean coast near Béziers.

The people who knew the most about hydraulic engineering in uneven terrain were peasant women who lived in towns in the vicinity of former Roman bath colonies in the Pyrenees. Some of the classical infrastructure for the baths remained in use into the seventeenth century, spas frequented by nobles with health problems.[34] But in

[32] Giorgius Agricola, *De re metallica*, trans. Herbert Hoover (New York, 1950), 149, 166, 172–81, 190–96; Joseph de Lalande, *Des canaux de navigation, et spécialement du canal de Languedoc* (Paris, 1778), 26–29; Belidor, "De l'Usage des Eaux a la Guerre," in *Architecture Hydraulique seconde partie qui comprend l'Art de diriger les eaux de la Mer & des Rivieres à l'avantage de la défense des places, du Commerce & de l'Agriculture par M. Belidor, Colonel d'Infanterie . . .* (Paris, 1753), vol. 2, especially, 233.

[33] Andreossy and the commission did such studies, but they were not considered really trustworthy. That is why Riquet was required to make a *rigole d'essai* for the water supply to prove that enough water could be brought from Mt. Noire to the main canal to keep refilling the canal where it crossed the continental divide. "Relation particuliere de la rigolle dessay," 02-14, ACM; Rolt, *From Sea to Sea* (cit. n. 13), 35–37; Louis de Froidour, *Lettre à M. Barrillon Damoncourt contenant la relation & la description des travaux qui se font en Languedoc pour la communication des deux mers* (Toulouse, France, 1672), 9–10; François Gazelle, "Riquet et les eaux de la Montagne Noire: L'idée géniale de l'alimentation du canal," in Bergasse, *Canal du Midi* (cit. n. 5), 4:143–67; and Michel Adgé, "Chronologie des Principaux Événements" (cit. n. 10), 176–77.

[34] Louis de Froidour, *Mémoire du Pays et des États de Bigorre,* introduction and notes by Jean Boudette (Paris, 1892), 20–22. For the survey of French mineral waters ordered by Colbert and Riquet's role in this project, see Lettre de Colbert à Riquet, 12 Jan. 1670, 23-3, ACM; and his letter of 15 Feb. 1670, 23-11, ACM.

many towns of the Pyrenees, the baths lost their usefulness, and locals applied classical principles of hydraulics to new problems, producing a living tradition of water engineering with Roman provenance. The sophistication of the tradition was described
in detail by Colbert's forestry expert, Louis de Froidour.[35] Women from these towns
worked on the systems in the summer when the men were away with their flocks of
sheep and other livestock in the high meadows, doing the essential job of making
cheese for winter. Women also sought seasonal employment as agricultural laborers
on either side of the mountains, so they were a good target for Riquet's recruitment
drives,[36] and they appeared in large numbers, first to work on the dam, hauling dirt
up the mountain to fill the voids between the masonry walls. Although hired for unskilled work, they nonetheless demonstrated skills in hydraulics that proved essential
for finishing the water supply and, later, for taking the canal through the mountains
by Béziers to the sea.[37]

The women knew methods for controlling wild rivers, working with rocky soil,
and keeping debris out of a water supply. They could measure and maintain water
volumes in a set of locks, control silt, manage drainage, tap water sources, and in
other ways negotiate meeting places of water and land. Most important for getting the
Canal du Midi through uneven terrain, the women could cut contours or maintain the
incline of the canal in complex topography.[38] Romans had controlled inclines for their
aqueducts, using gravity alone to bring fresh water to their cities and towns. This was
apparently a skill learned and reproduced over the centuries by Pyrenean peasants
who could make use of the abundance of water in the mountains to serve their communities.[39] After the fall of Rome, these once-formal techniques were maintained in
the mountains, but their provenance was lost.[40]

Threading the canal through the mountains while keeping the water flowing with
a gentle incline was a particular problem for engineers in the seventeenth century
because there were limited techniques for calculating elevation with precision. Elevation measures were so misleading that they were even officially abandoned for
the design of lock staircases, in which poor calculations resulted in locks containing
different volumes of water.[41] Working with subtle shifts in elevation was hard enough
in any case, but maintaining the incline became even harder when surveyors could
not see around the next set of hills. Experience with the terrain and knowledge of the
trails that ran through it provided better clues about how to take a channel unerringly

[35] Chandra Mukerji, "Women Engineers and the Culture of the Pyrenees," in *Knowledge and Its Making in the Early Modern World*, ed. Pamela Smith and Benjamin Schmidt (Chicago, 2008), 19–44;
Froidour, *Mémoire du Pays et des États de Bigorre* (cit. n. 34), 20–21.

[36] On October 20, 1668, Riquet wrote to Colbert that he was staying on in Perpignan to recruit
large numbers of workers for the canal; 21-18, ACM. He wrote later about finding them in Bigorre,
30-65, ACM.

[37] Women of the Pyrenees and their seasonal migrations in work are described in Isaure Gratacos,
Femmes Pyrénéennes, un statut social exceptionnel en Europe (Toulouse, France, 2003), 105–15. For
women working on the canal, see Rolt, *From Sea to Sea* (cit. n. 13), 89; 13-3, ACM, and the accounts
from the Somail region, 1071, ACM, and 1072, ACM.

[38] Mukerji, "Women Engineers" (cit. n. 35); and for descriptions of the water systems, see Froidour,
Mémoire du Pays et des États de Bigorre (cit. n. 34), 20–21.

[39] Lesley Adkins and Roy Adkins, *Handbook to Life in Ancient Rome* (New York, 1994), 135–36;
Mukerji, "Women Engineers" (cit. n. 35); 13-3, 1071, and 1072, ACM.

[40] Mukerji, "Women Engineers" (cit. n. 35); Froidour, *Mémoire du Pays et des États de Bigorre* (cit.
n. 34), 20–21.

[41] Riquet à Colbert, 24 Dec. 1669, 22-38, ACM; "Memoire, 1 March 1672," 5, 13-15, ACM.

down the topographically complex Mediterranean watershed. Indigenous women engineers were disproportionately used in the mountains, presumably because they provided a rich set of skills for doing this work based on hydraulic methods used in the Pyrenees for Roman baths.

Still, by bringing classical methods of hydraulics on site, women laborers could do this work and add to the range of skills with classical provenance available to the state for building the New Rome. In this way, they made France seem more Roman and a construct of the people, not just a dream of the king.

COLLECTIVE REPRESENTATION AND COLLABORATIVE PRACTICES

Making the canal worthy of Rome required political work as well as engineering. The evocation of the ancients began at the opening ceremony for the canal near Toulouse in the fall of 1667. The intendant of Languedoc at that time, M. Bezons, organized the event to celebrate the successful indemnification of the first parcels of land for the canal. Père Matthieu de Mourgues, *inspecteur du canal*, described the event with these words:

> On the 17th of November, the notables of the town, the ancient and new Capitouls [city officials] dressed in red and black, the clergy and Parliament in grand attire with all the attributes of their rank, paraded to the walls of the capital city of Languedoc. They met there the workers for the canal, without whom nothing would have been accomplished. The *chefs d'atelier* [stood] in front of their brigades of workers; there were close to six thousand *terrassiers* set out in battle order, drums beating. It was a powerful sight: the [notables of the town were] lined up behind the cross [parading] along the still-dry basin of the canal. The procession of authorities . . . flowed like a current into the middle of the enthusiastic crowd massed on the open banks of the channel, the people of Toulouse and workers mixed together.[42]

The New Romans marched like soldiers and were assembled in brigades. They consisted of laborers as well as their supervisors, working in tandem. The canal itself was described in Mourgues's account as an arena like those of the ancients, where nobles and peasants assembled to bless and inaugurate this start of the New Rome:

> They shouted out cries of joy, "Vive le Roy," or in the words of the author of the *Annales de Toulouse*, "[they] formed a kind of amphitheater and provided a sense of the spectacles of the ancient Romans." [This] perfect assembly of the powerful and honorable members of society . . . came to the place for the foundation of the locks [to link the canal to the Garonne]. The archbishop of Toulouse took the first two stones in his hand. He blessed them, giving one to the president of Parliament [the legal system that constitutes part of the importance of Toulouse] and the second to the Capitouls. A little mortar was taken with a trowel of gold from a silver plate, and the stones were placed. To the joy of the people, commemorative medals were thrown into the crowd, and Riquet had wine and liquor distributed, as the artillery massed on the banks of the Garonne fired as did

---

[42] Matthieu de Mourgues, *Relation de la seconde navigation du Canal Royal* (1683), in Phillippe Delvit, "Un Canal au Midi," in *Canal royal de Languedoc: Le partage des eaux*, ed. Conseil d'Architecture, d'Urbanisme et de l'Environnement de la Haute-Garonne (Caue, France, 1992), 204–24, especially, 205–6. Mourgues's position as *inspecteur du canal* when he wrote this "relation" is described by François de Dainville, *Cartes Anciennes du Languedoc XVIe-XVIIIe Siècle* (Montpellier, France, 1961), 53.

the musketeers. God was present. In this ceremony marking the opening of the canal, it was the 17th of November, but it was like a spring day, which people took as a good sign for the project.[43]

The event was appropriately commemorated with plaques written in both Latin and French, equating the accomplishments of the moderns with those of the ancients.[44]

The ritual efforts to celebrate the Canal du Midi were in part a response to the opposition to the waterway and the technical problems that were plaguing early engineering in Languedoc. There was little progress and many embarrassing failures in 1667–70. Near Toulouse, long sections of the ditch were dug up and stood empty, cutting across roads and orchards or standing between fields and a mill. In rainy periods, the walls would erode, and mud accumulated on the bottom. The canal excavations had to wait because the water supply was not ready and the locks needed to carry the canal across the continental divide were falling down.[45] Dreams of the New Rome, in this context, were sustained more in compensatory propaganda than in the engineering.

If the triumph of the New Romans was slow in coming in the 1660s, the work sped up in the 1670s. Part of the canal near Toulouse was opened to traffic in 1670, proving to be particularly useful to the dyers of the city.[46] Riquet wrote to Colbert on the 27th of March of that year, expressing his growing confidence in his own abilities and the canal's progress:

> God has inspired me with thoughts that I call sacred, since with them my works will be advanced, and I will not lose [my way]: it is in truth, Sir, a philosopher's stone which gives to all and misses no one. . . . [F]rom dreams, I've found such miraculous expedients for the advancement of the canal.[47]

Riquet spoke of a philosopher's stone—an object that could make everyone smart. This aptly described the character of the collaborative work on the canal and the miraculous process by which ordinary men and women could together achieve extraordinary things.

In spite of the collaborative nature of the work, Riquet began to position himself in this letter, and in many others, as the sole author of the canal. He imagined he was a solitary figure who dreamed up all the innovations himself. Riquet's self-conception as a lonely hero was clearly not sustained by the work in Languedoc, but his surprise

---

[43] Mourgues, *Relation de la seconde navigation* (cit. n. 42), 206.

[44] The inscription was a list of the august individuals associated with the presumed great work. These included "Pierre de Riquet, INVENTEUR de ce gd. Ouvrage." Quoted in "Notes sur l'Histoire du Canal de Languedoc par les Descendents de Riquet," 1803, 66, uncataloged papers, ACM.

[45] "Memoire de mes remarques du Canal M. de Seguelay," Oct. 1670, 13-05, ACM. For Riquet's critics and Colbert, see Murat, "Colbert et Riquet," in Bergasse, *Canal du Midi* (cit. n. 5), 3:105–22.

[46] For the profits of the dyers, see Lettre de Riquet à Colbert, 14 Jan. 1668, 21-1, ACM. In a letter from Bordeaux in May 1669, Clerville offered to come look at whatever Riquet wanted him to see and asked the entrepreneur to pay his expenses; 31-25, ACM. One of the verifications he did make after M. d'Aguesseau became intendant for Languedoc in 1673 is a detailed inventory of all the major worksites, commentary on their status given to M. d'Aguesseau by Riquet, and Clerville's review of the work completed in these areas (13-12, ACM). For an example of a different sort of verification, less like an account book, there is a 1672 review of work on the water supply system. New and deeper *rigoles* were being dug then, and some of the connections to rivulets and rivers were being fixed. Drains were being installed to get rid of excess water and return it for use by locals; 13-07, ACM.

[47] Riquet to Colbert, 27 March 1670, 23-16, ACM.

and pleasure at his newfound abilities were in fact real effects of collaborative work; the interactions among diverse participants stimulated all of them to think in new ways and made questions of attribution problematic.

Colbert not only did not believe Riquet's claims to personal powers—particularly in light of the technical problems and mistakes—but also feared Riquet's hubris in attributing his successes to God rather than to the king. He wrote a chilling letter in 1677 to the newly appointed intendant of Languedoc, d'Aguesseau, expressing his views of the entrepreneur:

> Although it might be best to treat [Riquet] as ill, we must, nevertheless, apply ourselves with care in order that the course and strength of his imaginings does not bring on us a final and grievous end of all his works. . . . This man does as great liars do who, after telling a story three or four times, persuade themselves it is true. . . . It has been said to him so many times, even in my own presence, that he is the inventor of this great work that in the end he has believed he is in fact the absolute author. And, on the greatness of the work he has founded [the idea of] the grandeur of the service he renders to the state and the greatness of his fortune. It is for this that he has made his son a Master of Requests, and which has given to his spirit, regarding the establishment of his children, a vast career and an inflation which hasn't any proportion or relationship with what he is or with what he has done.[48]

Rather than the hero he wanted to be, Riquet was a "visionary" in the derogative sense used in the period.[49] He was delusional in his failure to understand his place in patrimonial politics, recognizing neither his low standing nor his obligations to the king.

The minister was also frustrated with the slow progress and high costs of the canal because they threatened his own career. Again he wrote d'Aguesseau in 1679:

> It has been a long time since I wrote you on the subject of the canal for the communication of the seas and port of Cette even though there have often been letters from you on the subject. I admit that often the embarrassment of work stops one from giving everything equal application, and as I see, from your last letters and frequent ones from the sieur Riquet, the work that he owes is not advancing due to lack of funds although he is almost completely paid [what is owed]. . . . I swear that the end of this affair is annoying me for the reason that I have never seen enough resolve in Riquet's spirit to come out well from such a weighty affair as this.[50]

The New Rome was not the grand success it was supposed to be. In 1679, the Canal du Midi was an expensive burden on the state, and Riquet was perceived as a dangerous crazy man. Both were threats to the administration. Still, Colbert needed to make the king's New Rome a reality, so he found ways to try to get the job done. This entailed both technical support and political work that multiplied in 1680, when Riquet died before the canal was completed.

Colbert had been increasingly worried in the late 1670s about the slow pace of the work, in part because of Riquet's deteriorating health. He believed that, if the canal

---

[48] Letter from Colbert to d'Aguesseau in 1677, quoted in Rolt, *From Sea to Sea* (cit. n. 13), 91–92.

[49] *Avis a messieurs les Capitouls de la Ville de Tolose, par Arquier, Doyen de anciens Capitouls. Et Response a cet Avis, Article par Article par Jean de Nivelle, Ancien Capitaine Chasseauant du Canal dans l'atelier de Mr. Sagadenes,* 1667, 01-16, ACM.

[50] Lettre de Colbert à d'Aguesseau, 6 Sept. 1679, in Pierre Clément, *Lettres, Instructions et Mémoires de Colbert* (1867; Nendeln, Liechtenstein, 1979), 4:386.

was not in working order when Riquet died, the local nobles who had opposed the canal could finally stop the project. The Canal de Briare had been stalled under similar circumstances, and Colbert did not want to face the same outcome for the Canal du Midi.[51] Colbert was also sure that Riquet had been stealing treasury money and that his graft would be revealed on close examination of his estate, giving opponents of the project reasons to stop it.[52] The only way to protect the canal was to make it too great a political asset in the public imagination to be assailed locally. So, Colbert and d'Aguesseau began a campaign to make the waterway a public marvel and make Riquet a hero for building it.

The first steps were modest, but crucial. As Riquet's son arrived in Languedoc to take up the work his father had left undone, Colbert emphasized to d'Aguesseau the importance of asserting to local leaders the king's ongoing support for the enterprise and the family responsible for it. Colbert wrote:

> M. de Bonrepos, *maistre des requestes*, [Riquet's son] having left for Languedoc for the continuation of the work on the canal for the communication of the seas, I pray you to give him all the assistance that you can to achieve this grand enterprise and to let it be known in all the province that the king still honors his family with his protection, it being very important and necessary that all the province, convinced of this truth, provide the sieur de Bonrepos and his family, including the sieur Pouget and all those who are involved in their affairs, the credit and help that are necessary to finish the grand enterprise that the deceased sieur Riquet made, and in the execution of which he died.[53]

The king also forgave Riquet's debt to the treasury for taxes he had contracted to pay but failed to deliver before his death. This meant that the Etats du Languedoc was left with no excuse for withholding funds for the Canal du Midi that came from the treasury. It helped, too, that Riquet never did defraud the treasury. This made it easier than Colbert expected to represent the entrepreneur on his death as an honest man who had given his life for a project of great worth to the king and kingdom.[54]

The Canal du Midi itself also was quickly endowed with political dignity. The minister wanted d'Aguesseau to place a celebratory description of the Canal du Midi in the *Mercure*, a French publication that circulated widely, to gain some international interest in it, and he pushed for the canal to be officially opened as soon as possible. Early in 1681, "the king sent an order to the sieur d'Aguesseau, *intendant* of Languedoc, to inspect the canal dry, begin to fill it with water, and try the first navigation."[55] Verification of the dry channel began in May 1681. The canal was deemed ready, empty parts of the channel were flooded, and the inauguration of the canal took place starting in Castelnaudery on the 19th of May, 1681—less than a year after Riquet's death. Even though the canal still needed work after the navigation and was drained in part to make repairs, the Canal du Midi nonetheless had carried boats from one end to the other. That made it ready to become an icon of the New Rome.

[51] Pierre Pinsseau, *Le Canal Henri IV, ou Canal de Briare. 1604–1943* (Orléans, 1944), 98–105.

[52] Lettre de d'Aguesseau à Colbert, 5 Oct. 1680, and de d'Aguesseau à Colbert, 19 Oct. 1680, in Clément, *Lettres* (cit. n. 50), 4:598–99.

[53] Lettre de Colbert à d'Aguesseau, 1 Oct. 1680, in Clément, *Lettres* (cit. n. 50), 4:389–90.

[54] Lettre de d'Aguesseau à Colbert 5 Oct. 1680, and de d'Aguesseau à Colbert 19 Oct. 1680 (cit. n. 52), 4:598–99.

[55] "La première navigation sur le canal du Langudedoc" (cit. n. 10).

Figure 1. *Carte du Canal Royal de Communication des Mers en Languedoc, 1681. (Archives Historiques de la Marine, Vincennes.)*

In 1681, François Andreossy represented this visually on an official map he made of the Canal du Midi. The map was drawn from a bird's-eye view, with the canal and the province of Languedoc far below the clouds and dwarfed by the king, represented as Apollo. He dominated the heavens, pointing with his finger at the canal below or drawing it with his hand. This was the monarch's New Rome—a direct artifact of his power.

Andreossy's 1681 map of the Canal du Midi presented the waterway as a superhuman accomplishment, something far too grand for political squabbling or earthly concerns. Its technical faults were invisible from this vantage point. The waterway connected the two seas almost without technical measures. It was a product of the heavenly powers of the king.

To make the canal on the ground fit this heroic political imagery, the famous military engineer Vauban was sent to Languedoc in 1684 to correct the problems that still plagued the project. He brought the dignity of his lofty reputation to the work and, in praising the canal, placed the quality of the engineering beyond dispute—if not beyond fault. The waterway was officially a marvel of France's New Rome, an object of admiration for people all over Europe. And with the repairs made, it did indeed become a reliable form of transport through the region.

By the 1690s, the Canal du Midi was being portrayed as a jewel of the province as well as an icon of Rome. A map by Nolin made in 1697 intertwined images of technical finesse and patrimonial power. The map was framed doubly along its borders

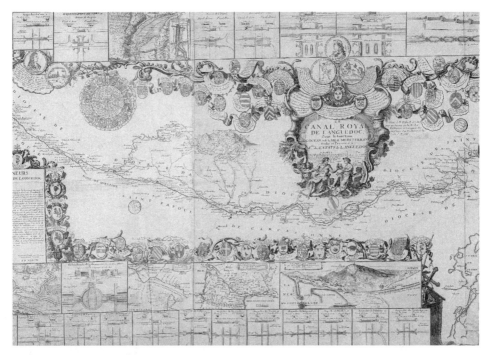

Figure 2. *Left side of the Canal Royal de Languedoc by Nolin, 1697. (Archives Historiques de la Marine, Vincennes.)*

with insets cataloging the major engineering features and with the coats of arms of Languedoc's noble families attached with ribbons, laurels, and medals to the king. Louis XIV himself was shown in the top center of the map in profile on a coin, like a Roman emperor. On Nolin's map, then, patrimonial politics and the New Rome were visually aligned, indirectly crediting the provincial nobility with the success of the project. The Canal du Midi might have been willed by the king, but it also stood for a region, a people—New Romans.[56]

The maps by Andreossy and Nolin of the Canal du Midi depicted the waterway as a work of Roman-style efficacy, embodying a greatness coded as masculine, military, and noble. Books from Belidor, Lalande, Vallancey, and others echoed this view, tracing the technology back to Rome and providing a cultural profile for New Romans.

> The military Genius of the Romans made them trouble very little about Commerce, because all the Nations they had conquered traded to the profit of the Conquerors, who thus enjoyed every Thing useful and agreeable that the World produced; yet [the empire] produced a great number of magnificent Works [to uphold] its Advantage; how many Ports did they improve on all the Coasts of the Empire? On the other hand, the grand Bridges they made were not less designed for Commerce than for the Passage of the Armies and

---

[56] In the eighteenth century, this program of fostering regional pride in the canal was so successful that the États du Languedoc even considered taking the waterway from the Riquet family, claiming it as an asset of the province. "Histoire du Canal par les descendents de Riquet," 162. Unpublished typescript of the original. Uncataloged papers, ACM.

warlike Stores; nor did they only make those great Roads for the Communication of all Parts of their vast Empire, but also rendered most of the Rivers navigable, and joined them by Canals.[57]

Writings on engineering that carried celebratory descriptions of the Canal du Midi often started with stories of the Romans, placing the canal in a genealogy of technique that presented it as a natural accomplishment of descendents of Gaul. Responsibility for the success of the canal was at moments ceremoniously assigned to the king—much as in Andreossy's map. But books that tried to trace the genealogy of canal building and hydraulics also began to detach the technical story of the New Rome from the power of the king. The authors wrote about the intelligence of the canal builders that allowed them to surpass both the work of the ancients and the canals of the moderns in the Netherlands and Italy.

By the eighteenth century, Riquet and Vauban were viewed as models of French ingenuity. In some tellings of the history of the Canal du Midi, Riquet was heralded as the lone genius he always claimed to be.[58] In other accounts, Vauban was given more credit for the canal, even though most of the improvements he made were based on precedents from earlier work on the canal itself. As a scholar and warrior, he was an appropriate descendent of the ancients.

> [T]hus *Vauban* had the honour of bringing this Canal to perfection. A Canal which all the World acknowledges to be the greatest piece of Hydraulic Architecture, that ever was undertaken, and which is of infinite Consequence to the finest Princes in France, thro' which a great Trade is carried on from Sea to Sea.[59]

Over time, Riquet became the more important historical figure. His story of personal authorship and struggles against local nobles resonated with readers of the eighteenth and nineteenth centuries, who saw in him an ordinary Frenchman with extraordinary abilities. For some authors, such as Lalande, who studied the canal through the entrepreneur's letters, Riquet was a man of science in an age of absolutism, a man who was misunderstood but nonetheless persevered to realize his work of genius. In this way, the entrepreneur came to stand for the very people his legend erased—the artisans and peasants so central to the enterprise.

Through maps and writings, then, the Canal du Midi was raised in cultural stature from a muddy scar in the landscape of Languedoc to a triumph of French engineering. Artisans and women laborers who were not lofty enough to fit the king's New Rome disappeared from these histories, and did not reappear even after the French Revolution. Riquet was sanctified as a native French genius; the collaborative abilities of those who worked on the waterway simply became his.[60] Riquet became a national hero and an icon of national identity in the nineteenth century—demonstrating the historical connection between the New Rome and French nation building.[61]

---

[57] Belidor, translated and quoted in Charles Vallancey, *A Treatise on Inland Navigation . . .* (Dublin, 1763), 100.

[58] Ibid., 109.

[59] Ibid., 116–17.

[60] Chandra Mukerji, "Cultural Genealogy: Method for a Cultural Sociology of History or Historical Sociology of Culture," *Cultural Sociology* 1, no. 1 (2007): 49–71.

[61] Bergasse, "Le 'culte' de Riquet en Languedoc au XIXe siècle" (cit. n. 8), 1:217–31.

CONCLUSION

The emergence of a national identity in France has been viewed as a product of political maturation and the growth of mass communication in the nineteenth century. But
there was also an itinerary of the imagination that led to French national identity. The
kingdom itself first had to become Roman before it could be imagined as a political
site defined less by the king than by his people. National identity linked places and
persons in political terms and had roots in the infrastructural engineering of the seventeenth century, such as the Canal du Midi. Territorial engineering demonstrated the
New Romans to be "true" descendents of ancient Gaul, and thus they were described
in engineering books. The New Rome may have begun as masculine, military, and
a dream of Louis XIV—appropriate to patrimonial politics. But imagery of Rome
could serve republican longings, too, so the New Romans could become French, capable of taking on the mantle of Gaul and becoming historical actors of consequence
on the world stage.

Projections of the Revolutionary Nation:
French Expeditions in the Pacific, 1791–1803

By Carol E. Harrison[*]

ABSTRACT

Revolutionary France's two Pacific expeditions, under the command of Jean-Antoine Bruny d'Entrecasteaux (1792–94) and Nicolas Baudin (1801–1804), demonstrate the importance of scientific inquiry to the newly sovereign nation. France's scientific community adapted to the changed circumstances of revolutionary upheaval by describing its work in terms of national priorities. Individuals on board the expeditions, both naval and scientific personnel, behaved as scientific citizens, intent on composing an encyclopedic body of knowledge about the Pacific. Disputes over whose science mattered more and how credit should be assigned through publication, however, broke down the consensus that science should be a national project.

INTRODUCTION

On January 2, 1791, the naturalist and deputy Louis-Augustin Bosc d'Antic presented the French National Assembly with a petition from the Society for Natural History, of which he was a member, calling for a dramatic rescue of Frenchmen left stranded, Robinson Crusoe–like, in the Pacific. The sailors in question had left France with Jean-François de Galaup, comte de La Pérouse, in 1785, placing "their lives in danger for the service of their *patrie* and the benefit of all peoples." Even now, the petition claimed, La Pérouse might be "on an island in the South Seas, stretching his arms out toward his *patrie*, whence he hopes . . . for a liberator." Saddest of all was that La Pérouse and his men knew nothing of "the astonishing revolution that has regenerated this empire." The petition continued, "If only he could know of your laws, if he could know that the French nation has reclaimed its sovereign rights, imagine his anguish at missing these fortunate changes." The petitioners were confident, however, that La Pérouse would not be abandoned. The composition of France's new National Assembly, which, if La Pérouse only knew, included "those names most dear to literature, to philosophy, and to humanity," would reassure him: men dedicated to the pursuit of knowledge now governed France. The "generous nation" that sent sailors

[*]Department of History, University of South Carolina, Gambrell Hall, Columbia, SC 29208; ceharris @mailbox.sc.edu.

Much of the research for this essay was conducted in collaboration with Danielle Clode, Department of Zoology, University of Melbourne, and with the support of the University of South Carolina and the History Department of the University of Melbourne. In addition, I thank the history faculty of the University of Adelaide, who heard an early version of this paper, and Tom Brown, Ann Johnson, and *Osiris*'s anonymous readers.

© 2009 by The History of Science Society. All rights reserved. 0369-7827/09/2009-0003$10.00

on scientific and humanitarian missions would not abandon its "brothers" to "the fury of the seas or the rage of the cannibals."[1]

The members of the Society for Natural History were right; the regenerated nation would not forget its lost brothers or the "philosophical navigator" who had led them into uncharted regions of the Pacific. Instead of marking the end of Old Regime France's interest in scientific exploration, La Pérouse's disappearance led revolutionary France to expand the program of Pacific discovery. The National Assembly's committee report announced that the naturalists' petition had "torn the veils from the eyes" of ordinary citizens and brought the cries for help of La Pérouse and his men into the very chamber of the assembly. Naturally, France would send a rescue mission; the National Assembly would "show the world the great value [Frenchmen] place on those who devote their talents to [France]. . . . By the concern that France shows for her children, the French will recognize what it is to have a *patrie* and they will dedicate themselves even more fully to its service. . . . This national gratitude, this offering to humanity will inspire heroism and acts of civic virtue."[2] The assembly applauded the report enthusiastically, including the estimate that a two-year voyage would cost 700,000 livres, and voted to carry out the plan the same day. Expeditionary science went from being a project of royal absolutism to a statement of the philosophical ambitions of the revolutionary nation.

The National Assembly's emotional and enthusiastic embrace of La Pérouse was surely France, in Greg Dening's phrase, "acting out [its] scientific, humanistic [self]" against a Pacific backdrop.[3] La Pérouse's expedition had been a project of the Old Regime, closely connected to Louis XVI's personal interest in the Pacific.[4] Finding La Pérouse and, especially, pursuing his encyclopedic program of Pacific discovery, however, were easily nationalized and quickly became goals of the revolutionary state. In the midst of the political, social, and military upheaval of the revolutionary years, France sent two major voyages of discovery to the south Pacific. The first, under the command of Rear Admiral Jean-Antoine Bruny d'Entrecasteaux, departed in 1792 with two ships, the *Recherche* and the *Espérance*, and a full scientific staff as well as instructions to look for the lost La Pérouse. D'Entrecasteaux's expedition broke up in 1794 in the Dutch East Indies as its survivors learned of the dramatic events that had taken place in France. On the Revolution's Pacific periphery, French sailors and scientists who believed that they served a constitutional monarch learned of his execution, and their sense of common national purpose disintegrated. Some embraced the new republic while others rejected it; all of them had to cope with the Pacific ramifications of France's European war—notably their hostile Dutch hosts—as they trickled back to France over the next few years.[5]

[1] Text of the petition and report on the debate published in the *Journal des Etats Généraux* 20, 22 Jan. 1791.

[2] Committee report, *Journal des Etats Généraux* 21, 9 Feb. 1791.

[3] Greg Dening, *Performances* (Chicago, 1996), 109.

[4] Louis XVI was a great reader of Captain James Cook and arranged for a French edition of his voyage. The Bibliothèque Mazarine also has a copy of La Pérouse's instructions annotated in the king's hand. John Dunmore and Maurice de Brossard, eds., *Le voyage de La Pérouse, 1785–1788: Récits et documents originaux*, 2 vols. (Paris, 1985), 1:5–7, 13–19. See also Catherine Gaziello, *L'expédition de Lapérouse, 1785–1788: Réplique française aux voyages de Cook* (Paris, 1984).

[5] Hélène Richard, *Une grande expédition scientifique au temps de la Révolution française: Le voyage de d'Entrecasteaux à la recherche de Lapérouse* (Paris, 1986); Frank Horner, *Looking for La Pérouse: D'Entrecasteaux in Australia and the South Pacific, 1792–1793* (Carleton, Australia, 1995); and Edward Duyker, *Citizen Labillardière: A Naturalist's Life in Revolution and Exploration (1755–1834)*

Scientific expeditions as expressions of national ambition remained priorities for regimes that followed the Reign of Terror and attempted to establish a republic that would represent the sovereign people without opening the door to democratic excess. As Jean-Luc Chappey has observed, the "savant" served as the model republican of the period after the Terror; the scientific expert represented "a return to the reign of Reason . . . the reestablishment of the principle of human progress that had made the Revolution possible and that would now bring it to a close."[6] Even as the stragglers from the d'Entrecasteaux expedition were returning home, the successful young general Napoleon Bonaparte was recruiting scientists to participate in his conquest of Egypt. When, in a 1799 coup, Napoleon seized power and engineered a new constitution that granted him considerable executive power, France's scientific establishment was quick to lobby the new regime for, among other things, another Pacific expedition. In 1801, the *Géographe* and the *Naturaliste* along with an extraordinary scientific staff under the command of Captain Nicolas Baudin left France with orders to explore the south Pacific, paying particular attention to the still largely uncharted continent of Australia.[7]

The Pacific, which had been a focus of cosmopolitan Enlightenment inquiry for much of the eighteenth century, increasingly became a site of national ambitions at the turn of the nineteenth century.[8] British interest in the region quickly came to focus on the colony at Port Jackson, New South Wales, the convict settlement that, from 1788 on, both "rid Britain of its criminals" and "dramatically increased Britain's advantage over its rivals in the Pacific."[9] Supplying this Pacific outpost absorbed considerable resources in the final decade of the eighteenth century. French national interest in the Pacific, however, was different; it focused on scientific investment. In self-conscious contrast to the British, revolutionary French regimes mounted major expeditions that established a French presence in the region by means of maps, charts, collections, and taxonomies.

La Pérouse, d'Entrecasteaux, and Baudin served, respectively, France's last absolute monarch, the new and experimental National Assembly, and a revolutionary general well on the way to declaring himself emperor. Their careers suggest that France's scientific community was remarkably adaptable in its ability to craft relationships with rapidly changing regimes. The absolutist model of close ties between science and the state transferred readily to the new regime: the Old Regime relationship of the royal patron and the scientist-client quickly nationalized under revolutionary circumstances. Indeed, the reorganization of institutions for scientific research

---

(Melbourne, Australia, 2003). In the 1820s, Peter Dillon discovered the wreck of La Pérouse's ships on the reefs off the island of Vanikoro: J. W. Davidson, *Peter Dillon of Vanikoro: Chevalier of the South Seas*, ed. O. H. K. Spate (New York, 1975).

[6] Jean-Luc Chappey, *La Société des Observateurs de l'homme: Des anthropologues au temps de Bonaparte (1799–1804)* (Paris, 2002), 52.

[7] Frank Horner, *The French Reconnaissance: Baudin in Australia, 1801–1803* (Melbourne, Australia, 1987); Edward Duyker, *François Péron, an Impetuous Life* (Carlton, Australia, 2006).

[8] Kapil Raj, "Eighteenth-Century Pacific Voyages of Discovery, 'Big Science,' and the Shaping of an European Scientific and Technological Culture," *History and Technology* 17, no. 2 (2000): 79–98; and Lorraine Daston, "Nationalism and Scientific Neutrality under Napoleon," in *Solomon's House Revisited: The Organization and Institutionalization of Science*, ed. Tore Frängsmyr (Canton, Mass., 1990), 95–119.

[9] Alan Frost, *Voyage of the Endeavour: Captain Cook and the Discovery of the Pacific* (St. Leonard's, Australia, 1998), 83.

and education was a major accomplishment of the revolutionary period.[10] Thanks to ongoing state sponsorship of science and the creation of public institutions, French scientists claimed to represent the best ambitions of the sovereign people.

The French Revolution, as David Bell has recently argued, revealed the "inescapably paradoxical" nature of nationhood: "it [made] political claims which [took] the nation's existence wholly for granted, yet it propose[d] programs which treat[ed] the nation as something yet unbuilt."[11] French science participated in both aspects of the nation. On the one hand, Frenchmen *were already* rational thinkers who excelled in the mastery of nature, and revolutionaries did not hesitate to present themselves as the heirs to the philosophical spirit of the Enlightenment's science of society. Lazare Carnot emerged as a revolutionary hero precisely because of what was understood as his particularly French combination of scientific inquiry and practical political activism.[12] On the other hand, however, Frenchmen had to *become* enlightened citizens; they needed inducements to abandon superstition and embrace their rational potential. The revolutionary state needed to establish a system of public instruction and invent festivals and rituals that would replace Catholic observance in peasant lives.[13] Science, in this scenario, was a tool—even a weapon—for creating the new revolutionary man; according to the title of one tract of the period, science needed to be "*sansculottisée*"—radicalized like a sansculotte—so that it might serve a nation of truly free men.[14]

The identification of the scientist's interest with the national interest was fundamental to the origins and conduct of France's Pacific missions. Frenchmen such as the poet and deputy André Chénier could speak of "friendship, science, love, and French glory"[15] combining to plead for La Pérouse's rescue because he believed that the Revolution had assembled these attributes in the citizens of the regenerated nation. Chénier's Frenchmen understood and shared the love of humanity and of nation that sent men such as La Pérouse into danger; members of the expeditions that followed La Pérouse into the Pacific conceived of themselves as philosophical navigators as well. The organization of science in the d'Entrecasteaux and Baudin expeditions demonstrates this commitment to a notion of the Frenchman-scientist. Scientific activity permeated the ships, encouraging all the men on board—sailors as well as civilian scientific staff—to participate in the creation of an encyclopedic body of knowledge about the Pacific. The identification between the nation and its scientific citizens was not perfect, however, and it broke down in the Pacific as the polymath nature of

[10] Patrice Bret, *L'état, l'armée, la science: L'invention de la recherche publique en France (1763–1830)* (Rennes, France, 2002).

[11] David A. Bell, *The Cult of the Nation in France: Inventing Nationalism, 1680–1800* (Cambridge, Mass., 2001), 5, 14–15.

[12] Nicole Dhombres and Jean Dhombres, *Naissance d'un pouvoir: Sciences et savants en France (1793–1824)* (Paris, 1989), 54–63.

[13] Mona Ozouf, "La Révolution française et la formation de l'homme nouveau," in *L'homme régénéré: Essais sur la Révolution française* (Paris, 1989).

[14] Dhombres and Dhombres, *Naissance d'un pouvoir* (cit. n. 12), 32–34; and Michael A. Osborne, "Applied Natural History and Utilitarian Ideals: 'Jacobin Science' at the Muséum d'histoire naturelle," in *Re-creating Authority in Revolutionary France*, ed. Bryant T. Ragan and Elizabeth Williams (New Brunswick, N.J., 1992), 125–43. On the Revolution's largely unrealized plans for overhauling education, see Isser Woloch, *The New Regime: Transformations of the French Civic Order, 1789–1820s* (New York, 1994).

[15] André Chenier, 7 Thermidor year 2 (25 July 1794), quoted in Richard, *Une grande expédition scientifique* (cit. n. 5).

the national project set individuals at odds over whose science mattered more. The question of publication similarly revealed the fault lines within the national mission as individual scientists sought to allocate credit by transforming a national project into proprietary, individual results.

FRENCH SCIENCE BETWEEN MONARCHY AND REVOLUTION

The science upon which revolutionary regimes built appears poised between Old Regime systems of patronage and anticipation of the modern world of professional expertise. Patronage shaped the practice of eighteenth-century science; access to the monarchy and the ability to define science as a form of royal service opened doors to scientific success. Science's entrenchment within Old Regime society made possible, for instance, scientific dynasties such as that of the Jussieus, father, son, and uncles, all of whom derived their livelihood from royal service as botanists.[16] French science's relationship to the absolutist state also produced distinctively modern features, however. The Jussieus and many other French scientists did, after all, owe their livings to their expertise, an achievement largely possible because of the institutional organization of Old Regime science. An institution such as the Jussieus' Jardin du Roi channeled royal patronage toward specific disciplinary ends, in this case, the natural sciences, and also toward experts in those fields. As Roger Hahn has observed of the members of the Academy of Sciences, "[B]y accepting government pensions, [they] tacitly indicated their willingness to devote their efforts to the 'national good' and to forgo the freedom and pleasures of dilettantism."[17] Affiliation with an institution came to function as a guarantor of scientific integrity; receiving payment, instead of calling the scientist's good faith into question, solidified his credentials.

Although Charles Coulston Gillispie has recently posited a "long half century of French scientific predominance, from the 1770s through the 1820s" in which revolutionary political transformations and scientists' roles in them were merely surface events, adaptation to revolutionary change was nonetheless crucial.[18] Centralized state scientific institutions and the elite associated with them had their origins in the Old Regime; nevertheless, these institutions and their denizens modified their rhetoric and practice to accommodate themselves to their new masters. The enduring success of the French scientific community across the Revolution was not simply because Old Regime institutions were already modern, professional, scientific bodies. Science practiced by individuals beholden to the government and gathered together in government-sponsored institutions could easily have been a casualty of the revolutionary impulse to sweep away the abuses of monarchy. The 1793 abolition of the Royal Academy as part of a wholesale elimination of the privileged corporations of the Old Regime aimed at that very goal. Emma Spary's recent analysis of the transformation of the Jardin du Roi into the Muséum d'Histoire Naturelle charts this practice of accommodation. Rhetorically, the garden's naturalists represented themselves as the guardians of the natural order to which a regenerated France aspired: the "gar-

---

[16] Emma Spary, *Utopia's Garden: French Natural History from Old Regime to Revolution* (Chicago, 2000), chap. 1.

[17] Roger Hahn, *The Anatomy of a Scientific Institution: The Paris Academy of Sciences, 1666–1803* (Berkeley, Calif., 1971), 10.

[18] Charles Coulston Gillispie, *Science and Polity in France: The Revolutionary and Napoleonic Years* (Princeton, N.J., 2004), 4.

den would be the abridgement of the physical world as regenerated France [would] be that of the moral world."[19] They simultaneously reformulated their administrative practice, acknowledging (though also repressing) the desire of the sovereign people to pick the garden's flowers. Adaptation was in many ways a process of nationalization, of articulating the ways in which scientific projects served the nation.

Pacific ventures exemplify French scientists' strategic adaptation to new revolutionary regimes, particularly to the notion of a specifically French capacity for scientific humanitarianism. The Society for Natural History's petition calling for a mission to rescue La Pérouse, for instance, suggests that its authors understood their audience well; their emphasis on sensibility as a motive force is reminiscent of plans for national regeneration in the early Revolution.[20] The creation of the Society for Natural History was itself a canny move; it included virtually all of the staff of what had been the Jardin du Roi, but it was a "free" society, one of many established in 1790 when knowledge was released from the monopoly that royal charter granted to privileged institutions. The 1793 reconstitution of the Jardin as the Muséum d'Histoire Naturelle marked the successful conclusion of naturalists' strategies. Within a few years, scientists based at the Muséum were putting forward proposals for the ambitious Pacific venture that would become the Baudin expedition. This time their arguments drew on what promoters referred to as the "successes" of Bonaparte's Egyptian campaign. The establishment of an institute for arts and sciences in Cairo, modeled on the Institut National, newly organized from the remains of the Royal Academy, suggested that French science could conquer and transform remote societies in the national image. The Muséum and French science had already enjoyed tremendous success "thanks to our victorious armies and the enlightened protection of the government"; a voyage around the world (as the Baudin expedition was originally conceived) would be a suitable next venture.[21] As Baudin asked the scientists of the Institut National, "Think of how glorious it would be for the Nation and the Institut to add to all of the trophies that surround us a voyage undertaken for the increase of human knowledge in the midst of a war whose like has never been seen in the annals of history."[22]

BRITAIN AND FRANCE IN THE PACIFIC

A comparison of British and French approaches to the Pacific illustrates the impact of different relationships between science and the state. A comparative approach is also useful because the figure of Captain James Cook looms so large over the eighteenth-century Pacific that he establishes the British experience as the model for exploration and colonization—a measure by which the efforts of other nations inevitably appear as failures. The British scientific establishment lacked the government-sponsored institutions that characterized France, and the role of British science in Pacific exploration was correspondingly different. British interest in the Pacific was fundamentally

[19] Spary, *Utopia's Garden* (cit. n. 16), 227.

[20] Jessica Riskin, *Science in an Age of Sensibility: The Sentimental Empiricists of the French Enlightenment* (Chicago, 2002).

[21] Jussieu and Lacépède on behalf of the Muséum to the naval minister, Marine BB 995, 12 Thermidor year 6 (Aug. 1798), microfilm reel 1, Papers of the Baudin Expedition, National Library of Australia, Canberra (hereafter cited as NLA Baudin).

[22] Baudin to the Institut National, letter read in the meeting of 7–8 March 1800, in Nicolas Baudin, *Mon voyage aux Terres Australes: Journal personnel du commandant Baudin*, ed. Jacqueline Bonnemains (Paris, 2001), 32.

geostrategic; it originated in the British Admiralty and aimed to create naval bases and safe sailing routes for a commercial empire.[23] Scholars looking for similar motivations for revolutionary France's Pacific expeditions face a near-total dearth of documentary evidence, a startling contrast to the abundance of documentation of the voyages' scientific equipment, personnel, and purpose.[24]

The tradition of gentlemanly autonomy and the virtuoso remained stronger in Britain than in France, and the figure of Joseph Banks dominates British expeditionary science in the period. John Gascoigne has recently argued that publicly funded science was a casualty of demands for cheaper government in the wake of Britain's loss of the North American colonies; in the 1790s, British science depended more heavily than ever on men such as Banks, who worked as a sort of unpaid consultant, bringing gentlemanly expertise in botany, agriculture, commerce, and colonization to the conduct of affairs of state. Banks could broker exchanges between government departments, the Royal Society, the East India Company, and the Royal Botanic Gardens at Kew precisely because he was beholden to none of them. The virtuoso-naturalist remained the norm in Britain, closely connected with the aristocracy and his ties to the state mediated primarily by a sense of noblesse oblige.[25]

The first Cook voyage was the most explicitly "scientific" of Britain's eighteenth-century Pacific voyages, its ostensible purpose to observe the transit of Venus. The Bureau of Longitude and the Royal Society sent geographers and astronomers, and the young Joseph Banks—making his really grand Grand Tour with a few naturalists, friends, servants, and dogs—organized the expedition's natural history work. Banks, for instance, insisted that a scientific expedition should have an official artist and commissioned Sydney Parkinson.[26] Banks had hoped to accompany the second Cook expedition but quarrelled with the Admiralty about accommodations on board. In his stead, the naturalists George Forster and Johann Reinhold Forster, their assistant, the Linnaean-trained Anders Sparrman, the artist William Hodges, and the astronomer and meteorologist William Wales were the "experimental gentlemen" on board. Cook disliked the Forsters, and his third voyage included no nonnaval scientific personnel; William Anderson, surgeon's mate, was responsible for most of the natural history. The only nonnaval expert on board was a draftsman, John Webber.

[23] See, e.g., Frost, *Voyage of the Endeavour* (cit. n. 9); and Frost, "Shaking off the Spanish Yoke: British Schemes to Revolutionise Spanish America, 1739–1807," in *Science and Exploration in the Pacific: European Voyages to the Southern Oceans in the Eighteenth Century*, ed. Margarette Lincoln (Woodbridge, UK, 1998), 19–37. But see David Mackay, *In the Wake of Cook: Exploration, Science, and Empire, 1780–1801* (Wellington, New Zealand, 1985), who agrees the Admiralty tried to do science cheaply but presents British approaches to the Pacific as less coherent and coordinated.

[24] The only text from the Baudin expedition that suggests an imperial vision of Australia comparable to contemporary ideas either of a "British Pacific" or of a Napoleonic Europe is a memo written by François Péron, one of the surviving naturalists, to Governor Decaen of the French colony of Île de France, where the expedition stopped on the trip back to France. There is no evidence that Péron wrote the memo under orders or that it reflects anyone's ideas but his own. Duyker, *François Péron* (cit. n. 7), 204–6. The archival void on French territorial ambitions is so complete that it leads Pacific scholar Jean-Paul Faivre to declare that "in spite of the almost total absence of documents, I still persist in believing in the political character of the Baudin expedition." Jean-Paul Faivre, preface to Nicolas Baudin, *The Journal of Post Captain Nicolas Baudin, Commander in Chief of the Corvettes Géographe and Naturaliste*, ed. and trans. Christine Cornell (Adelaide, Australia, 2004), xiii.

[25] John Gascoigne, *Science in the Service of Empire: Joseph Banks, the British State, and the Uses of Science in the Age of Revolution* (New York, 1998). See also Richard Drayton, *Nature's Government: Science, Imperial Britain, and the "Improvement" of the World* (New Haven, Conn., 2000), especially, 72–80.

[26] Bernard Smith, *European Vision and the South Pacific*, 2nd ed. (New Haven, Conn., 1985), 36–37.

Collections from the Cook voyages circulated among networks of patrons, individual and institutional, and virtuosos. Johann Reinhold Forster presented some of his materials to Oxford, Anders Sparrman to the Swedish Academy of Sciences, James Patten (surgeon on the *Resolution* in 1772) to Trinity College Dublin, John Webber to the Museum of Bern. Joseph Banks had tremendous influence over patterns of distribution of scientific materials from all three Cook voyages. His ability to "activate" the expertise of a network of savants across Europe proved crucial to transforming disparate collection items into scientific knowledge.[27] Although Banks's networks included public institutions such as the Royal Society and Kew Gardens, he was deeply committed to a "gentlemanly culture with . . . clublike connections" and a common sense of public service as an obligation of rank. The best way to produce new knowledge from Pacific curiosities, Banks believed, was to put them in the correct private and public-spirited hands.[28]

The French governments that supported Pacific expeditions in the revolutionary period, in contrast, funded science lavishly and maintained strict controls over scientific production. Both expeditions carried a full civilian scientific staff; even the hydrographers and geographers were civilian experts.[29] On Baudin's ships, for instance, the twenty-two savants outnumbered senior officers; each ship had an astronomer, a geographer, two or three zoologists, a botanist, and an artist. There was also a mineralogist with two assistants and a gardener with three. Scientists with both expeditions owed their positions to nominations from Parisian scientific institutions.

In addition to the civilian staff, naval personnel also often possessed some scientific competence as it was standard practice on voyages of exploration to appoint men with multiple competencies. The chaplain's position, for instance, was not generally wasted on individuals with merely theological expertise. Louis Ventenat, chaplain on board d'Entrecasteaux's *Recherche*, was typical: he was an amateur naturalist and probably owed his appointment, at least in part, to the recommendation of his brother, Pierre-Etienne, a member of the Society for Natural History. The chaplain on the *Espérance*, similarly, was the expedition's junior astronomer. The artwork from Baudin's expedition was produced not by its official artists, who left the expedition in Mauritius, but by two gunners, Charles-Alexandre Lesueur and Nicolas Petit, whom Baudin selected at least as much for their artistic ability as for their accuracy of aim.[30] Men such as Petit and Lesueur understood that the nation might well command them to produce scientific illustration as well as to fulfill their ordinary naval duties.

The French government was also far more possessive about the scientific results of the expeditions it sponsored than its British counterpart. Instructions from the minister of the navy always indicated that all notes, journals, collections, charts, and maps to result from a voyage were public property. Representatives of the naval ministry and of the Muséum d'Histoire Naturelle met returning ships at the dock to take possession of this material. Officials advised both D'Entrecasteaux and Baudin to

[27] Ibid., 127; David Philip Miller, "Joseph Banks, Empire, and 'Centers of Calculation' in late Hanoverian London," in *Visions of Empire: Voyages, Botany, and Representations of Nature*, ed. David Philip Miller and Peter Hannes Reill (New York, 1996), 21–37.

[28] Gascoigne, *Science in the Service of Empire* (cit. n. 25), 5.

[29] E. de Rossel, *Voyage de D'Entrecasteaux envoyé à la recherche de La Pérouse* (Paris, 1808), 1:xlvi; Baudin, *Journal of Post Captain Nicolas Baudin* (cit. n. 24), 579–82.

[30] See the catalog to an exhibition of their work: Susan Hunt and Paul Carter, eds., *Terre Napoléon: Australia through French Eyes, 1800–1804* (Sydney, Australia, 1999).

remind men not to "oppose the government in its property rights over the voyage results."[31] This more centralized approach concentrated scientific influence in Parisian establishments such as the Muséum, whose members were guaranteed immediate access to all of the data. The scattering of British collections, in contrast, sometimes slowed publication and created confusion. While most of the type specimens from the Baudin voyage are still in the Muséum's collections, British dispersal of specimens resulted in a great deal of taxonomic confusion, such as that produced by the three kangaroo skulls, skins, and partial skeletons (now all lost) that James Cook brought back from Australia in 1790. Although described as all belonging to a single species, the specimens in fact appear to have been mixed up, and the classification had to be retrospectively reapplied in the twentieth century. Similarly, Banks's Australian *Florilegium*, though available to other naturalists, was not actually published until the 1980s.

On British Pacific expeditions, science was either a naval matter conducted by naval personnel, such as charting, or it was a matter of personal interest to men of sufficient influence, such as Joseph Banks, that their private pursuits could shape public policy. By the 1790s, the purpose of British presence in the Pacific had, according to Glyndwr Williams, "change[d], to put it crudely, from exploration to exploitation."[32] British voyages represented the Admiralty or various commercial interests, and their purposes were correspondingly narrowly defined. Following his death in Hawaii, Cook emerged as a national hero, his laconic prose and imperturbable demeanor representing a new model of British manhood, but later expeditions were "more workaday affair[s]" that offered little scope for anyone, officer, sailor, or scientist, to rival Cook's achievement.[33] By contrast, France in the same period sent expeditions of sailors and scientists who understood that their collective purpose was discovery and that each individual on board possessed the ability to analyze and understand nature. The scientific aims of the voyages were encyclopedic; Frenchmen aspired to know all there was to know about the Pacific. This breadth of ambition made it possible for men of d'Entrecasteaux's and Baudin's expeditions to conceive of themselves as citizen-scientists; they could all be "philosophical travelers,"[34] participating in the creation of French knowledge of the Pacific.

PHILOSOPHICAL NAVIGATORS: THE CULTURE OF SCIENCE ON FRENCH SHIPS

A culture of doing science pervaded French ships as they explored the south Pacific; everyone was engaged in measuring, sketching, calculating, and recording data. These expeditions produced a striking number of journals—a full third of the men on the Baudin expedition wrote journals of some sort, and some twenty journals survive

[31] Baudin's orders of 9 Vendemiaire year 9 in AN Marine BB 995, reel 1, NLA Baudin.

[32] Glyndwr Williams, "'To Make Discoveries of Countries Hitherto Unknown': The Admiralty and Pacific Exploration in the Eighteenth Century," in *Pacific Empires: Essays in Honour of Glyndwr Williams*, ed. Alan Frost and Jane Samson (Melbourne, Australia, 1999), 13–31, on 27.

[33] Ibid., 28. On Cook's reputation, see Kathleen Wilson, *The Island Race: Englishness, Empire, and Gender in the Eighteenth Century* (New York, 2003); Sujit Sivasundaram, "Redeeming Memory: The Martyrdoms of Captain James Cook and Reverend John Williams," in *Captain Cook: Explorations and Reassessments*, ed. Glyndwr Williams (Woodbridge, UK, 2004), 201–29; and Frost, *Voyage of the Endeavour* (cit. n. 9), 113–14.

[34] The phrase is from the anthropological instructions for the Baudin expedition: Joseph Marie Degérando, *The Observation of Savage Peoples*, ed. and trans. F. C. T. Moore (Berkeley, Calif., 1969), 63.

from the d'Entrecasteaux voyage.[35] From the most literate savant to the seaman with dubious penmanship and worse spelling, men on board obviously felt that they had cause to keep a regular record of their activities. Many men simply copied data on position and weather from the ship's log and observed that "nothing" happened on most days. Others made their journals into a space for privacy, reflected on their relationships with their shipmates, mulled over grievances, and replayed confrontations, seeking better resolutions. Officers were expected to keep some basic records, but many expanded their journal writing well beyond the necessary minimum. As Lieutenant Jacques de Saint-Cricq observed in October 1801, "a nautical journal is . . . the most boring thing in the world to read or to write" since it just repeated the details officially recorded in the ship's log. Instead, he proposed to offer "a purely narrative account of [his] voyage," with began with a description of the "pleasant accord between the Naturalists and [the officers],"[36] who were delighted to find how much they had in common. Saint-Cricq, like many of his shipmates, constructed his journal to address a putative reader and proposed to offer that reader an interpretation of Pacific discovery, not merely raw data.

Many of the journals reveal their authors' pursuit of their own scientific interests alongside their official duties. Louis Ventenat, chaplain with d'Entrecasteaux, composed an extensive travelogue, which includes some of the most sustained ethnographic material of the voyage.[37] Pierre Gicquel, who sailed on both expeditions, took a particular interest in language and compiled comparative vocabularies of the peoples with whom the expedition came into contact.[38] Gicquel's job certainly did not involve linguistic research, but he was obviously fascinated with the problem of communication with the natives of the south Pacific islands that the d'Entrecasteaux expedition visited. François Michel Ronsard, engineer on board the *Géographe*, took an active interest in naval hygiene and the organization of the ship, as one might expect of an ambitious young officer, but his journal also overflows with enthusiastic reports on the work of the expedition's naturalists. Ronsard often accompanied them on shore visits, delighted that he could participate in their mission "to gather knowledge of . . . New Holland that is as yet unknown to Civilized Nations."[39] Ronsard filled his journal with speculations on the nature of property ownership among Australian aboriginals and lists of the plants and shells that he collected at every opportunity.[40]

Scientific education was also common on board these expeditions, with formal instruction in mathematics and navigational technique offered at all levels. The botanist Jacques Julien Houtou de Labillardière, sailing with d'Entrecasteaux, was impressed that all of the officers on board possessed Borda's reflecting circle, which he considered superior to the sextant, and they all learned to use it over the course of the voyage. He also noted that the expedition's mineralogist and astronomer offered

[35] Paul Carter, "Looking for Baudin," in Hunt and Carter, *Terre Napoléon* (cit. n. 30); Horner, *Looking for La Pérouse* (cit. n. 5), xii.

[36] Saint-Cricq journal, AN Marine 5 JJ 48, 1–2, reel 15, NLA Baudin.

[37] Ventenat journal, AN Marine 5 JJ 4, reel G 24,564, Papers of the d'Entrecasteaux expedition, National Library of Australia, Canberra (hereafter cited as NLA d'Entrecasteaux).

[38] Gicquel notes, AN Marine 5 JJ 1 n°10, reel G 24,649, NLA d'Entrecasteaux; and journal AN Marine 5 JJ 6, reel G 24,657, NLA d'Entrecasteaux. Gicquel's stay with the Baudin expedition was brief; he hated the commander and arranged to leave at Mauritius on the voyage out.

[39] Ronsard journal, AN Marine 5 JJ 28, p. 36, reel 10, NLA Baudin. See also pp. 10–12 for his remarks on naval stores and pp. 20–24 for naval hygiene.

[40] Ibid., 70–72.

instruction in mathematics, some of which seems to have been relatively advanced and suitable for ambitious young officers, but some quite simple.[41] One journal of an anonymous sailor includes pages and pages of simple arithmetic problems; in particular, he spent a lot of time doing sums with nondecimal currency.

Most of the drawing, writing, and figuring that occurred on these expeditions did not lead directly to the production of new knowledge. Indeed, the pervasive will to record the journey was probably more useful as a disciplinary tool than as a scientific one; journals kept men busy during the long and frequently boring days when nothing much happened. Math lessons and journal writing were not, however, the only means of maintaining discipline; indeed, they form a stark contrast to the British navy, which relied far more on the threat of the lash.[42] References to any form of corporal punishment are relatively rare in the records of the French expeditions; the more common confinement to quarters, if anything, encouraged the habit of journal writing. However, even as a means of keeping bored men busy, record keeping had its disadvantages, as Baudin observed when he scolded his lieutenant for drawing during his watch or complained that his ship's doctor preferred to write pamphlets rather than monitor the health of the crew.[43] We will return to Baudin's sense that his expedition had too many scientists—or at least too many members who aspired to be scientists—but for the moment the key point is that the aspiration to produce science was widespread.

Léon Brèvedant, helmsman on Baudin's ship the *Naturaliste*, offers in spite of himself particularly compelling evidence of the pervasive culture of record keeping on board ship. Brèvedant's journal regularly parodies naturalists' accounts of the voyage. Thus on one shore excursion he notes that he observed rocks and a lot of trees, "which I won't describe because I don't know anything about them." He began one entry: "To my readers. I have as much right as anyone else to bore you, but I'm not going to take advantage of it, instead I'll leave that to the botanists and the others." For Brèvedant, "savant" was a common and extremely flexible term of abuse, which he also applied to naval officers whom he disliked. Thus Lieutenant Freycinet was one of those men "who thinks he's a savant because he has a job that he doesn't know how to do and who prefers to let an injustice stand rather than to retract an order often delivered without paying attention." Brèvedant did not like scientists, by which he seems to have meant most men in a position to give him orders—science, for Brèvedant, was what distinguished the higher ranks from the lower. He clearly thought that the trappings of science surrounding him—meticulous descriptions of perfectly ordinary rocks and trees—were ridiculous. Even the disaffected Brèvedant kept a journal, however, with his disdain manifesting itself as satire rather than a simple rejection of the expedition's scientific culture.[44]

As Baudin and Brèvedant's complaints suggest, the conduct of science aboard ship revealed the ways in which sailors and savants threw themselves into the projection of a scientific and revolutionary nation, but it also uncovered fault lines in that project. Members of the expeditions certainly embraced the notion that they formed

---

[41] Labillardière, *Voyage in Search of La Perouse, performed by order of the Constituent Assembly . . .* (1800; repr., New York, 1971), 50.

[42] On British naval discipline, see Greg Dening, *Mr. Bligh's Bad Language: Passion, Power, and Theatre on the Bounty* (New York, 1992).

[43] Baudin, *Journal of Post Captain Nicolas Baudin* (cit. n. 24), 65, 68.

[44] Brèvedent journal, AN Marine 5 JJ 56, reel 21, NLA Baudin.

part of the collective production of scientific knowledge and its generous distribution to humanity. Quarrels, jealousy, and vindictiveness—not a consensus around noble scientific goals—are equally characteristic of the narrative accounts of both the d'Entrecasteaux and Baudin voyages, however. Scholars have explained the ill will so abundant on board these voyages as a consequence of political divisions generated by the French Revolution (particularly for d'Entrecasteaux) or of tensions between naval officers and scientific personnel (especially Baudin). What was often at stake in the battles that raged aboard ship, however, was who got to be the scientist and whose science mattered more. Given that French ambitions in the Pacific were encyclopedic—no realm of knowledge was outside of the mandate of France's explorers—there were many opportunities for debates between branches of science and their practitioners.

Nicolas Baudin particularly liked to stage episodes in his journal in which his men's naïve enthusiasm for science erodes the patience of their more experienced commander and results in setbacks for the expedition as a whole. For instance, when the ship first sighted land in the Canaries, having barely left France, Baudin observed caustically that "all the scientists and even most of the officers were so overjoyed that they behaved like madmen. . . . [E]veryone went off to get his portfolio and his pencils and, to fore and aft of the ship, there was not a soul to be seen who was not busy sketching."[45] Running sketches of the cost of the Canaries were hardly likely to contribute much to science as the area had been well charted for centuries and the habitat of its flora and fauna posed few questions. The willingness of sailors and savants to participate in recording the journey left Baudin unimpressed. Similarly, Baudin enjoyed recording stories of squabbling among scientists. His journal, for instance, indicates that the first shark available for dissection became the scene of conflict between the ship's zoologist, François Péron, and its surgeon, François Lharidon. According to Baudin, Péron, dripping in blood, sought out his commander to complain that the surgeon had snatched the shark's heart to dissect it himself; Baudin promised Péron that he should have the next shark all to himself. Four days later, Baudin was adjudicating between the anatomists, who wanted a porpoise on its back so that they could investigate its innards, and the artists, who wanted the same animal on its belly in a lifelike pose.[46] Baudin clearly enjoyed reporting these instances of childish and competitive behavior, ill suited to a national scientific project.

Baudin's disdain for his men did not stem from conflicts between naval officers and scientists but rather from disputes about whose competence was greater and whose science mattered more. Baudin was convinced that he was as qualified a naturalist as any of the men on board in an official natural history capacity. Like many of the savants, he owed his appointment to his connections at the Muséum d'Histoire Naturelle, whose director, Antoine-Laurent Jussieu, had been particularly impressed with Baudin's earlier work as a scientific collector. Baudin had already made significant botanical contributions to French national glory, Jussieu explained to the minister for the navy in his letter recommending Baudin for command of the Pacific expedition; until Baudin's "campaign" in the Caribbean, French national collections of tropical plants had been decidedly inferior. Indeed, Baudin's strictly *naval* qualifications for his command were dubious at best, as he had spent the 1790s working for

[45] Baudin, *Journal of Post Captain Nicolas Baudin* (cit. n. 24), 20.
[46] Ibid., 46, 49.

the wrong side, supplying the gardens of the Hapsburg emperor.[47] When the French nation had been in greatest danger, Baudin had been augmenting the scientific treasures of an enemy nation, a fact he awkwardly explained away with protestations of his neutral commitment to the "progress of Science" and vague hints that some Old Regime injustice had driven him from France.[48]

In theory, the aims of these French expeditions had no boundaries; they aspired to total knowledge of the Pacific and to that end staffed the voyages with representatives of every branch of science. In practice, however, the encyclopedic ambitions of this expeditionary science often set representatives of specific disciplines at odds. Most obviously, natural history and geography/cartography made for poor shipmates. The geographers were happiest in motion; traveling along a coastline was their ideal situation. The naturalists, in contrast, wanted to spend as much time at anchor as possible so that they could go on shore to pursue their investigations. Squabbling over a shark or a porpoise was almost certainly, in part, symptomatic of the naturalists' boredom at sea. Naturalists fought with geographers over access to landing boats, and tensions reached their peak on the not-infrequent occasions when naturalists failed to make a rendezvous. Naturalists described getting lost and wandering in the interior as indicators of bravery and willingness to put the cause of science above everything else; sailors understood the same episodes as evidence of the naturalists' complete lack of common sense and regard for the expedition as a whole.[49] For all the resources that the Muséum, the Institut, the National Assembly, and the First Consul invested in these voyages, there simply was not enough to go around all of the scientists, who each wanted more time, more storage space, more assistance, and more resources. Indeed, given the unbounded nature of the project, there could never have been enough resources to satisfy all of these discoverers.

DISSEMINATION AND DISSENT: PUBLICATIONS AND COLLECTIONS

The general consensus that all members of the expedition were participants in a national scientific project broke down most completely over the question of publication of results. Because scientific investigation was a national venture, it was also potentially a means of individual advancement. But if scientific production were to further individual careers, then there had to be some way of distinguishing the merit of one savant from the efforts of the group. The revolutionary principle of careers open to talent entered the calculations, and individual scientists, each intent on making his own career, sought to define research as proprietary. In spite of agreement that the expedition represented French genius for humanitarian discovery, assigning credit for particular discoveries led to conflicts over the disposition of collections and the publication of results.

Ideally, these state-sponsored voyages were to result in a single publication, authored by the captain, which would be both a narrative of the expedition and an

---

[47] Jussieu to Lacépède, 12 Thermidor year 6 (Aug. 1798), AN Marine BB⁴995, reel 1, NLA Baudin.

[48] Baudin to the minister of the navy, 15 Nivose year 4 (5 Jan. 1796), AN Marine BB⁴995, reel 1, NLA Baudin.

[49] See, e.g., Labillardière, *Voyage in Search of La Perouse* (cit. n. 41), 98, on the issue of boats. On naturalists failing to meet boats, see Baudin, *Mon voyage* (cit. n. 22), 287–88, 293; and François Péron, *Voyage de découvertes aux terres australes, executé par ordre de sa Majesté l'Empereur et roi sur les corvettes le Géographe, le Naturaliste, et la goëlette le Casuarina* (Paris, 1807), 1:122–23.

account of the research that took place on board. The official collection of journals and specimens at the end of the voyage was to facilitate the production of this volume. The state's position, then, did not change appreciably between the Old Regime and the Revolution with regard to intellectual property. Previously, absolutist patronage had chartered the Academy of Sciences, and that institution controlled the publication of scientific knowledge. After the Revolution, scientific knowledge, now the result of the investment of the sovereign people, was still supposed to have a single outlet, and a major project such as an expedition to the South Seas was supposed to produce a single, authoritative statement of national achievement. The National Assembly's approach to La Pérouse's text exemplified this attitude; while discussing the rescue mission, the assembly also decreed that the proceeds from the sale of his narrative (which runs up to the expedition's last stop at Port Jackson and was returned to Europe on a British ship) would go to the lost explorer's wife. The assembly's gesture did not imply that Madame de La Pérouse enjoyed any rights over the narrative composed by her husband; rather, she benefited from the generosity of a grateful nation. Where La Pérouse's work had once belonged to the king, his patron, now it belonged to the National Assembly as representative of the sovereign people.[50]

Again, the contrast with the Cook voyages is enlightening. The British government would have preferred a single account of Pacific exploration, and Cook did gather up his men's notes and journals, but the British government never really abandoned the notion that journals were the private property of their authors. Cook's first voyage was published in an officially commissioned edition with the heavy-handed literary and philosophical editing of John Hawkesworth, who recounted Cook's travels in a larger compendium that included John Byron, Samuel Wallis, and Philip Carteret's expeditions as well. Disliking Hawkesworth's added commentary, Cook took care to prepare his own editions of his second and third voyages. When the heirs of artist Sydney Parkinson published his journal, however, Hawkesworth and Banks, attempting, unsuccessfully, to stop the publication, argued that the Parkinson book would damage Hawkesworth's legitimate profits, not that the Parkinson text was government property. Johann Reinhold Forster also published his account of the second voyage, again against the opposition of Banks and the Admiralty.[51] To a much greater extent than in France, British voyages were available for literary use; their narratives were not closely guarded public property.

The notion that the fruits of a national project belonged exclusively to the nation did not win unanimous assent among members of revolutionary France's Pacific expeditions. Both d'Entrecasteaux and Baudin prepared narratives in the neutral, authoritative voice that would announce all the results—geographic, zoological, or botanical—of the collective endeavor of the expedition's scientists. However, both commanders died before returning to France, and neither expedition produced the kind of univocal account that would stand as a monument to French achievement. Many of the individuals on board saw no reason why there should be a conflict between participation in the national project and the talented individual making his own career. Inevitably on these ships full of scientists, many men had their own notes, thoughts on nature or society in the Pacific, and collection materials, and back in Eu-

[50] Dunmore and Brossard, *Le voyage de La Pérouse* (cit. n. 4), 1:91.
[51] See the discussion in J. C. Beaglehole, *The Life of Captain James Cook* (Stanford, Calif., 1974), 456–71.

rope there was a significant market for the literature of discovery of which they hoped to take advantage.

Journals, such an important part of life to so many on board, not surprisingly became a locus of conflict. Louis Ventenat, the chaplain-naturalist on d'Entrecasteaux's *Recherche*, evidently planned to publish his journal, and his text includes a preface in which he debates with himself whether the state had "the right to take from individuals the fruits of their work." He concluded that he could "in good conscience publish his research," which he had conducted "to enrich his fellow citizens with the discoveries made during a long and painful voyage."[52] He did his best to hang on to his journal, but his commanding officer eventually located it; it is in the National Archives with others of the expedition. The journal of Hyacinthe de Bougainville, midshipman on the Baudin expedition, remained in his possession at the end of the voyage and only later joined the collection in the National Archives. The fact that he was the son of the famous Pacific explorer Louis-Antoine de Bougainville (who by 1800 had become a member of the Institut National committee that recommended in favor of Baudin's expedition) almost certainly meant that the young officer's privacy received greater consideration than that of his colleagues. Hyacinthe's colleague, François-Michel Ronsard, also an officer on the *Géographe*, conducted an extended battle with Baudin over Ronsard's journal. Ronsard insisted on a receipt for his journal, which Baudin refused to provide. The two disliked each other and were bound to disagree over something, but it is nonetheless significant that their enmity focused on a journal. Baudin, refusing to acknowledge any transfer of property, denied Ronsard a receipt and insisted that a notation in the logbook to the effect that every man had handed over his papers should suffice.[53] Louis Freycinet, like Ronsard, handed over his journal during the final stay in Dutch Timor, but he made a point of sealing it, adding a note to that effect to Baudin's order.[54] Freycinet and Ronsard both used their journals to nurse their grievances, many of them against their captain, and they undoubtedly did not want Baudin to read them. The disaffected Brèvedant took more resolute steps to keep his journal private. The fragment that exists informs the reader defiantly that he tossed most of it overboard at Île de France, precisely the moment when he should have handed it over to form part of the official record.[55]

Ownership of collections, like texts, was disputed. D'Entrecasteaux's naturalist Riche complained that "shell mania" had overtaken the entire ship to the extent that the "passion for collecting on the part of persons who are not naturalists" prevented him from acquiring any good specimens for the official collection.[56] Much of the enthusiasm for natural history that expedition journals reveal was associated with individuals' creating their own collections. Stanislas Levillain, zoologist with the Baudin expedition, had a private collection of ten boxes of shells, two boxes of birds, and a single box of mineral samples that, according to naval tradition, were auctioned off among the crew when he died at sea.[57] François Péron noted that he was obliged to

[52] Ventenat, "Discours préliminaire, divisé en trois articles principaux," AN Marine 5 JJ 4, reel G 24,564, NLA d'Entrecasteaux.

[53] Ronsard journal, 17 Thermidor year 11, AN Marine 5 JJ 28, reel 10, NLA Baudin.

[54] Louis Freycinet journal, 23–24 Thermidor year 11, AN Marine 5 JJ 49, reel 16, NLA Baudin.

[55] Brèvedent journal, AN Marine 5 JJ 56, reel 21, NLA Baudin.

[56] Riche to d'Entrecasteaux, 29 July 1792, AN Marine 5 JJ 4, reel G 24,564, NLA d'Entrecasteaux.

[57] Naval ministry to Hamelin, 26 Brumaire year 12 (inquiring about the proceeds of the sale); Hamelin to the ministry, 12 Fructidor year 12 (denying that there was any money because the collection, being government property, could not have been sold). AN Marine BB 995, reel 1, NLA Baudin.

re-create much of Levillain's Shark Bay collection when the expedition returned to western Australia; he was particularly indignant that some of the original specimens had fallen into English hands in Port Jackson.[58] One of Baudin's botanists, Louis Leschenault, ended up in Java in 1808, sick and in debt, with a miscellaneous collection that included thirty-four mammals, an assortment of birds, 714 insects, 900 Javanese plants, 200 types of seeds, and assorted reptiles preserved in alcohol. He proposed to sell the collection to the French government for a sum that would allow him to pay his debts and return to Europe. The ministry refused to pay for what it considered it already owned but did offer Leschenault a stipend for meritorious service that would allow him to bring the materials back to France.[59]

The collections of the d'Entrecasteaux voyage, which broke up in leaderless confusion before returning to France, raised particularly knotty questions about legitimate ownership. The expedition had left France in the service of the king and the National Assembly; it arrived in the Dutch East Indies in 1793 to learn that Louis XVI was dead and that France was a republic at war with the Netherlands. These radically changed circumstances produced competing property claims over research materials. The surviving senior officer, a nobleman of royalist sympathies willing to cooperate with the Dutch authorities, demanded that all members of the expedition hand over their journals. Those whose loyalties remained with the National Assembly objected and in some cases subverted the order by hiding or destroying their notes.[60] The collection materials ultimately fell into the hands of the British navy; back in London the authorities treated them as the property of Louis XVIII, then living in British exile. He was not interested in them and in turn offered them to his hostess, Queen Charlotte. Joseph Banks naturally entered the picture at this moment to choose specimens for the queen's collection and to send others to the British Museum. The French Directory appealed for return of the collection, but more effectively, Labillardière contacted Banks directly and appealed for the return of his materials. Banks apologized to the queen but gave his opinion that "it was necessary for the honor of the British nation and for the advancement of Science that the right of the Captors to the Collection should on this occasion be wav'd and that the whole should be returned to M de Billardière."[61] The exchange between Banks and Labillardière is cited as an instance of scientific internationalism, but it is equally significant as a miscarriage of French national science policy.[62] Labillardière succeeded not only in holding on to his own notes but also in having the collection defined as his private property—as it might well have been had he been in Banks's position on a British expedition.[63]

[58] Duyker, *François Péron* (cit. n. 7), 184.

[59] See the correspondence in AN F 17 3930, dossier 14, reel 24, NLA Baudin.

[60] At least one sailor's journal went overboard at this point. The chaplain-naturalist Ventenat tried unsuccessfully to smuggle his back to France. The naturalist Labillardière, the geographer (Willaumez), and one of the senior officers (La Motte du Portail) succeeded in retaining their journals. For a further account of the political tensions in Surabaya, see Horner, *Looking for La Pérouse*, 203–22; and Duyker, *Citizen Labillardière*, 189–99. (Both cit. n. 5.)

[61] Joseph Banks to Major William Price, 4 Aug. 1796, in G. R. de Beer, "The Relations between Fellows of the Royal Society and French Men of Science When France and Britain Were at War," *Notes and Records of the Royal Society* 9 (1952): 255–56.

[62] Ibid., and de Beer, *The Sciences Were Never at War* (London, 1960).

[63] Because of this designation as private property, the collection of the d'Entrecasteaux expedition was sold off after Labillardière's death and, unusually for material from a French expedition, is widely dispersed. See Joan Apfelbaum, "Australian Collections of Labillardière in the Herbarium of the Academy of Natural Sciences of Philadelphia," *Taxon* 26, nos. 5/6 (1977): 541–48.

Labillardière's acquisition of the expedition's most important scientific work guaranteed that the d'Entrecasteaux expedition would never result in the single authoritative narrative that French authorities expected. Labillardière's own *Relation du voyage à la recherche de La Pérouse*, which appeared in 1799, was the most significant publication to emerge from the voyage. Rumors that Labillardière had his own publication plans contributed significantly to tensions aboard ship. Why should the botanist's career advance independently while his shipmates saw their scientific contributions effaced in the larger national project? From Labillardière's point of view, however, the question was why botanical knowledge—and his own specialist's role in creating that knowledge—should be submerged in a general account of French scientific prowess. Although heavy on botany, the *Relation* was in fact a narrative account of the voyage, and Labillardière was able to draw on some of his colleagues' nautical work. The expedition's artist, Jean Piron, offered Labillardière copies of his work, so the *Relation* contains images of various Pacific peoples, their homes, and their tools that support Labillardière's extensive anthropological speculations.[64]

Labillardière's *Relation* infuriated Lieutenant Paul-Edouard de Rossel, the senior surviving officer of the expedition, whose own official history (which incorporated d'Entrecasteaux's journal) did not appear until 1808. Labillardière's book was a great success, running through several French editions in its first few years of publication as well as four English editions and two German translations before 1804. In 1804–1805 his multivolume catalog of the plants of New Holland appeared.[65] The botanist completely upstaged Rossel's official account, and the lieutenant was outraged by this breach of discipline. Labillardière's book, he claimed, was "totally sterile" on every subject except botany, and in particular the book was completely lacking in geographical information.[66] Rossel missed the point here, since Labillardière saw himself primarily as a naturalist producing a work of natural history, not an official account of French scientific achievement. Nonetheless, Labillardière's *Relation* ensured that Rossel's own book was something of an anticlimax. Rossel could only publish two volumes: one with d'Entrecasteaux's narrative and the second with hydrographic and astronomical results. Labillardière had already satisfied the general audience for tales of overseas discovery, and Rossel sold poorly. Only Rossel's second volume—with the charts and hydrography—was ever translated into English. Clearly, there was not room for both private accounts of the work of individual scientists and official accounts of national achievement.

Baudin's expedition similarly failed to produce a single, authoritative account that assembled various forms of scientific expertise into a single narrative of French achievement. As with the d'Entrecasteaux voyage, the leaderless disarray of the expedition's end allowed individual scientists to pursue their own projects. In this case, the Muséum d'Histoire Naturelle intervened with the minister of the navy to request that, following Baudin's death, the naturalist François Péron be entrusted with the responsibility for the volume. The staff of the Muséum lobbied the ministries of the navy, the interior, and the colonies to pay Péron a salary for a further six years; it would be "glorious," they insisted, for the narrative of this French voyage to appear

[64] Labillardière, *Voyage in Search of La Perouse* (cit. n. 41), xi, xii.

[65] See the reprint edited by Frans Stafleu: *Novae Hollandiae Plantarum Specimen* (New York, 1966); Duyker, *Citizen Labillardière* (cit. n. 5), 230–33.

[66] "La Relation du voyage à la Recherche de M de la Peyrouse," draft letter in Rossel's hand, n.d., AN Marine 2 JJ 14 dossier 18, reel G 24,493, NLA d'Entrecasteaux.

before the publication of an English account of the contemporaneous voyage of Matthew Flinders. The English, Jussieu reminded his superiors, "despite their Establishment [Port Jackson] and resources in this new land, [had] not yet published anything that can compare to the work of our naturalists."[67]

Naming Péron to write up the voyage was a remarkable step, since it completely removed the official account from naval jurisdiction and, moreover, handed it to an individual who had embarked as a "pupil zoologist." Certainly, Péron asserted his allegiance to the idea of the voyage as a national project. Isolated, individual travelers produced contradictory information, he asserted, which could only frustrate "those who seek exactitude and truth." "Great national enterprises" such as the expedition with which he sailed were the only way to produce reliable information.[68] Péron, however, had his own career to make as well as grievances left over from the voyage to pursue. His *Voyage de découvertes aux terres australes* not only focused on natural history but also effaced most of his colleagues' contributions.[69] In particular, Péron's account of the voyage devoted considerable attention to the inadequacies of Nicolas Baudin, whom Péron despised. Baudin's name never actually appears in the text at all—a remarkable break with naval tradition—but the incompetence and malice of the unnamed commander are themes that run through the book.[70] Péron was certainly not interested in composing an account of the voyage that depicted science and the navy united in the project of French discovery, and the completed text makes it perfectly clear that individual savants succeeded in producing scientific knowledge in spite of the expedition's command structure, not because of it.[71]

Expeditions to the Pacific after Baudin were far more successful in producing authoritative accounts of French achievement written in the anonymous and collective voice of the commander. In effect, later voyages simply eliminated the tensions between the pursuit of individual careers and of national achievement by drawing all scientific personnel from naval ranks. The increasing importance of national institutions and laboratory science encouraged scientists to remain in Paris and leave the field collecting to well-disciplined travelers.[72] French scientists increasingly saw the work of expeditions and the work of science as divergent, and they were happy to leave each with its own chain of command. Thus the 1837 expedition led by Jules

---

[67] Jussieu's report on the collections of the Baudin expedition, 8 Messidor year 12 (27 June 1804), Archives du Muséum d'histoire naturelle de Paris, reel 24, NLA Baudin.

[68] François Péron and Charles-Alexandre Lesueur, "Observations sur le tablier des femmes hottentotes," *Bulletin de la Société Zoologique de France* 8 (1883): 22. Although not published until 1883, the paper was originally presented in 1805.

[69] For instance, Péron altered the taxonomic name of a newly discovered species, replacing the name of the deceased zoologist René Maugé with that of his friend Lesueur: Michel Jangoux, "L'expédition du Capitaine Baudin au Terres Australes: Les observations zoologiques de Francois Péron pendant la première campagne (1801–1802)," *Annales du Muséum du Havre* 73 (2005): 1–35.

[70] To the extent that reassessing Baudin's reputation has become a major concern of recent work on the expedition: Horner, *French Reconnaissance* (cit. n. 7); Jean Fornasiero, Peter Monteath, and John West-Sooby, *Encountering Terra Australis: The Australian Voyages of Nicolas Baudin and Matthew Flinders* (Kent Town, Australia, 2004).

[71] Péron enlisted Lieutenant Louis Freycinet to assemble the expedition's cartographic work in a second volume, which Freycinet completed and published in 1816 after Péron's early death.

[72] Richard W. Burkhardt, "Naturalists' Practices and Nature's Empire: Paris and the Platypus, 1815–1833," *Pacific Science* 55, no. 4 (2001): 327–41; Marie-Noëlle Bourguet, "La Collecte du monde: Voyage et histoire naturelle (fin XVIIe siècle–début XIXe siècle)," in *Le Muséum au premier siècle de son histoire*, ed. C. Blanckaert et al. (Paris, 1997), 163–96.

Dumont d'Urville, which landed on the Antarctic continent, succeeded in producing precisely the kind of official account its predecessors missed.[73] Dumont d'Urville survived the voyage and returned to France to write a twenty-three-volume work that integrated his own narrative account with the scientific research of the expedition's geographers, astronomers, and naturalists, none of whom challenged his textual authority. Crucially, all of Dumont d'Urville's savants were naval officers, and usually they were doing double duty, like the expedition's medical staff, René-Constant Quoy and Paul Gaimard, who produced the most significant natural history of the expedition. Both of these philosophical navigators were making careers in the navy, and they had no incentive to hold back specimens or to rush to publication.

By the time that French interest in the Pacific had moved on to Antarctic discovery, the revolutionary notion that Frenchmen had a particular capacity to produce or at least to appreciate science had developed into more limited nineteenth-century ideas about the citizen-scientist. Revolutionaries were confident about the philosophical bent of all Frenchmen, who needed only to be freed from corporate oppression and given appropriate training in the expansive educational system that revolutionaries imagined but never established. Pacific voyages reflected this optimism about the unbounded capabilities of liberated Frenchmen. Postrevolutionary observers were less sanguine about the capacities that Frenchmen shared, however, and postrevolutionary regimes were anxious to limit the political nation to those who could demonstrate their suitability. By the first decades of the nineteenth century, discussions of how to allocate suffrage focused on scientific capacity, among other attributes such as wealth, that, far from being universal, were specific to those individuals who, in the government's view, merited full citizenship. Evidence of scientific abilities such as membership in a learned society testified to an individual's suitability for political responsibilities such as participation in a jury or voting.[74] Just as the French navy concluded that ships full of philosophical discoverers were illusory—or at least undesirable—the French state retreated from revolutionary ideals about universal rationality and democratic citizenship.

National investment in the Pacific suggests that Tocquevillian questions about continuity and transformation in French science across the revolutionary period need to be investigated in layers; new practices and rhetorics often underlay what appeared to be a continuous tradition. Certainly the d'Entrecasteaux and Baudin voyages showed affinities with Old Regime science and Enlightenment ideas about the Pacific; in particular they relied on the public funding and institutional support that had characterized science under France's absolute monarchy. Their ambitions were encyclopedic, ranging from practical desires to discover, transfer, and cultivate useful species to more amorphous aspirations to identify humans in a state of nature. These voyages were revolutionary ventures, however, and confidence in the limitless capabilities of regenerated Frenchmen inspired both their conception and their conduct. When

---

[73] Jules Dumont d'Urville, *Voyage au Pôle Sud et dans l'Océanie, sur les corvettes l'Astrolabe et la Zélée*, 23 vols. (Paris, 1841–54); Danielle Clode and Carol E. Harrison, "Savants, Sailors, and Surgeons: Natural History Publishing and French Expeditions to Australia" (unpublished paper presented to the Colloque Terres Australes, Le Havre, France, December 2007).

[74] Pierre Rosanvallon, *Le Sacre du citoyen: Histoire du suffrage universel en France* (Paris, 1992); and Carol E. Harrison, "Citizens and Scientists: Toward a Gendered History of Scientific Practice in Postrevolutionary France," *Gender and History* 13, no. 3 (2001): 444–80.

the centralized institutions of French science such as the botanical garden and the
academy ceased to depend on royal patronage and became expressions of the will of
the sovereign people, scientists who depended on them had to alter their rhetoric and
their practice. The experience of French science in the Pacific—both the early enthusiasm for discovery and the later jealousies and jockeying for precedence—reveals
the adjustments demanded by a Revolution that claimed science as part of its project
of national regeneration.

Material Experiments:
Environment and Engineering Institutions in the Early American Republic

By Ann Johnson[*]

ABSTRACT

In nineteenth-century America, strength of materials, an engineering science, focused on empirical research that yielded practical tools about how to predict the behavior of a wide variety of materials engineers might encounter as they built the nation's infrastructure. This orientation toward "cookbook formulae" that could accommodate many different kinds of timber, stone, mortar, metals, and so on was specifically tailored for the American context, where engineers were peripatetic, materials diverse, and labor in short supply. But these methods also reflected deeper beliefs about the specialness of the landscape and the providential site of the American political experiment. As such, engineers' appreciation of natural bounty both emerged from and contributed to larger values about exceptionalism and the practical character of Americans.

INTRODUCTION

In the overview to his edited collection of articles on American engineering, historian Terry Reynolds argues that a distinctly American style of engineering emerged in the nineteenth century. American engineering had two parents, according to Reynolds, one French and institutional, the other British and empirical. He writes, "As these traditions blended in the 1800s, however, they yielded a distinctly American engineering tradition, better adapted to American economic and political conditions and American social values than either of its predecessors."[1] Rather than disputing the distinctiveness or parentage of American engineering, in this paper I will present a third element that shaped American engineering practice. This factor also played a critical role in the shaping of American identity and culture, yet it has often been ignored there as well. This factor is the environment, what nineteenth-century engineers

[*] Department of History, University of South Carolina, Columbia, SC 29208; annj@sc.edu.

I particularly wish to thank the Charles Warren Center for Studies in American History at Harvard, led by Laurel Thatcher Ulrich and organized in 2001–2002 by Joyce Chaplin and Charles Rosenberg, on "Exceptional by Nature? American Science and Medicine, 1600–1900." Conversations with the other fellows, Jorge Cañizares-Esguerra, Cornelia Dayton, Mordechai Feingold, Marina Moskowitz, Katherine Pandora, Susan Scott Parrish, and Alice N. Walters, challenged and substantially deepened my thoughts on the role of science, technology, and medicine in the construction of national identity. Special thanks to Jay Richardson for his dedication, enthusiasm, and research skills.

[1] Terry S. Reynolds, "Overview: The Engineer in 19th Century America," in *The Engineer in America: A Historical Anthology from Technology and Culture* (Chicago, 1991), 7.

© 2009 by The History of Science Society. All rights reserved. 0369-7827/09/2009-0004$10.00

would have called "nature." Both engineering practices and American identity drew from the unique physical experience of the continental nation, gained through efforts to conquer, settle, improve, shape, engineer, and develop that land.

The role of environment in shaping values and perceptions through the construction of infrastructure has come under close scrutiny by environmental historians since the 1970s. Still, the project remains incomplete. Environmental historian Ted Steinberg has argued that while earlier historians such as Frederick Jackson Turner, Samuel Hays, Roderick Nash and the Annales historians did integrate nature and the environment into their studies, the development of the professional subfield of environmental history has actually marginalized studies of nature. Environmental history has served to remove these concerns from social, cultural, political, and economic history.[2] A similar claim could be made for the history of science and technology; their maturity as professional subfields has led to their relative isolation from "mainstream" history. As a result, the history of science, technology, and environment are rarely closely examined in histories of national identity, even when the nations in question developed crucial scientific elements.[3]

Nowhere is the development of a scientific character as a dimension of national identity clearer than in the United States, where a particular vein of science—practical science—was closely aligned with Americans' self-perception by the late nineteenth century. Americans saw themselves as a practical people—this characterization even elicited pride—and science that served practical ends distinguished such a people. However, this alignment of scientific and cultural values has not attracted much attention from either historians of science or from American historians, in part because of the way American science has been studied. Following the lead of I. Bernard Cohen, historians have spilled far more ink lamenting Americans' failure to develop "proper" theoretical and basic science (so-called pure science).[4] The absence of theoretical science represented a scientific lacuna—especially following the nation's promising beginning under the intellectual and political leadership of Benjamin Franklin and Thomas Jefferson; according to this argument, nineteenth-century America was an unscientific, even antiscientific, nation. America's emergence as a scientific power in the twentieth century appeared to be a relatively late development spurred by an increasingly industrial economy, the World Wars, and the decline of competitors such as Germany and Britain. However, this assumption that science equals theoretical science has obscured the scientific knowledge that Americans were furiously producing in the antebellum period; much of that production would be considered "ap-

---

[2] Ted Steinberg, "Down to Earth: Nature, Agency, and Power in History," *American Historical Review* 107 (June 2002): 798–820.

[3] See Daniel J. Boorstin, *The Americans: The National Experience* (New York, 1967); Perry Miller, *The Life of the Mind in America from the Revolution to the Civil War* (New York, 1965); John Kasson, *Civilizing the Machine: Technology and Republican Values in America, 1776–1900* (New York, 1976); John R. Stilgoe, *The Common Landscape of America, 1580–1845* (New Haven, Conn., 1982); William Goetzmann, *Exploration and Empire: The Explorer and Scientist in the Winning of the American West* (New York, 1966). All of these books integrate the history of science and technology into more general historiographical concerns in U.S. history in ways that are much less common today than they were a generation ago.

[4] I. Bernard Cohen, "Science and the Growth of the American Republic," *Review of Politics* 38 (1976): 359–98. For a contrary view, see Nathan Reingold, "American Indifference to Basic Research: A Reappraisal," in *Nineteenth Century American Science*, ed. George H. Daniels (Evanston, Ill., 1972), 38–62. For details of the pure science ideal in America, see David Hounshell, "Edison and the Pure Science Ideal in the 19th Century," *Science* 207, no. 4431 (1980): 612–17.

plied" science—that is, engineering, agriculture, and public health.[5] In fact, none of these fields relied on applications of basic science; all produced their own corpus of knowledge, driven by the social, political, and commercial needs of building a nation.[6] Examining these kinds of science yields an account of American scientific development that can and should be integrated into questions of American national identity. It is notable that the sciences of engineering, public health, and agriculture are all predicated in some way on environmental conditions, thus setting the stage for American claims about the exceptional nature of the physical environment or landscape flowing freely from science to broader culture and vice versa.

The environment played a critical role in shaping American engineering practices, and those practices yielded views about development and internal improvement that were, at least partially, constitutive of early American identity. Engineers found themselves in a mediating role, both between land and landscapes and between citizens and government. Engineers were often closely tied to government, as many of the projects they designed were government funded.[7] Even on projects not financially dependent on the state, engineers and their work served to reconfigure political, economic, and natural landscapes in ways that the state had to accommodate. The projects detailed in this paper were government funded and of national significance in the process of literal nation building. As such, these engineering efforts symbolized the government's dynamic vision for the nation and acted as a conduit for the transfer of perceptions of the land back to metropolitan politicians.

This paper focuses on the processes and activities through which engineers came to know the landscape and particularly came to know the resources American land could offer. Americans knew land; national identity was dependent on what Martin Brückner calls "a geographic imagination."[8] The perception that the American landscape offered a uniquely abundant array of resources was a widespread and common assumption of eighteenth- and nineteenth-century Americans, citizens and leaders alike. As Samuel Smith claimed in his address to the American Philosophical Society in 1797, "Opportunities for research on this continent could not fail to elevate the United States far above other nations."[9] Jared Mansfield, surveyor general under Jefferson and a West Point professor, wrote his wife during the survey, "Nature has done enough for this country."[10] Mansfield's nephew Joseph Totten, the main character of the story told here, claimed that "the providential diffusion over our territory of an admirable material" created ideal conditions for the production of both new knowledge and a novel nation-state.[11] No one saw this phenomenon more clearly than did

---

[5] On nonengineering applied sciences, such as public health, medical, and agricultural sciences, in eighteenth- and nineteenth-century America, see Charles Rosenberg, *No Other Gods: On Science and American Social Thought*, 2nd ed. (Baltimore, 1997); and Joyce Chaplin, *An Anxious Pursuit: Agricultural Innovation and Modernity in the Lower South* (Chapel Hill, N.C., 1993).

[6] See Edwin Layton Jr., "Mirror-Image Twins: The Communities of Science and Technology in 19th-Century America," *Technology and Culture* 12 (1971): 562–80.

[7] For comparative accounts of the interactions of engineers and nation-states, see the fall 2007 issue of *History and Technology* focusing on engineering and national identity.

[8] Martin Brückner, *The Geographic Imagination in Early America: Maps, Literacy, and National Identity* (Chapel Hill, N.C., 2006).

[9] Smith, quoted in Miller, *Life of the Mind in America* (cit. n. 3), 284.

[10] Jared Mansfield to Elizabeth Mansfield, 27 Oct. 1804, folder 2, box 1, collection 68, Jared Mansfield Papers, Ohio Historical Society, Columbus (hereafter cited as Mansfield Papers).

[11] Lt. Col. J. G. Totten, "On Hydraulic and Common Mortar," *Journal of the Franklin Institute* 20 (Oct. 1837): 230.

engineers and technicians—especially the nation's surveyors—working to design the nation's infrastructure.

Surveying was an important scientific activity of the antebellum period.[12] Surveys were essential for defining the boundaries of the nation, understanding its resources, and moving European Americans onto western lands. Although historians of science have often overlooked surveys as scientific endeavors, in the period between the late sixteenth and early nineteenth centuries, it was quite common to see reports of new methods in surveying in the scientific journals, particularly the *Philosophical Transactions of the Royal Society.*[13] Surveying, alongside navigation, was applied astronomy, complicated by the need to inscribe a two-dimensional map onto three-dimensional land. Historians of cartography have commented at length about the difficulties and techniques of representing three-dimensional land on two-dimensional maps, but the problems of inscribing that map back onto land have, by and large, been overlooked.

In eighteenth- and nineteenth-century America, surveying activities often overlapped engineering; many engineers did surveying work as a way to make more money. Many engineers rose from positions in surveys to construct key engineering works such as the Erie Canal and the Baltimore and Ohio Railroad. According to a New York comptroller in the 1870s:

> I know that the commissioners thought they must send for some great engineer from the Duke of Bridgewater's canal, to teach us how to build a canal, fearing to trust our commonplace Americans. At length they settled upon that plain, unsophisticated, and unpretending land surveyor[—]nothing but a land surveyor—Benjamin Wright for the engineer on Erie and James Geddes on the Champlain. From this "school" arose nearly all the canal engineers who have lined the country with their works of internal improvements.[14]

The organization of surveying provided an apprenticeship through which men with ambition, but little formal education, could advance. Furthermore, although surveyors were usually quite well paid, surveying also provided opportunities to make (and lose) great fortunes through land speculation.

To some extent, particularly among federal politicians, the voices of surveyors and engineers helped establish the view that America's resources were uniquely rich. This perception of richness had religious overtones that converged with Protestant American beliefs about Providence.[15] Although providential thinking and engineering may seem to be orthogonal, in fact, in the minds of many Americans, they converged. For example, in an 1851 sermon at Boston's Hollis Street Meeting House, Rev. Thomas King claimed:

> Providence had another and a higher use for those iron tracks and flying trains. After the mercantile heart had devised and secured them, God took them for His purposes:

[12] See William H. Goetzmann, *Army Exploration in the American West, 1803–1863* (New Haven, Conn., 1959); see also Goetzmann, *Exploration and Empire* (cit. n. 3).

[13] See, e.g., Charles Mason and Jeremiah Dixon, "Observations for Determining a Degree of Latitude in the Provinces of Maryland and Pennsylvania, in North America," *Philosophical Transactions* 58 (1768): 274–328.

[14] Charles Stuart, *Lives and Works of the Civil and Military Engineers of America* (New York, 1871).

[15] Nicholas Guyatt, *Providence and the Invention of the United States, 1607–1876* (New York, 2007).

without paying any tax for the privilege. He uses them to quicken the activity of men; to send energy and vitality where before were silence and barrenness . . .Without any vote of permission from legislatures and officials,—even while cars are loaded with profitable freight and paying passengers, and the groaning engines are earning the necessary interest,—Providence sends without charge its cargoes of good sentiment and brotherly feeling . . . entwining sea-shore and hill-country, mart and grain-field, forge and factory, wharf and mine, slowly prepares society to realize, one day, the Saviour's prayer, "that they may all be one."[16]

American providence aligned with views of America as a divinely authorized political experiment; both claims were closely related to early American scientific endeavors. Albert Barnes, speaking to Hamilton College in 1836, concluded, "It seems almost as if God, in favor to science and the enlargement of the human mind, had reserved the knowledge of the western world, until almost the last felicitous investigations that could be made had been made in the old world."[17]

Through surveys and peripatetic surveying careers, early American engineers came to realize that the country possessed a wide array of resources out of which the nation would be built. One question came to the fore for most engineers: how to best utilize resources, particularly given the labor shortage that most frontier locations suffered from. Overbuilding, or requiring more material than structurally necessary, strained already scarce labor. Ideally, for completely practical reasons, engineers needed to figure out how to use materials efficiently. But different materials clearly had different capacities; some might be stiffer, others stronger but more brittle, others prone to change significantly with temperature or humidity fluctuations. When the materials encountered were unfamiliar, how could one know how a material might perform?

In response to these questions, engineers invented new ways of performing engineering design, in large part to accommodate the broad array of materials they encountered but also to make engineering design accessible to individuals without formal training or well-established modes of apprenticeship. American engineers had to design and build structures from the materials they found in the field; as a result, they needed to know how to predict the various physical properties of those materials. For American engineers, these were not theoretical issues but rather empirical ones. Starting in the 1830s, engineers began writing large numbers of inexpensive and mathematically accessible handbooks to provide information on various materials to other engineers working in the field; the numbers had exploded by the 1850s. The title of an early handbook explains what it is for: *A Compendium of Mechanics or Textbook for Engineers, Millwrights, Machine-makers, Founders and Smiths containing practical rules and tables connected with the Steam Engine, waterwheel, pumps and mechanics in general; Also examples for each rule, calculated in decimal arithmetic, which renders the treatise particularly adapted for operative mechanics.*[18]

[16] Thomas Starr King, *Two Discourses Delivered in Hollis-Street Meeting-House, Sunday, September 21, 1851* (Boston, 1851), 17–18. Thanks to Philip Jay Richardson for locating this source.

[17] Quoted in Miller, *Life of the Mind in America* (cit. n. 3), 284.

[18] Robert Bruton, *A Compendium of Mechanics or Textbook for Engineers, Millwrights, Machine-makers, Founders and Smiths containing practical rules and tables connected with the Steam Engine, waterwheel, pumps and mechanics in general; Also examples for each rule, calculated in decimal arithmetic, which renders the treatise particularly adapted for operative mechanics* (New York, 1830). There are literally hundreds of similar books, including W. E. Worthen, *Appletons' Cyclopedia of Drawing designed as a textbook for the Mechanic, Architect, Engineer and Surveyor* (New York,

This orientation toward empiricism is something that Alexis de Tocqueville focused on in *Democracy in America*.[19] Chapter ten is titled "Why Americans are more attracted to practical rather than theoretical aspects of the Sciences," and Tocqueville wrote, "You may be sure that the more a nation is democratic, enlightened, and free, the greater the number of those self-interested champions of scientific genius and the more profit will result from discoveries immediately applicable to industry."[20] Later he added, "Those same Americans who have not laid eyes upon a single general law of mechanics have changed the face of the world with a new machine for navigation."[21] However, by reducing practical science to invention and machinery, Tocqueville also devalues American contributions to the engineering sciences. Clearly, there exists a middle ground between general laws of mechanics and machine invention, and although Americans did excel at invention, it is also important that they pioneered the development of engineering science. Eda Kranakis shows this middle ground in her study by comparing the "ideologies of practice" of American and French engineers.[22] Her conclusion is Tocquevillean: American engineers were more empirical, their research practices more deeply rooted in craft traditions, and the rural and frontier experiences more central to understanding such ideologies of practice.[23]

In this paper, the middle ground between science, on the one hand, and invention and craft tradition, on the other, is represented by the strength of materials, an engineering science developed through the rigorous and systematic empirical testing of American materials at several sites. Strength of materials has a long and distinguished pedigree; it was one of Galileo's "Two New Sciences." However, until the Industrial Revolution, strength of materials was not a practical science; therefore, the development of an empirical science of strength of materials was a critical stage in the development of scientific, professional engineering. The history of strength of materials shows how engineering knowledge and practices were constructed. American scientists and engineers in the 1830s began to use and extend the science of strength of materials to help design and build a wide variety of artifacts, from fortresses to steam boilers. These engineers addressed practical problems in a new way by performing experiments on local materials, mimicking European practices but developing authentically American knowledge stores.

1857); Simeon Borden, *A System of Useful Formulae adapted to the practical operations of locating and constructing railroads* (Boston, 1851); Charles Davies, *Elements of Surveying and navigation with Descriptions of Instruments and their Necessary Tables* (New York, 1854); Davies, *First Lesson in Geometry with practical applications in Mensuration and Artificers' Work and Mechanics* (Hartford, Conn., 1840); C. S. Cross, *Engineer's Field Book* (New York, 1855).

[19] See Alexis de Tocqueville, *Democracy in America*, trans. Gerald E. Bevan (London, 2003), vol. 2, chap. 10, 529–37.

[20] Ibid., 534.

[21] Ibid.

[22] Eda Kranakis, *Constructing a Bridge: An Exploration of Engineering Culture, Design, and Research in Nineteenth-Century France and America* (Cambridge, Mass., 1997).

[23] Ibid., 308.

JOSEPH TOTTEN: LEADER OF AMERICAN MATERIALS RESEARCH
IN THE NINETEENTH CENTURY

Joseph Gilbert Totten was born in 1788 in New Haven, Connecticut.[24] His mother died when he was three years old, and his father subsequently moved to the West Indies to become a vice-consul. Consequently, Totten's upbringing fell to his uncle, Jared Mansfield. Although this sequence of events hardly seems joyful, it did put his care in the hands of a man who set him on a scientific path. A Yale graduate, Mansfield in 1801 published *Essays Mathematical and Philosophical*, the first original treatise on mathematics written by a native-born American.[25] Mansfield belonged to the scientific elite in the late eighteenth and early nineteenth centuries, corresponding with European natural philosophers and mathematicians, as well as with other elite American intellectuals such as Thomas Jefferson and Jefferson's secretary of the treasury, Albert Gallatin.

In 1802, Mansfield earned an appointment as the first professor of mathematics and natural and experimental philosophy at the newly opened United States Military Academy (USMA) at West Point.[26] In 1803, fourteen-year-old Totten enrolled at the academy and was commissioned in 1805 as a second lieutenant in the Army Corps of Engineers. While at the academy, he took a leading role in establishing the United States Military Philosophical Society (USMPS), a group of cadets and professors, who, under the leadership of Superintendent Jonathan Williams, gathered to discuss scientific work and research. Mansfield was also a founding member. The society's technical studies represented a diverse group of subjects, but a dedication to experiment tied the researches together. Sidney Forman's article on the USMPS emphasizes this, claiming "Experiment was employed as a method of teaching the cadets who were immersed in an environment of technical learning. No branch of scientific study was beyond their interest."[27] The central place of experiment in Totten's education stayed with him throughout his career, as would eclectic scientific interests.[28]

Totten resigned from the army shortly after his commission to assist Mansfield, who had taken a position as surveyor general of the Ohio and Western territories. Mansfield, with Totten as his assistant, was responsible for extending the federal rectangular survey west into the Indiana Territory and for correcting the mathematical mistakes of Rufus Putnam, the first surveyor general and a crony of George Wash-

[24] For Totten's biographical details, see his entry in the *Dictionary of American Biography* and the United States Military Academy Register of Graduates, available at http://www.aogusma.org/pubs/ register/totten.htm; John Gross Bernard, "Biographical Notice of Joseph Gilbert Totten," in *Annual of the National Academy of Sciences 1865* (Cambridge, Mass., 1866); Mary Margaret Thomas, "Science Military Style: Fortifications, Science, and the U. S. Army Corps of Engineers" (PhD diss., Univ. of Minnesota, 2002).

[25] It should be added that Spanish Americans were working far in advance of Anglo-Americans, producing many mathematical texts by the turn of the nineteenth century.

[26] Peter Michael Molloy, "Technical Education and the Young Republic: West Point as America's École Polytechnique, 1802–1833" (PhD diss., Brown Univ., 1975); Terry S. Reynolds, "The Education of Engineers before the Morrill Act of 1862," *History of Education Quarterly* 32 (Winter 1992): 459–82.

[27] Sidney Forman, "The United States Military Philosophical Society, 1802–1813," *William and Mary Quarterly*, 3rd ser., 2 (1945): 278.

[28] Although not a topic of this paper, Totten performed important work in natural history, particularly in conchology, in which he collected and classified several shells, two of which have been named for him.

ington's.[29] Totten was hired specifically to make astronomical observations and to complete calculations to determine a new meridian—that is, a true north-south line that the rectangular survey would use as a baseline. Totten's work was data gathering, although such a term did not exist in the vocabulary of early American surveyors. This mission was also empirical in its nature, and Totten saw more than a thousand miles of the nation—enough to become convinced that the variety of landscapes, tree species, and other natural resources were both an example of God's great bounty and the great scientific and engineering challenge of the American nation.

After this project and initial exposure to the American landscape, Totten returned east in 1808. Upon his return, he received reappointment to the army in the same commissioning ceremony as future West Point Superintendent Sylvanus Thayer. Jonathan Williams, as chief of the Corps of Engineers, then sent Totten to oversee the construction of fortresses guarding New York Harbor. During this period, Totten also worked with Thayer under the supervision of Joseph G. Swift, the first graduate of West Point and future chief engineer of the U.S.A., in the inspection and repair of New England coastal forts.[30] The network of West Pointers at work constructing the nation's defenses helped create a privileged position for these half dozen or so men. They had a disproportional effect on nation building and on the development of particularly American ways of knowing in engineering. The first project of this network was to oversee the development of a national system of coastline defense—and although this essay is less concerned with fortification design than with experimental research on materials, the two projects came together in Totten's longest single assignment: the design and construction of Fort Adams in Newport, Rhode Island.

NARRAGANSETT BAY AND THE FIRST FORT ADAMS

Narragansett Bay provides Providence and Newport, Rhode Island, two important eighteenth- and nineteenth-century ports, with deepwater harbors and access to the Atlantic. The bay has the distinction of being the only American harbor on the Atlantic coast that vessels can enter with a northwesterly wind; consequently, it has been called the "best naval refuge in the United States."[31] Newport Harbor, a part of the bay, is protected from the Atlantic by islands and spits that create three passages into the harbor. Fort Adams sits on the southwestern entrance to the harbor on a neck of land called Brenton's Point. An earthen fort was originally constructed on the site in the summer of 1776.[32] Nevertheless, the British took Newport in December of that year, since Brenton's Point protects only one entrance to Newport Harbor. During the war, the American colonists, the British, and the French each successively held Newport, emphasizing how difficult it proved to hold this precarious location, with its multiple approaches.[33] Given the commercial importance of Providence and Newport, the new

---

[29] Jared Mansfield to William Lyon, 20 Feb. 1804, folder 2, box 1, collection 68, Mansfield Papers.

[30] For biographical details of Swift, see his Register of Graduates entry, http://www.aog.usma.edu/pubs/register/swift.htm.

[31] Aubrey Parkman, *Army Engineers in New England: The Military and Civil Work of the Corps of Engineers in New England* (Waltham, Mass., 1978), 9.

[32] Theodore L. Gatchel, "The Rock on Which the Storm Will Beat: Fort Adams and the Defenses of Narragansett Bay," at http://www.fortadams.org/history.htm. See also Parkman, *Army Engineers in New England* (cit. n. 31), 9–10.

[33] Bvt. Major-General George W. Cullum, *Historical Sketch of the Fortification Defenses of Narragansett Bay* (Washington, D.C., 1884), 8–11.

American government had to concern itself with a plan for protecting Narragansett Bay. In 1792, Lt. Col. Louis Tousard, a French engineer who once served the marquis de Lafayette, arrived in the United States, fleeing the French Revolution. By 1798, Tousard had planned out a new system of defense for Narragansett Bay, and the fortress on Brenton's Point composed only one piece of the plan. Tousard's plan detailed the need for four forts to fully protect the bay; only two were ever completed. Tousard completed one fort on Brenton's Point in 1799, naming it Fort Adams, after the sitting U.S. president, John Adams.

Between 1799 and 1808, Fort Adams, along with most coastal forts, fell into disrepair due to federal neglect and underfunding. An effort begun by Swift in 1808 to inspect and repair Atlantic fortresses soon became entangled in disagreements between Swift and the War Department.[34] In examining Fort Adams, Swift argued it should be replaced, while the War Department preferred to first repair all existing structures before considering any new projects. Despite his conviction that Fort Adams would be ineffective in defending Newport, Swift worked with the assistants assigned to him, Totten and Thayer, to restore Fort Adams to Tousard's design.

Although Swift's doubt did not prove prescient in the particular case of Fort Adams, war in 1812 did validate his concern over the porousness of the coastal defense system. The British blockade of New England made the defense of ports in the area irrelevant. The invasion of Washington, however, drove home the need for better coastal and harbor defense. Although Totten and other West Point–trained engineers proved themselves indispensable to the war, the long Atlantic coast proved impossible to defend effectively. In assessing Revolutionary War–era forts, Totten's later assistant J. G. Barnard wrote, "[M]ost of the works were . . . defective in design, small, weak, and, being built for the present economy of cheap materials and workmanship, very perishable."[35] Still none of the forts constructed by West Pointers fell to the enemy, a point used after the war to argue for new fortresses rather than repairs to Revolutionary War–era structures.[36] Henry Adams even argued that "had an engineer been employed at Washington by Armstrong and Winder, the city would have been saved."[37] Totten, as the successful chief engineer of the fortifications at Plattsburgh, New York, was representative of the West Point engineers to whom Adams was referring.

TOTTEN AND THE NEW FORT ADAMS

The convergence of Totten's success and the older forts' failures meant that when government attention turned to restructuring coastal defenses after the war, the newly promoted Major Totten earned an appointment to the Board of Engineers. The board consisted of a small group of engineers who recommended and prioritized internal improvements important to commerce and defense. Overlooking the success of West Pointers in the War of 1812, the War Department decided in 1818 to hire French engineer Simon Bernard to head the board. Immediately, Swift and William McRee, another West Pointer and USMPS member, resigned in protest over the elevation of a foreigner. Totten agreed that Bernard's appointment was "deplorable." He wrote in an 1823 letter to Secretary of War John C. Calhoun that Bernard's appointment

---

[34] Parkman, *Army Engineers in New England* (cit. n. 31), 13.
[35] Bernard, "Biographical Notice of Joseph Gilbert Totten" (cit. n. 24), 61.
[36] Forman, "United States Military Philosophical Society" (cit. n. 27), 284.
[37] Henry Adams, *History of the United States* (New York, 1889–91), 9:235.

and continued employment represented the "injustice of excluding native talent" and that Bernard's expertise was not applicable to the unique challenges of defending "a country, the like of which he has never seen; and where, the character of the war, derived from the nature of the country, must be such as to entirely exclude the benefits of his experience."[38] Yet Totten did not follow his comrades in resigning from the board, and as a result, Totten was initially one of only two board members, and as the author of the reports emanating from the board, he wielded considerable power. In these reports to Congress and the War Department, Totten repeatedly illustrated the weaknesses of coastal defense and particularly drove home the point that Tousard's Fort Adams would not protect commercial interests in either Newport or Narragansett Bay. Newport needed a modern masonry fortress.[39] Totten declared, "Unless the main works be competent to withstand a siege of a few days, they will not fulfill their trust and will be worse than useless."[40] In 1824, the Monroe Doctrine led to increases in the defense budget, and the government razed the old fort to clear space for the new Fort Adams.

In the summer of 1825, Totten moved to Newport to oversee the construction of the new Fort Adams. The design of the new fort proved innovative; taking ideas from Marc René, marquis de Montalembert and Sébastien Le Prestre, marquis de Vauban, Totten added his own ideas drawn from inspections of dozens of less formal defenses. Totten's design was massive, covering twenty acres. The fort was irregularly shaped, with bastions on the seafront and Vaubanian earth-filled ramparts on its rear to protect against a land attack. To deter an attack from higher ground, Totten added a redoubt.[41] The resulting design stood as the most innovative fort of the so-called Third System, planned by Totten and Simon Bernard.[42]

FORT ADAMS AS A "RESEARCH SCHOOL"[43]

While construction on Fort Adams began with a $50,000 allocation, nearly half of that was used to purchase land to enlarge the facility, and subsequent funding proved irregular and inadequate.[44] Totten did not sit by idly and accept the continual underfunding of fort construction, particularly on his pet project at Fort Adams. Totten urged more spending on fortification as a critical matter of a four-part interdependent national defense system, which included the navy, fortifications, an interior communications network, and the army and militias. He wrote,

---

[38] Joseph Totten to John C. Calhoun, 18 Dec. 1823; Totten Letterbooks, entry 146, record group 77, National Archives Building, Washington, D.C.

[39] Gatchel, "Rock on Which the Storm Will Beat" (cit. n. 32).

[40] "Report of the Board of Engineers," quoted in Cullum, *Historical Sketch* (cit. n. 33), 29.

[41] For descriptions of Fort Adams, see Cullum, *Historical Sketch* (cit. n. 33); Bernard, "Biographical Notice of Joseph Gilbert Totten" (cit. n. 24); Parkman, *Army Engineers in New England* (cit. n. 31); and Gatchel, "Rock on Which the Storm Will Beat" (cit. n. 32).

[42] See Robert S. Browning, *Two if by Sea* (Westport, Conn., 1983); Emanuel Raymond Lewis, *Seacoast Fortifications of the United States* (Washington, D.C., 1970); Bernard, "Biographical Notice of Joseph Gilbert Totten" (cit. n. 24), 79.

[43] For historiography of research schools see, particularly, Gerald Geison and Frederic L. Holmes, eds., *Research Schools: Historical Reappraisals, Osiris* 8 (1993); Geison, *Michael Foster and the Cambridge School of Physiology* (Princeton, N.J., 1978); and Kathryn Olesko, *Physics as Calling* (Ithaca, N.Y., 1991).

[44] The $50,000 figure comes from Cullum, *Historical Sketch* (cit. n. 33), 30. On irregular funding, see Parkman, *Army Engineers in New England* (cit. n. 31), 17–18.

It is important to notice the reciprocal relation of these elements of national defense; one element is scarcely more dependent on another than the whole system is on each one. Withdraw the Navy and the defense becomes merely passive. We expose ourselves the more to suffer the evils of war, at the same time we deprive ourselves of all mean of inflicting them. . . . Withdraw fortifications and interior communications are broken up and the Navy is left entirely without collateral aid.[45]

Addressing the funding problem would take years of reports and requests to Congress, and to a great extent, Congress never would provide the allocations required for the system of forts envisioned by Totten.[46] Then again, facing funding difficulties simply redirected Totten's attention to problems in fortress design and construction that he did control.

Over the course of a decade of inspections, Totten had come to the conclusion that fortress design in the United States was flawed. He felt that the designs needed to take into consideration the nature of the materials from which American forts were constructed and the unique sites of the forts. The land-based orientation of Vauban and Montalembert simply did not suit the American condition. Totten was not satisfied with duplicating French plans; he wanted to produce new French-style military science on the proper design of forts for the American locations. Vauban and Montalembert's works were not mere cookbooks of "off-the-shelf" designs—they were not useful in that sense—but rather examples of the kind of science Totten expected to produce. Consequently, given a situation in which he lacked the immediate and regular financial allocations to build the fortress he felt the site called for, Totten instead developed a major research program investigating the question of how to design and build the best fortification for Brenton's Point. Over the thirteen years he spent in Newport, Totten did design and construct Fort Adams, but ultimately the research into design and materials testing he performed there proved to be more important work.

Totten planned to make Fort Adams, as a place for experimental research, a site for learning, a kind of a postgraduate laboratory in fort design and construction. Totten knew through his own experience under Mansfield, Williams, and then Swift that engineers, even West Point graduates, learned on the job, and he wanted to help fulfill this requirement.[47] In Fort Adams, he had a long-term, complicated design and construction project where he could take advantage of his connections with Thayer and West Point. Thayer would send him the best West Pointers, and after giving them practical experience, Totten would send the now more experienced and qualified engineers on to other projects. The USMA connection was important to ensure that Totten's assistants were the kind of officers who wanted to engage in research. This notion of fortress as training ground, coupled with funding difficulties that slowed construction, led directly to Totten's research program in strength of materials. Furthermore, the engineers who passed through Fort Adams between 1825 and 1838 experienced firsthand the place engineering research could have in the actual work of building. Many continued this type of work elsewhere.

When Totten started researching materials at Fort Adams, he initially focused on

---

[45] Joseph Totten, "Report to Congress," in *American State Papers, Military Affairs* (Washington, D.C., 1836), 6:377.

[46] Browning, *Two if by Sea* (cit. n. 42), 42.

[47] Bernard, "Biographical Notice of Joseph Gilbert Totten" (cit. n. 24), 83.

three problems: timber, stone, and mortar. He undertook many more research projects that did not focus on material properties, including, among many others, projects on excavating rock, reducing the size of casements, determining the thrust of arches, and measuring the impact force of ordnance. All of his materials projects, and many of his other researches, yielded important publications in the two major American scientific outlets of the antebellum period: the *American Journal of Science and Arts* and the *Journal of the Franklin Institute*. Reports of Totten's projects also appeared in British and French scientific journals, including the *Journal of the Royal Institution* and the *Annales de chimie et de physique*. Interestingly, only Totten authored the work on mortar, which consisted largely of translations of French treatises with comments on American materials. His junior officers authored the other studies.

Totten's impulse for starting a program on materials research can be located in a contradictory aspect of his character. As a West Pointer, his intellectual orientation was toward French science, on the École Polytechnique model.[48] But his early personal experiences in surveying Ohio and inspecting forts between 1808 and 1809 demonstrated to him that French sources, as much as he respected their rigor, failed to address the distinctly American landscape he encountered. In Totten's eulogy, John Barnard, one of Totten's assistants at Fort Adams, wrote:

> The art of the civil engineer was yet in its infancy in this country. Our resources in building materials were almost unknown, their qualities and adaptabilities to different purposes of construction undeveloped. Thus far the matter had excited little attention; the building material, whether brick or stone, lime or timber, nearest at hand was indiscriminately used, and its aggregation left much to the skill of the mechanic.[49]

Totten's interest in unique properties of American materials was further explained in the conclusion to one of the reports in the *Journal of the Franklin Institute*. Lt. T. S. Brown, another one of Totten's assistants at Fort Adams, wrote:

> In Tredgold's Carpentry and other similar works may be found the constant numbers (*a*) and (*c*) for nearly all the kinds of wood useful in the arts; but besides that these numbers are in many instances calculated from insufficient experiments, most of the specimens used were of European growth and, of course, the results obtained are inapplicable to American timber, though bearing the same name. It is much desired that numerous and accurate experiments be made, in this country, by those having the requisite zeal and opportunities; our architects will then know with certainty the qualities of different kinds of wood they are using and instead of working at hazard and in the dark, as they so often do now, they will be guided by the sure light of practical science to certain and definite results.[50]

---

[48] Todd Shallat, *Structures in the Stream: Water, Science, and the Rise of the U.S. Army Corps of Engineers* (Austin, Texas, 1994), chap. 3.

[49] John G. Barnard, "Eulogy on the Late Joseph G. Totten," *Historical Papers Relating to the Corps of Engineers*, Engineer School Occasional Papers, no. 16 (Washington, D.C., 1904).

[50] Lt. T. S. Brown, "On the Strength of Pine and Spruce Timber," *Journal of the Franklin Institute* 7 (April 1831): 238. Brown refers to Tredgold's *Elementary principles of carpentry; a treatise on the pressure and equilibrium of timber framing, the resistance of timber, and the construction of floors, centres, bridges, roofs; uniting iron and stone with timber, etc.*, which had been excerpted in the *Journal of the Franklin Institute* in 1829, prior to its American publication in 1837. The variable *a* refers to a constant of stiffness used in the equation $BD^3/aL^2 = W$ and *c* to a constant of strength used in $BD^2c/L = W$.

For American engineers to emulate polytechniciens they needed to construct their own practical science on the French model. For Totten, American engineers did not need to reinvent the wheel but rather pick up where the French had left off and investigate materials that the French had not encountered. Consequently, American engineers first had to be made aware of the state of the art, and for Totten this justified translating French works and adding the findings of his own research in a coda. He made that claim in a footnote to the first article on mortar, writing:

> In this country we have been led, within a few years, to some improvements in the practice of mortar making, by the actual necessity imposed by extensive hydraulic works, and by the providential diffusion over our territory of an admirable material, still we are, for the most part, the slaves of an antiquated routine, elsewhere known to be radically wrong. The works here translated are recent, and among the best issued from the French press . . . To most American readers they will present many new results, highly important, and of a character to be applied with great advantage to our own operations.[51]

The articles that Totten and his staff wrote are modern scientific articles: they begin with methods and move on to present novel results of experiments. Totten's intent was to create a blueprint for strength of materials research that other engineers could follow for further investigations of other problems and in other locations. This coupled with the dozens of engineers trained at Fort Adams effectively created a community of researchers who worked in very similar ways—that is, a research school. Although Totten and the men of the Corps of Engineers obviously identified themselves as engineers, the frequency and form of their publications suggest that Totten's research program may deserve more attention than it has previously received as a stage in the professionalization of science in America.

TOTTEN'S STRENGTH OF MATERIALS RESEARCH

In 1830, Totten began a series of experiments to determine the rate and extent of the expansion and contraction of building stone. Totten and Lt. William Bartlett, a future professor of natural philosophy at West Point and Totten's assistant on this project, wanted to determine how the changes in the stone itself affected the cement in the joints and therefore the overall strength of the composition.[52] Bartlett, in his report to the *American Journal of Science and Art* in 1832, wrote:

> In the progress of this work, we have had occasion to use considerable quantities of coping stones, taken from different localities, with all of which it has been found impossible to obtain tight joints. The walls, on which these stones were placed, have not undergone the slightest change; and, notwithstanding they were laid with the greatest possible care, yet at the expiration of a few weeks, these joints were broken up by fissures, which extended from the top to the bottom of the coping. These fissures were supposed to have arisen from a change in the dimensions in the coping stones, in consequence of the ordinary variations of atmospheric temperature.[53]

[51] Totten, "On Hydraulic and Common Mortar" (cit. n. 11), 230.

[52] After his work at Fort Adams, Barlett returned to West Point and spent his career from 1836–71 as a professor of natural philosophy at the USMA.

[53] Lt. William H. C. Bartlett, "Experiments on the Expansion and Contraction of Building Stones by Variations of Temperature," *American Journal of Science and Art* 22 (1832): 136.

Bartlett and Totten tested three types of stone: granite, limestone, and sandstone. The granite used came from Buzzard's Bay, the limestone from the Sing Sing (Ossining) prison quarries in New York, and the sandstone from Chatham, Connecticut. Bartlett describes each sample in detail—its color, grain, and method of extraction from each quarry or, in the case of the granite, boulder. Bartlett fit similarly sized samples of each stone into frames that held pine-measuring rods connected to the stone with copper elbows. A thermometer was inserted into a groove in each stone sample. The project became mathematically complicated, since to calculate accurately the expansion of the stone, Bartlett had to also account for the expansion of the wood and copper measuring rod. Bartlett knew that the copper and wooden components would change size at different rates for different temperatures, so their rates of change had to be calculated independently. The stones were subjected to temperatures ranging from 6 degrees to 109 degrees Fahrenheit.[54] Bartlett's goal was to devise a scale showing the average expansion or contraction per degree for each material. Taking stones approximately 94 inches long, Bartlett found that granite changed the least, .0004538 inch per degree; marble changed .0005297 inch per degree; and sandstone changed .0008965 inch per degree, or nearly twice the rate of granite. For application to construction, Bartlett's findings became more significant. Given a fluctuation of 96 degrees over the course of a year, the sandstone would expand and contract .05 inch. A crack of this size would be quite visible, and water would easily leak through it. Bartlett concluded his report writing, "The mischief does not stop here: by this constant motion back and forth in the coping, the cement of whatever kind the joints be made, would be crushed to powder, and in a short time be totally washed by the rains from its place, leaving the whole joint open."[55] But as the report noted, the choice of stone determined the severity of the problem. For Totten, this research project intersected with his work on mortar, an effort to find a cement or method of masonry construction that would further minimize the effects of expansion and contraction.

Concurrent with Bartlett's research on stone, Lt. T. S. Brown undertook research on the relative stiffness and strengths of American timber. Brown published reports of his research in both the *American Journal of Science and Art* and the *Journal of the Franklin Institute*. Brown tested three types of timber, northern white pine, southern pine, and northern spruce. For each variety, Brown tested three samples. Each sample measured 9.389 feet long and was 2.75 inches wide and 5.5 inches deep (i.e., a nominal 3 by 6). Placing the beam on supports, the total span was 7.1 feet. These measurements and dimensions represented a real structural problem that Totten faced in the design of Fort Adams. Brown used bricks, of a predetermined weight, and added them to a carriage that Brown suspended from the center of the beam. The carriage hung from the beam by means of an iron collar 3 inches wide and 1/2 inch thick, so it would not cut into the wood. Brown also attached an index to the bottom of each sample as a way to measure deflection, but Brown was careful in his design of the apparatus so the measuring device did not add any weight to the beam. Brown's scale was divided into increments of .177 inch, which represented 1/40 of an inch per foot

[54] Presumably, Newport's temperature never reached 109 degrees; instead the effect of the stone's sitting in the sun drove the temperature of the stone to 109 degrees. Only the sandstone reached this temperature. The highest temperature measured in the granite was 102 degrees and 99 degrees for the limestone (which is alternately referred to as marble in the report).

[55] Bartlett, "Experiments" (cit. n. 53), 140.

of beam. According to Brown, the figure of 1/40 of an inch per foot was well established as the maximum deflection allowed in practice.[56]

Brown had men add the bricks to the carriage and recorded the number of bricks each time the index moved down to a new division. Each beam was tested to the breaking point. Brown dismissed readers' anticipated concerns that the weights were not allowed to act on the beams over time by arguing that in one case they allowed weight totaling 2,662 pounds to remain for more than two hours and saw no further deflection. In addition, Brown's goal was to determine the relation between the different varieties, and the effect of time would be similar on each species.[57] Using his empirical findings, Brown then calculated constants for the stiffness of each species, which took into account the amount of deflection. Brown's chart looked like this:

Species	deflection (inches/foot)	weight (#)	constant (a)
White Pine	1/320	121	not given
	1/40	777	0.0116811
	2/40	1,485	0.0061119
	3/40	2,108	0.00436056
	4/40	2,701	0.0033603
Spruce	1/320	134	not given
	1/40	892	0.0101751
	2/40	1,613	0.0056269
	3/40	2,291	0.0039617
	4/40	2,966	0.0030601
Southern Pine	1/320	218	not given
	1/40	1,175	0.0077244
	2/40	2,218	0.0040921
	3/40	3,192	0.0028434
	4/40	4,111	0.0022078

Brown included two more tables, one of which showed the amount of weight that broke each species, giving a constant for strength. The last table compared the stiffness of the three species, taking white pine as one and showing that spruce was slightly stiffer, while southern pine was nearly twice as stiff.[58]

The third and most significant materials research project undertaken at Fort Adams was Totten's mortar project. Totten detailed this project in a series of eleven articles in the *Journal of the Franklin Institute* between October 1837 and September 1838, although translations of French treatises by Treussart, Petot, and Courtois compose most of the pages of these reports. These works really did represent the current state of the art, all bearing French publication dates in the 1830s. The last two articles were dedicated to explications of Totten's empirical work on mortar. Totten's experiments on mortar focused not only on the three constituent ingredients (lime, cement, and

[56] Brown cites Tredgold as the source for the 1/40 inch per foot deflection standard.

[57] Brown, "On the Strength of Pine and Spruce Timber" (cit. n. 50), 231.

[58] The ratios vary slightly depending on the amount of deflection, since the deflection curves are not identical.

sand or aggregate) of different kinds of mortar but also on the processes of making that mortar. Totten examined three kinds of lime—that from Smithfield, Rhode Island; Thomastown, Maine; and the local lime found at Brenton's Point. Totten tested three different methods of slaking, or dissolving, the lime—by sprinkling it with water, "drowning" it in water, and leaving it out to absorb atmospheric water. Totten investigated three different kinds of cement, two from New York and one imported from England. Totten used four different kinds of sand, all local but differing in particle size.

Totten's report carefully detailed the methods of mortar preparation and the amount of time each sample was allowed to cure. Totten specified in detail how the samples had been assembled:

> The strength of mortars as regards tenacity, was determined by measuring the force required to separate bricks that, having been joined by the mortar had been left, for the desired length of time, in some place safe from frost or accident. The bricks were joined in pairs, being crossed at right angles, so that supposing each brick to be 4 inches wide, the surface of contact would be 16 square inches. The real surface or surface of effectual contact, was, in every case found by actual measurement. The mortar joint separating the bricks was made about 3/8 of an inch thick: and in order that this mortar should in all cases be equally consolidated, each pair of bricks was submitted to the pressure of 600 lbs. for 5 minutes, immediately after being joined.[59]

Totten referred to the ability of the mortar to resist forces pulling the bricks apart as the tenacity of the mortar. To determine the tenacity of the forty-one different combinations of lime, cement, and sand, Totten and his assistants constructed an apparatus for pulling the samples apart. A drawing plate of the apparatus appeared in the *Journal*. On one end of an iron beam was a clamp that grabbed the top brick. On the other end of the beam was a bucket to which sand could be added. The experimenter added weight, in the form of sand, to the bucket until the bricks were torn apart, after which the bucket plus the sand was weighed. The variation in the tenacity of the forty-one samples was enormous; the strongest mortar required 61.9 pounds to tear the sample apart, while the weakest withstood only 5.6 pounds, demonstrating a tenfold variation. Totten tested multiple samples of most combinations; the value given for tenacity was a mean of all tests on that sample.

After pulling the bricks asunder, Totten tested the hardness of each mortar sample. Using the brick on which the mortar was still stuck, the hardness-testing machine used a similar setup of a beam and a bucket to press an iron point into the mortar. The apparatus determined the amount of weight the sample could withstand before cracking under pressure. Again, the variation was about tenfold, but the hardest sample (supporting 1,055 pounds) often coincided with the most tenacious. The weakest cracked with only 98 pounds of weight. Totten's tests showed that a mortar's resistance to tension (the tenacity test) and compression (the hardness test) would be related to one another—the strongest mortar was the best performer in both categories.[60]

---

[59] Totten, "Brief Observations on Common Mortars, Hydraulic Mortars, and Concretes; with some Experiments made therewith, at Fort Adams, Newport Harbor, R. I., from 1825 to 1838," *Journal of the Franklin Institute* 22 (July 1838): 5.

[60] This is not to say that the mortar resisted tension and compression equally well. The numbers bear this out. The strongest sample bore more than 1,000 pounds in compression, but not even 62 in

Totten's mortar tests continued for several years, and samples of many different ages were tested. Totten's reports added to the knowledge he disseminated in the translations of Treussart, Petot, and Courtois. However, by the time Totten finally completed the work on mortar, the construction of Fort Adams was nearly complete. Totten had little time to celebrate the completion of either the test or the fort, as he received reassignment to Pensacola in 1838 to plan the Navy Yard there.[61] That same year, he was appointed chief of the Corps of Engineers and remained chief until his sudden death from pneumonia in 1864 at the age of seventy-six.

TOTTEN'S INFLUENCE

Totten consciously designed the program at Fort Adams as a training ground; therefore, he expected his charges at Fort Adams to take away certain values and behavior patterns from their work with him, values consistent with broader American values such as practicality and empiricism. These were also the values that characterized Totten's career and disposition; Totten certainly held his own work and personal behavior up as a standard to be followed by his junior officers. In looking at his career, we see three clear themes. First, Totten believed in observation and inspection. Whenever Totten received a new assignment, his initial action was to take an inventory and assess the problems and resources at hand. This is clear in his nearly obsessive interest in fort inspection; he tried to visit every fort once every two years.[62]

The second theme of his career centers on the primacy of experimentation. Totten thought the first job of the scientist was to test the world. Drawing from his vast knowledge of European science, Totten addressed engineering problems by first categorizing them, seeking what knowledge had already been developed, and then creating a program to produce more knowledge through experiment. In the process, Totten seemed particularly interested in developing new techniques and machines for measurement and assessment. For all three strength of materials projects from Fort Adams, as well as his later work at West Point on casement embrasures and the impact of ordnance, Totten focused on experiments and experimental apparatus.

Third, Totten's career and research were delineated by committee work. From his position on the Board of Engineers in 1818 to the work at Fort Adams during the 1820s and 1830s to his position on Winfield Scott's staff in the 1840s to his leadership of the Lighthouse Board in the 1850s, including his involvement in the early Smithsonian and National Academy of Science, Totten's career forms a catalog of successful collaborations. While some in the early corps, including Swift and Thayer, had numerous difficulties working with others, Totten seemed to cooperate well with a diverse group of engineers, politicians, and scientists. As a result, the number of people who interacted with Totten in one project or another is much greater than the circle which can be drawn around any other engineering figure in the antebellum

---

tension. It was not news to anyone that mortar was incredibly weak in tension, but it was news that the best scores for tenacity occurred on the samples that also scored the highest in the hardness tests.

[61] Still Totten remained officially in charge of Fort Adams until his death, often to the discomfort of the officer in place. Thomas, "Science Military Style" (cit. n. 24), 194.

[62] Barnard, "Eulogy" (cit. n. 49); George W. Cullum, *Biographical Register of the Officers and Graduates of the U.S. Military Academy, from 1802 to 1879* (New York, 1897); and *Dictionary of American Biography* (New York, 1996).

period, save perhaps John Jervis. One can only speculate on the reasons for Totten's successful collaborations, but it is worth noting that the projects Totten found the most engaging were not narrowly focused, requiring instead diverse and wide-ranging vision. It is also clear from his career that Totten kept to the work of engineering and often shied away from making the kind of political waves that made others, such as Thayer or Swift, more difficult to work with.

TOTTEN'S PROGRAM AT FORT ADAMS AS A MODEL FOR ENGINEERING RESEARCH

One of the young officers who passed through Fort Adams was Alexander Dallas Bache. In 1825 Bache graduated first in his class at the USMA, and after working for a year as a professor of mathematics and natural philosophy at West Point, he moved on to Fort Adams, where he spent two years with Totten. After his time at Fort Adams, Bache resigned his commission to accept a position as professor of natural philosophy and chemistry at the University of Pennsylvania. Bache is best known for his work to professionalize the American scientific community and generate government support for science; he was the self-styled leader of America's elite scientific clique of the mid-nineteenth century, the Lazzaroni.[63] However, most studies of Bache focus on his work at the Franklin Institute and the University of Pennsylvania or as the chief engineer of the Coast and Geodetic Survey in the 1850s and 1860s, where he was the most significant patron of basic scientific research in the nineteenth century. Looking at his work with the Franklin Institute in light of his work with Totten sheds some new light on the kinds of science he supported. Although Bache is the larger figure in the history of American science, he should also be seen as a disciple of Totten's.

By 1831, Bache had been in Philadelphia for three years. Hugh Slotten, in his biography of Bache, claims that even at this early stage of his career, Bache

> led a clique of like-minded men of science who gained control of the Franklin Institute and restructured it according to their particular ideals. . . . As chairman of the Committee on Instruction, Bache raised the quality of the lectures at the Franklin Institute in accordance with his own European-oriented standards and stressed the importance of the physical sciences. . . . Bache also spearheaded an effort to transform the Institute into a resource for experimental research.[64]

In other words, Bache saw in the Franklin Institute an organization that could support and promote the type of research that he performed at Fort Adams under Totten's direction. Nathan Reingold writes that, for Bache, "routine work and the research were ideally related to a theory in mathematical form—the theory giving direction to the research and the routine work, while the resulting data confirmed, extended, or refuted the theory."[65]

Bache expressed his scientific values most effectively through the *Journal of the Franklin Institute*. Bache had gained control of the *Journal* in 1831 and quickly

---

[63] On Bache, see Hugh Slotten, *Patronage, Practice, and the Culture of American Science* (Cambridge, Mass., 1994); Robert Bruce, *The Launching of American Science, 1846–1876* (New York, 1987); Nathan Reingold, "Alexander Dallas Bache: Science and Technology in the American Idiom," *Tech. Cult.* 11 (1970), 163–77; Bruce Sinclair, *Philadelphia's Philosopher Mechanics: A History of the Franklin Institute, 1824–1865* (Baltimore, 1974).

[64] Slotten, *Patronage* (cit. n. 63), 39.

[65] Reingold, "Alexander Dallas Bache" (cit. n. 63), 174.

changed its focus to publish original experimental research. Slotten argues that Bache's goal was to produce a journal that would "gain respect from Europeans."[66] Bruce Sinclair, in his history of the Franklin Institute, claims that Bache and his cronies equated research with publication—men of science produced public knowledge; therefore the process by which knowledge was distributed was as much a part of the work of science as anything that happened in the lab.[67] Totten would have agreed, and it is no surprise that after 1831 the reports of research at Fort Adams appeared nearly exclusively in the *Journal of the Franklin Institute*. That Bache sought to publish this work shows his respect for Totten's scientific standards. Totten's interest in contributing to international knowledge of the strength of materials and Bache's goal of international recognition converged nicely in the 1830s. Sinclair points to three goals of Bache's: to use the magazine to express his own research interests and standards; to link the *Journal* with important, contemporary scientific problems in America and elsewhere; and to encourage the talents of a new generation of men of science by opening a place for publication of their work.[68] One could make nearly the same statement about Totten's goals at Fort Adams.

Bache also led the most important research project of the Franklin Institute in this period: a federally funded investigation into the causes and prevention of steam boiler explosions. From its beginning in 1824, the Franklin Institute had a place for the on-going discussion of the explosion propensity of high-pressure steam boilers on ships. When the *Journal* began publication in 1826, articles began to appear investigating the causes and putting forth patents for devices that might offer relief.[69] Boiler explosions constituted a real public-safety threat; between 1825 and 1830, 273 people had been killed in boiler accidents. After half-hearted attempts by Congress and noncooperation from steamboat operators to determine the cause and remedies for the explosions, the Franklin Institute itself undertook a considerable research program. As luck would have it, Samuel Ingham, secretary of the treasury who had been similarly charged to investigate the problem, read about the institute's inquiry in a newspaper. Ingham approached the institute with a proposal to undertake the research collaboratively.[70] Because of this cooperation, the story of the investigations into boiler explosions is usually told as one of early federal support of scientific research. However, the steam boiler research can also be seen in light of the Fort Adams research, as an inquiry, performed in the interest of the nation, into the strength of materials.

By 1831, the boiler investigations at the Franklin Institute had been organized into two subcommittees: one to determine the causes for the buildup of excess pressure in the vessel; and a second to test the strength of materials used in boiler construction. The strength of materials project bears striking similarities to the Fort Adams work. Overseen by Bache, the strength of materials tests were performed by Walter R. Johnson, a Harvard graduate who was an instructor of Latin, Greek, and mechanics at the institute's high school. Following Totten's pattern of assessment and experimentation, Johnson spent the first year of his research designing and producing machines to perform the tests that he, Bache, and Benjamin Reeves, the third strength

---

[66] Slotten, *Patronage* (cit. n. 63), 39.
[67] Sinclair, *Philadelphia's Philosopher Mechanics* (cit. n. 63), 195.
[68] Ibid., 208–9.
[69] John G. Burke, "Bursting Boilers and the Federal Power," *Tech. Cult.* 7 (1966): 13.
[70] Sinclair, *Philadelphia's Philosopher Mechanics* (cit. n. 63), 176.

of materials committee member, had drawn up. Principally there were three purpose-built testing machines. They used one to determine the amount of longitudinal stress required to fracture the sample.[71] This machine was clearly similar to the one used at Fort Adams, showing visually the link between Bache's and Totten's research programs. Johnson devised another machine to test materials at high temperatures. It consisted of a device to test the tensile strength of metals raised to a certain temperature in either an oil or molten metal bath. The third machine was the most innovative; it was a steam pyrometer used to determine temperatures above those that could be measured using a mercury thermometer.[72] Johnson's pyrometer measured temperatures up to 1,300 degrees.[73]

Johnson's research—the plan for it set out by Bache—determined that temperature affected the strength of various metals significantly. Copper became much weaker when subjected to high temperatures, as it would be in a boiler vessel. Iron, by contrast, became notably stronger as temperatures rose to 600 degrees. The relation of structural properties to temperature harkened back to Totten's work on stone and mortar, which had been published in the years during which Johnson and Bache performed their work. The subcommittee published its findings of the strength of materials in a series of articles in the *Journal of the Franklin Institute* between February and August of 1837. Sinclair argues that Bache's report of this work

> was cast in near perfect form to secure legislative action. The report was complete and thorough. The problem and its solutions were clearly outlined and firmly supported. It was an authoritative document and self-contained, even to the point of including at the end a bill which would provide the necessary legislation. Experimental results were translated into three proposed categories of regulation: the establishment of an inspection and certification system for steam boilers; minimum standards for safety devices, operating procedures, and personnel qualification; and penalties for improper operation or noncompliance.[74]

Joseph Henry confirmed that the boiler research Bache oversaw gained Bache a scientific reputation in Europe, writing in Bache's eulogy, "No American has ever visited Europe under more favorable circumstances for becoming intimately acquainted with its scientific and literary institutions. His published researches had given him a European reputation, and afforded him ready access to the intelligent and influential classes of society which is denied the traveler whose only recommendation is the possession of wealth."[75] It cannot escape one's notice that Bache achieved Totten's goal, which was not merely to emulate European science but to become an equal contributor in the construction of new scientific knowledge and a participant in the discussion of science. It is also significant Bache exemplified the kind of American basic scientific research that began to emerge around the Civil War and that I. B. Cohen

---

[71] "Report of the Committee of the Franklin Institute of the State of Pennsylvania on the Explosions of Steam Boilers," *Journal of the Franklin Institute* 19 (Feb. 1837): 75–77.

[72] Sinclair, *Philadelphia's Philosopher Mechanics* (cit. n. 63), 167–68.

[73] Burke, "Bursting Boilers" (cit. n. 69), 13.

[74] Sinclair, *Philadelphia's Philosopher Mechanics* (cit. n. 63), 189.

[75] Joseph Henry, "Eulogy on Professor Alexander Dallas Bache, Late Superintendent of the United States Coast Survey," in *Annual Report of the Board of Regents of the Smithsonian Institution, showing the operations, expenditures, and condition of the Institution for the year 1870* (Washington, D.C., 1872), 97.

was looking for to define American achievement in science. As the example of Bache makes clear, it is neither separable from nor superior to the kind of practical science that characterizes the antebellum period, and both are aligned with the changing social and intellectual values of nineteenth-century America. The connective tissue between the two periods—ante- and postbellum—are the materials on which Bache, Totten, and others worked. American science, whether basic or applied, was experimentally based and depended on a supply of uniquely American materials.

CONCLUSION: THE ROLE OF MATERIALS IN STRENGTH OF MATERIALS RESEARCH

Seriously examining the role materials play in both the development of scientific knowledge and the moral economy of science is a recent trend in the history of science.[76] Still, most of the work in this area has concerned the role of organisms in life science, not construction materials in engineering science. Exploring how different materials constrained and shaped scientific ideas and practices can shed significant light on the development of scientific tools and ideas, but most studies of materials also emphasize the moral economies specific materials create. As such, materials lie at the intersection of the local and the universal in science. Claims about materials must be able to be detached from their context of discovery and invention, becoming universally accessible. But they always carry the fingerprints of their local origins, whether those be in a chemistry lab or in the field. Therefore, research on materials provides a probe into science's national, regional, and local identities.

In the past, the history of strength of materials has been written as a history of ideas and mathematical descriptions; a material context was oddly absent. Looking at the work of Totten and Bache retrieves the centrality of materials and the technologies used for testing them to the development of the practical science of strength of materials. In Steven Timoshenko's venerable *History of the Strength of Materials*, none of the researchers discussed here appears, because none of these men ever worked on the formal mathematics of strength of materials.[77] However, in investigating their work, especially in recovering the practice of materials research, one finds that the materials played an important role. When these men of science spend years designing and building new machines with which to test the various properties of these materials, it seems foolish to ignore the material world in which they work.

Totten's research program at Fort Adams played a critical role in opening the field of strength of materials to American experimenters. Totten's attention to the peculiarities of local materials and to the precise knowledge of their properties as manifest in the actual design of structures defined a research program that fit antebellum

[76] There is now a voluminous literature, particularly on biological materials. See Robert Kohler, *Lords of the Fly: Drosophila Genetics and the Experimental Life* (Chicago, 1994); Karen Rader, *Making Mice: Standardizing Animals for American Biomedical Research, 1900–1955* (Princeton, N.J., 2004); Gerald L. Geison and Angela N. H. Creager, "Introduction: Research Materials and Model Organisms in the Biological and Biomedical Sciences," *Studies in the History and Philosophy of Biology and the Biomedical Sciences* 30 (Spring 1999): 315–18; Adele Clarke and Joan Fujimura, *The Right Tools for the Job: At Work in the Twentieth Century Life Sciences* (Princeton, N.J., 1992).

[77] Stephen P. Timoshenko, *History of the Strength of Materials* (New York, 1983). The absence of the American empirical tradition in strength of materials is particularly ironic in the case of Timoshenko's work, as his claim for the importance of strength of materials is based on its practical application in construction. His history overlooks precisely the kind of strength of materials that had any bearing on the actual design of structures.

political, economic, social, and cultural values, diverse though they were. The research program at the Franklin Institute continued these priorities in investigations
of a completely different class of materials for a different kind of structure—steam
boilers instead of forts. But the seriousness with which both projects took the materials themselves is telling. In terms of transforming strength of materials into a practical science, as both T. S. Brown and General J. G. Barnard refer to Totten's subject,
the materials played a central role. Totten's interests were in the *use* of stone, timber,
and mortar, just as Bache and Johnson's interest was in the *use* of copper and iron in
steam boiler vessels. Without a focus on the materials themselves, the history of the
strength of materials does look a lot like ideas floating in context-free space. However, neither Totten nor Bache, nor any architect or engineer, ever referred to mathematical formalisms—their interest lies in the materials themselves.

Because practical applications were so tightly woven into the scientific enterprise
itself, Totten and Bache's research programs played a special role in mediating between the American landscape and Americans' identity as the people chosen to live in
such a bounty. Of course, the reality of life on the frontier was radically different, but
one of the compelling questions about national identity is how it is constructed from
experiences that in individual cases were quite opposed to the aggregate national perception of such episodes. Thus the Lewis and Clark expedition can be held up as a
national triumph, despite their disappointing failure not only to find a water passage
to the Pacific but also their "discovery" that crossing the Rocky Mountains was only
exceeded by the challenge of crossing the Bitterroot. The success of turning natural
materials into national infrastructure, while creating a uniquely American way of
engineering that was quickly disseminated by commercially available handbooks,
plays a similarly mythological role: engineers conquered nature for human benefit
under the American flag. Not only does this account of building the nation sui generis
obscure the already constructed landscape of Native Americans, it also exaggerates
the ease and rationality of the nation building. But it does so deliberately, to extend an
engineering identity that coheres with national proclivities.

The Children's Republic of Science in the Antebellum Literature of Samuel Griswold Goodrich and Jacob Abbott

By Katherine Pandora[*]

ABSTRACT

The antebellum years in the United States were marked by vigorous debates about national identity in which issues of hierarchy, authority, and democratic values came under intense scrutiny. During this period, a prime objective of indigenous authors writing for American children was educating the young so they would be ready to assume their republican responsibilities. The question of how depictions and discussions about nature and science were deployed toward this end is explored by examining key texts about nature and science from the era's two most prolific and popular children's authors—Samuel Griswold Goodrich (1793–1860) and Jacob Abbott (1803–79)—and highlighting assumptions within these works about what the proper relationship should be between the search for scientific knowledge and the larger polity.

INTRODUCTION

What did it mean to be a scientific American in the nineteenth century? One set of answers to this question can be found by examining the organizing of scientific societies, following the various paths that individuals took to earning higher degrees, and analyzing the practices and theories that contributed to the realization of disciplinary identities—foundational events that made possible the first volume of James McKeen Cattell's *American Men of Science* in 1906. In large measure, when historians of American science have studied the nineteenth century, it is this set of answers that they have pursued most diligently.[1] And yet studies of the making of American

[*]Department of the History of Science, University of Oklahoma, 601 Elm Avenue, Physical Sciences Bldg. 619, Norman, OK 73019; kpandora@ou.edu.

As a fellow at the Charles Warren Center for Studies in American History at Harvard University, I benefited greatly from the insights of Charles Rosenberg, Joyce Chaplin, and Laurel Thatcher Ulrich, the directors of our year-long investigation into the topic "Exceptional by Nature? American Science and Medicine, 1600–1900," as well as those of the other fellows: Jorge Cañizares-Esguerra, Cornelia Dayton, Mordechi Feingold, Ann Johnson, Marina Moskowitz, Susan Scott Parrish, and Alice N. Walters. I also appreciate the thoughtful commentary from the referee process for this paper and from the workshop that preceded this volume, which included the other authors as participants, as well as Michael Adas, John Krige, and Laura Dassow Walls.

[1] The interest of a previous generation in professionalization led to a vigorous research program that encompassed numerous scholarly articles, editions of collected essays, bibliographic resources, and biographical treatments. Examples of key texts include: A. Hunter Dupree, *Science in the Federal Government* (Cambridge, Mass., 1957); George Daniels, *American Science in the Age of Jackson*

© 2009 by The History of Science Society. All rights reserved. 0369-7827/09/2009-0005$10.00

scientists alone cannot answer this question. As the citizens of the former colonies turned new republic engaged in the dynamics of nation building over the course of the nineteenth century, their experiences with and interpretations of the natural world they inhabited—in its physical, faunal, floral, and human forms—were crucial resources they used in debates about what was at stake in the American experiment in self-government as sanctioned by the "Laws of Nature and of Nature's God."[2] Indeed, the turbulent era between Andrew Jackson and Abraham Lincoln was decisive in establishing the contours of argument about creating an American republic: it is in this formative period that the first generational cohorts who inherited the Revolution came of age and engaged in intense explorations about the meaning of national identity, with understandings of American "nature(s)" being central to key aspects of this ferment.[3]

Within vernacular discourse about the natural world, books for American children written by native authors have been unduly neglected by scholars as a means for exploring the intellectual, social, and political ramifications of ideas about nature and science in American culture during this period. Indeed, a push for a specifically American literature for youngsters that began in the 1820s had burgeoned by midcentury, stimulated by changing views of childhood itself and fostered by a worry that imported literature presented "a wrong direction to the minds of the young, modeled as they are on a condition of life, and on prevailing sentiments, civil, moral and social . . . varying from those which American children should early be taught to cherish." As Anne Scott MacLeod remarks, nineteenth-century Americans "wanted American

(New York, 1968); Howard S. Miller, *Dollars for Research: Science and Its Patrons in Nineteenth-Century America* (Seattle, Wash., 1970); Sally Gregory Kohlstedt, *The Formation of the American Scientific Community: The American Association for the Advancement of Science, 1848–1860* (Carbondale, Ill., 1976); and Robert Bruce, *The Launching of Modern American Science, 1846–1876* (New York, 1987). Although this work did discuss popular science to some extent in contextualizing the world from which professionalism diverged, this was not a key focus. Subsequent, more sustained engagement regarding popular science in this time by historians of science can be found in: Elizabeth Keeney, *The Botanizers: Amateur Scientists in Nineteenth-Century America* (Chapel Hill, N.C., 1993); Daniel Goldstein, "'Yours for Science': The Smithsonian Institution's Correspondents and the Shape of Scientific Community in Nineteenth-Century America," *Isis* 85 (1994): 573–99; and Sally Gregory Kohlstedt, "Parlors, Primers, and Public Schooling: Education for Science in Nineteenth-Century America," *Isis* 81 (1990): 425–45.

[2] The most interesting insights into the cultural meanings of natural knowledge in this period can be found in the interdisciplinary efforts of art historians, literary scholars, and those working in American Studies. As examples, see: Barbara Novak, *Nature and Culture: American Landscape and Painting, 1825–1875* (New York, 1981); Laura Dassow Walls, *Seeing New Worlds: Henry David Thoreau and Nineteenth-Century Natural Science* (Madison, Wis., 1995); Margaret Welch, *The Book of Nature: Natural History in the United States, 1825–1875* (Boston, 1998); and Christoph Irmscher, *The Poetics of Natural History, from John Bartram to William James* (New Brunswick, N.J., 1999).

[3] For issues such as nationalism, identity, and cultural debates in this time period, relevant texts are: Gordon Wood, "The Significance of the Early Republic," *Journal of the Early Republic* 8 (1988): 1–20; Charles Sellers, *The Market Revolution: Jacksonian America, 1815–1846* (New York, 1994); Anne C. Rose, *Voices of the Marketplace: American Thought and Culture, 1830–1860* (New York, 1995); David Waldstreicher, *In the Midst of Perpetual Fetes: The Making of American Nationalism, 1776–1820* (Chapel Hill, N.C., 1997); Lawrence Levine, *Highbrow/Lowbrow: The Emergence of Cultural Hierarchy in America* (Cambridge, Mass., 1998); Joseph A. Conforti, *Imagining New England: Explorations of Regional Identity from the Pilgrims to the Mid-Twentieth Century* (Chapel Hill, N.C., 2000); Joyce Appleby, *Inheriting the Revolution: The First Generation of Americans* (Cambridge, Mass., 2001); and Jill Lepore, *A Is for American: Letters and Other Characters in the Newly United States* (New York, 2002).

books for their children of destiny."[4] Nearly two decades ago, Sally Gregory Kohlstedt urged historians of science to "dig deeper into the national record" to better explain the "pervasive public curiosity about scientific subjects" in the nineteenth-century United States; she memorably argued for the relevance of "parlors, primers, and public schooling" in underwriting a variety of private and public responses to "cultural initiatives in the history of science in America, past and present." She suggested that "books for children particularly underscore the enthusiasm for science."[5]

In taking up scientific themes in instructing these children of destiny, what visions of the scientific enterprise did American authors convey? What aspects of natural lore and natural law did they believe were best suited to forming the growing mind and soul, and what were the ramifications of these assumptions? How did they place the middle-class American child within the natural and moral economy of the search for natural knowledge, and for what purposes? The generational cohort of Americans who worked at influencing their fellow citizens in the middle decades of the nineteenth century included such prominent individuals as Samuel Griswold Goodrich (1793–1860) and Jacob Abbott (1803–79), who developed and presented a vision of scientific Americanism to the young that emphasized natural knowledge as common property, acquired through mutual and united effort.[6] Adopting formats that favored

---

[4] William Cardell, *The Story of Jack Halyard*, 3rd ed. (Philadelphia, 1825), quoted in Anne Scott MacLeod, *A Moral Tale: Children's Fiction and American Culture, 1820–1860* (Hamden, Conn., 1975), 20. On American society and childhood and/or juvenile literature in this period, see Gillian Avery, *Behold the Child: American Children and Their Books, 1621–1922* (Baltimore, 1994); R. Gordon Kelley, *Mother Was a Lady: Self and Society in Selected American Children's Periodicals, 1865–1890* (Westport, Conn., 1974); Monica Kiefer, *American Children through Their Books, 1790–1835* (Philadelphia, 1948); Gail Schmunck Murray, *American Children's Literature and the Construction of Childhood* (New York, 1998); Jacqueline Reinier, *From Virtue to Character: American Childhood, 1775–1850* (New York, 1996); and Bernard Wishy, *The Child and the Republic: The Dawn of Modern American Child Nurture* (Philadelphia, 1968). On information circulation and the emergence of a reading public, essential starting points are: Richard D. Brown, *Knowledge Is Power: The Diffusion of Information in Early America, 1700–1865* (New York, 1989); and Ronald J. Zboray, *A Fictive People: Antebellum Economic Development and the American Reading Public* (New York, 1993).

[5] Kohlstedt, "Parlors, Primers, and Public Schooling" (cit. n. 1), 425, 436. In 1985, James Secord noted that "inquiry into the special historical problems of presenting science to a juvenile audience has barely begun"; there is much spadework yet to be done, particularly for the American context. Part of the issue is the assumption that "children's books sit securely on the bottom rung of historical significance, for here truth is in the most dilute form possible." See James A. Secord, "Newton in the Nursery: Tom Telescope and the Philosophy of Tops and Balls, 1761–1838," *History of Science* 23 (1985): 127–51, 127, 128. Among research that has begun to develop in this area is Aileen Fyfe, "Young Readers and the Sciences," in *Books and the Sciences in History*, ed. Marina Frasca-Spada and Nick Jardine (Cambridge, UK, 2000), 276–90; Barbara T. Gates, "Revisioning Darwin with Sympathy: Arabella Buckley," Pamela Henson, "'Through Books to Nature': Anna Botsford Comstock and the Nature Study Movement," and Greg Myers, "Fictionality, Demonstration, and a Forum for Popular Science: Jane Marcet's *Conversations on Chemistry*," in *Natural Eloquence: Women Reinscribe Science*, ed. Barbara T. Gates and Ann B. Shteir (Madison, Wis., 1997); Barbara T. Gates, *Kindred Nature: Victorian and Edwardian Women Embrace the Living World* (Chicago, 1998); Keeney, *Botanizers* (cit. n. 1); Greg Myers, "Science for Women and Children: The Dialogue of Popular Science in the Nineteenth Century," in *Nature Transfigured: Science and Literature, 1700–1800*, ed. John Christie and Sally Shuttleworth (Manchester, UK, 1989), 171–200; and Harriet Ritvo, "Learning from Animals: Natural History for Children in the Eighteenth and Nineteenth Centuries," *Children's Literature* 13 (1985): 72–93.

[6] Although Goodrich and Abbott are well known as popular juvenile authors in the history of children's literature, sustained analysis of their work and lives has been sporadic and extended secondary material surprisingly scarce. Concise entries in the *American National Biography* for each are useful starting points; fuller descriptions are contained in the *Dictionary of Literary Biography*. Other discussions of their places within the history of children's literature are contained in Avery, *Behold*

a welcoming tone and that featured unintimidating entry points into learning about a shared world of questions and answers, they helped to place the youngest citizens in the republic of science on a stable footing that, ideally, would allow the experiences of all to count in working out an understanding of American nature.[7] It was a plain style that eschewed constant reference to learned authorities, placed little emphasis on expensive instruments or complicated terminology, and promoted a belief in a democracy of learners, rather than an aristocracy of intellect.

The relaxed, inclusive, and optimistic approach of these authors to scientific education is one that was not premised on inculcating deference to learned authorities but was instead informed by a republican ethos that presumed the intellectual capacity of self-motivated learners to work independently toward goals of their own choosing. This cultural latitudinarianism on scientific topics is one that would find increasing disfavor among professionalizing scientific elites in the latter half of the nineteenth century. They would argue that popular approaches to science needed to be brought into closer alignment with the values and priorities that they identified as primary, particularly a more restrictive vision of intellectual authority compatible with the hierarchical norms of elite science current in Britain and Europe. In this sense, reading these children's authors can provide insight into views about the place of scientific knowledge within the larger American polity. Scholars have typically seen these texts as simply efforts to impart child-size bits of scientific information to the young for their edification, but they were more than this. They were also vehicles for forging readers' sensibilities toward the search for natural knowledge as properly personal, communal, and consequential in a mutual constitution of self and nation: a children's republic of science.

Why a *children's* republic of science? The emphasis on childhood was a necessity because of these authors' focus on the future. Goodrich and Abbott were less concerned with the republic-of-the-moment than they were with the republic-that-was-to-be when the children developing all around them would come to instantiate

---

the Child (cit. n. 4); and Cornelia Meigs, *A Critical History of Children's Literature* (New York, 1953). For Goodrich, a slight biography that offers a basic sketch is Daniel Roselle, *Samuel Griswold Goodrich, Creator of Peter Parley: A Study of His Life and Work* (Albany, N.Y., 1968). Goodrich did compose a memoir: *Recollections of a Lifetime, or Men and Things I Have Seen: In a Series of Familiar Letters to a Friend, Historical, Biographical, Anecdotal, and Descriptive*, 2 vols. (New York, 1857). There is no biography of Abbott, although two unpublished dissertations provide substantive overviews and analyses: Gregory Nenstiel, "Jacob Abbott: Mentor to a Rising Generation" (PhD diss., Univ. of Maryland, 1979); and Philip Kendall, "The Times and Tales of Jacob Abbott" (PhD diss., Boston Univ., 1968). Abbott's son Edward Abbott provided a biography in "Sketch of the Author" in the memorial edition of *The Young Christian* (New York, 1882); his son Lyman Abbott has a biographical chapter on his father in his book *Silhouettes of My Contemporaries* (New York, 1922). Pat Pflieger's Web site, "Nineteenth-Century American Children and What They Read" (http://www.merrycoz.org/) features both Goodrich and Abbott prominently and is an invaluable source of primary documents, contextual information, and helpful bibliographical material.

[7] In terms of history of science, published interest in these authors has been minimal. There are a few pages in Keeney on both Goodrich and Abbott in *Botanizers* (cit. n. 1) and in Welch on Goodrich and natural history in *Book of Nature* (cit. n. 2). Bruce Harvey explores Goodrich's geographies in his *American Geographics: U.S. National Narratives and the Representation of the Non-European World, 1830–1865* (Stanford, Calif., 2001). Abbott is the focus of Cheryl R. Smith, "*Learning about Common Things*: Conceptions and Uses of Science in the Juvenile Literature of Jacob Abbott" (master's thesis, Univ. of Oklahoma, 2000). Goodrich is touched on briefly in regard to a British work for children that fakes authorship from Goodrich, in James Secord's introduction to a reprint of *Peter Parley's Wonders of the Earth, Sea, and Sky* in vol. 3 of *Science for Children*, ed. Aileen Fyfe (Bristol, UK, 2003).

it as adults. The most crucial issue in this regard was how these later citizens would handle the question of where legitimate authority resided. And here is the point at which the significance of a children's republic of *science* enters in: for science was a new claimant in the bid for cultural authority, vying for a place alongside such venerable sources as religion, state power, inherited privilege, ancient traditions, and military strength. Reassessment of these extant forms of authority had been undergoing scrutiny and debate since the Revolutionary era, and the explosion of natural knowledge during the nineteenth century threw into the mix the question of what the proper place of science should be within the larger polity. Goodrich and Abbott thus placed particular value on nature and science within their children's works, because inculcating self-reliance in assessing and judging this new form of knowledge and its effects would be tasks that the coming generation would surely face.

Because those who had inherited the Revolution viewed the national experiment in which they were participating as yet open ended— rather than a playing out of events whose terms had been set by remote ancestral generations and required deference to traditional norms—cultural negotiations over the meaning of all forms of authority were inescapable. As literary theorist Philip Fisher has suggested in his book of essays, *Still the New World*, nineteenth-century authors such as Ralph Waldo Emerson and Mark Twain knew "that there are different truths when we know ourselves to be late in time near the completion of the world (or of our own country), ready to fit a few last details somehow onto a crowded page, and when, being near the beginning of time, we find ourselves still studying the sketch of what might someday be realized."[8] Within vernacular discourse about science and nature in this time period, Goodrich and Abbott likewise saw the nation, the nation's children, and science itself as being near the beginning of time, a sensibility at odds with American scientific leaders who looked to deeply rooted Anglo-European traditions for guidance in gaining membership in a transatlantic elite whose illustrious past was seen as having set the terms for its present achievements. While there were points of convergence between these two sensibilities, there were also differences in emphasis that could lead to differences in kind, as these children learned to become American by participating in a democracy of learners.

SAMUEL GRISWOLD GOODRICH AND JACOB ABBOTT: CULTIVATORS OF THE "NATIONAL NURSERY"

Goodrich began to publish for children in earnest in 1827, with *Peter Parley's Tales of America;* the character of Peter Parley—a benevolent old man who had much to tell of all he had seen, done, and heard about the world—was the purported author of the books, and he became a household name. The son of a Congregational minister, Goodrich was born into a Connecticut family of Yale graduates and Federalist officeholders, which included a United States senator; he, however, ended his schooling at twelve and began a set of apprenticeships at age fifteen. As a young man, Goodrich embarked on a course of self-study and was inspired to enter the publishing business in partnership with an older friend in 1816. The untimely death not long after of both Goodrich's partner and Goodrich's young wife initiated a time of uncertainty in his

---

[8] Philip Fisher, *Still the New World: American Literature in a Culture of Creative Destruction* (Cambridge, Mass., 1999), 2.

life; after a sojourn in Britain and Europe he returned home resolved to write books for children, with moralist Hannah More as his inspiration. Goodrich is credited with having written or edited 170 titles (130 for children) and to have sold seven million copies of his books in his lifetime; an 1884 estimate put the number at eleven million. He also founded two children's periodicals, *Parley's Magazine* and *Robert Merry's Museum*, which spread his influence even further.[9]

Goodrich began writing on scientific topics for young people early on, with *The Child's Botany* and *Peter Parley's Tales about the Sun, Moon, and Stars* in 1830; *Peter Parley's Tales about Animals* in 1831; *Peter Parley's Book of Curiosities, Natural and Artificial* in 1832; *Peter Parley's Farewell* in 1844 (on natural theology); *A Pictorial Natural History* in 1842; and *The Truth-Finder; or the Story of Inquisitive Jack* in 1845 (on natural history). In 1844–45, he published Parley's Cabinet Library, consisting of twenty volumes. Scientific fare in this set included *A Glance at the Physical Sciences—or the Wonders of Nature, in Earth, Air, and Sky*; *Wonders of Geology*; and *Anecdotes of the Animal Kingdom*. The Cabinet Library did not neglect consideration of human nature, especially in *Curiosities of Human Nature* and a trilogy on Indians: *Lives of Famous American Indians*; *History of the American Indians*; and *Manners, Customs, and Antiquities of the American Indians*.[10] That he capped his literary output in 1859 with the two-volume *Illustrated Natural History of the Animal Kingdom*, which contained more than 1,300 pages of text and 1,500 interspersed engravings, speaks to his sustained love of scientific topics.

The younger Abbott came fast on Goodrich's heels in the early 1830s, memorably creating a series of books following the development of a fictional creation named Rollo, a small-town rural New Englander who has been called "the first truly American child in fiction to become popular."[11] The *Rollo* series eventually numbered twenty-eight volumes, and characters in the *Rollo* books such as Lucy and Jonas received series of their own; together, the *Rollo*, *Lucy*, and *Jonas* volumes sold a million and a quarter copies in twenty-five years. All told, Abbott is credited with 211 books, the vast majority for the juvenile audience.[12] Abbott grew up on the northeastern frontier in Hallowell, Maine, where his father was a land trustee; he and his three brothers followed a common path to Bowdoin and Andover Theological Seminary. In

[9] The sales numbers are cited from Roselle, *Samuel Griswold Goodrich* (cit. n. 6), 53. Goodrich was an orthodox Congregationalist, and his work did contain religious commentary, in these books most usually in terms of natural theology.

[10] This later work, two massive tomes, was sold by subscription; its publisher indicates that it reached the sale of many thousand copies. (J. C. Derby, *Fifty Years among Authors, Books, and Publishers* [New York, 1884], 118.) Derby appears to be referring to the first editions brought out. After Goodrich's death, the A.J. Johnson & Son publishing company bought the stereotype plates and issued nine subsequent editions from 1868 to 1894 as *Johnson's Natural History: The Animal Kingdom Illustrated*. (Welch, *Book of Nature* [cit. n. 2], 255n12.) Goodrich's daughter recalled that the engravings alone cost thousands of dollars. Emily Goodrich Smith, "'Peter Parley'—as Known to His Daughter," part 2, *Connecticut Quarterly* 4 (1898): 399–407, 406.

[11] See Faye Riter Kensinger, *Children of the Series and How They Grew: Or a Century of Heroines and Heroes, Romantic, Comic, Moral* (Bowling Green, Ohio, 1987); Alice M. Jordan, *From Rollo to Tom Sawyer* (Boston, 1948), 73. Abbott also wrote books that featured children who were free blacks. On this, see Mary Quinlivan, "Race Relations in the Antebellum Children's Literature of Jacob Abbott," *Journal of Popular Culture* 16 (1982): 27–36; and Jeannette Barnes Lessels and Eric Sterling, "Overcoming Racism in Jacob Abbott's *Stories of Rainbow and Lucky* in Antebellum America," in *Enterprising Youth: Social Values and Acculturation in Nineteenth-Century American Children's Literature,* ed. Monika Elbert (New York, 2008).

[12] Frank Luther Mott, *Golden Multitudes: The Story of Best Sellers in the United States* (New York, 1947), 98.

1824, Abbott signed on to a position as lecturer and then a year later as professor of mathematics and natural philosophy at the newly founded Amherst College; when innovations that he had initiated for curricular reform foundered, he left after four years to run the Mount Vernon School, an academy for girls in Boston. Abbott first came to literary fame with his book *The Young Christian* in 1832, which was a bestseller at home and abroad, with a quarter million copies sold.[13]

Abbott introduced ideas about science in his earliest work, *The Little Philosopher* (a set of five short books that appeared in 1829 and then combined into a single volume in 1833), which consisted of queries, prompts, and examples of simple experiments meant to get children thinking about the nature of the common objects that surrounded them in their daily lives. Science-themed volumes within the *Rollo* series include *Rollo's Experiments* and *Rollo's Museum* (both from 1839), and in 1842 Abbott focused on science further in the four volumes of *Rollo's Philosophy* (on water, air, fire, and sky). Abbott also featured technology in his Marco Paul books, which took as topics visits by a young American to such places as the Erie Canal and the Springfield Armory in 1843 and 1844, respectively. Outside of series works, Abbott can be found addressing similar topics in other formats: the steam press in *The Harper Establishment, or How the Story Books are Made* in 1855; technical ingenuity in a Harper's storybook, *The Engineer, or How to Travel in the Woods* in 1856; and natural history in the nonfictional *Aboriginal America* in 1860 (the first volume of a chronological set on American history). A series for older children appeared near the end of Abbott's life, in 1871 and 1872, titled *Science for the Young: or The Fundamental Principles of Modern Philosophy Explained and Illustrated in Conversations and Experiments, and in Narratives of Travel and Adventure by Young Persons in the Pursuit of Knowledge.*

Both Goodrich and Abbott believed they could exert the greatest possible influence on the destiny of the republic by attending to the formation of the minds and souls of the rising generation. Abbott, for example, argued that because children worked so hard to build up a picture of the world from the resources they had at hand, childish perceptions did not simply fade away: "the nature and character of the images which the period of infancy and childhood impresses upon the mind" possess a vast "influence on the ideas and conceptions, as well as on the principles of action in mature years." Goodrich thought likewise. After a decade's worth of work, he would state that although the writing of children's literature "is regarded as a humble, and often a mean, vocation, yet it is not without the means of vindication, even in the light of philosophy," for while "hardened manhood" can defy the efforts of a genius to influence its course, young people can yield "to the slightest touch, [and] may be moulded, in hundreds and thousands" by an ordinary intellect such as his "into the image of God." Along these same lines, speaking to a group of his admirers in a speech sponsored by the People's Lyceum in New Orleans in 1846, Goodrich put the matter even more assertively, asking his audience: "What field so wide, so promising, in every point of

---

[13] Abbott's religious views fell well within Congregationalist orthodoxy, although his lack of support for such older theological views as infant depravity marked him as being of a more liberal bent than other orthodox clergy in this transitional time period. In fact, his downplaying of doctrinal differences and eschewal "of Biblical literalism and anthropocentrism" held him up to complaints of heresy, as when John Henry Newman, in the *Oxford Tracts for the Times* (no. 73), attacked the 1834 follow-up to *The Young Christian—The Cornerstone*—for Socinianism. Kendall, "Times and Tales of Jacob Abbott" (cit. n. 6), 173ff.

view so inviting, so worthy the attention of the patriot and statesman, as the *national nursery*, budding by millions into life and immortality?"[14]

Goodrich's and Abbott's works on nature and science for the young, therefore, were part and parcel of a larger mission. This fact influenced how they packaged and presented natural knowledge—not necessarily conforming to experts' views on how to structure a course in science, or on how to properly categorize scientific knowledge, but instead following their own lights on how to approach these subjects. It also influenced how their readers encountered science—as one facet of the Goodrich or Abbott brand-name enterprises that emphasized a community of learners taking an omnivorous approach to the world around them, accompanied by their own special guide, who was a companion rather than a tutor. They saw their works as being as adaptable to the schoolroom as to fireside education at home, and, in essence, they sought to build a foundation that could support the creation of a nation of lifelong learners who would be open to taking in all they could, confident in their skills of absorbing and assessing new information.[15] The goal was not to create scientists per se but to provide the means to recognize one's scientific citizenship and to access the rights and privileges attendant upon this status.

In adapting their work for young minds, Abbott and Goodrich employed similar techniques, including writing in an easy conversational tone and using vivid examples and striking imagery.[16] For example, in one story, rather than simply stating that the children in the chapter were walking over a deep snow, Abbott writes that "in one place, where the snow was very deep on the side of a hill, they went right over the top of a stone wall." Likewise, Goodrich presents our clinging to the earth in this manner: "The world, you know, is round, like a ball, or like the moon, and people go over its surface, and pass round it, just as flies creep round an apple or a pumpkin." In general, Goodrich had a loose, somewhat miscellaneous style, with the odd and the unusual likely to turn up at any moment (although his scientific texts were more sober in tone). Abbott's style was much quieter, and he depicted smaller-scale settings than did Goodrich. John Crandall, one of the few scholars to focus on both Goodrich and Abbott in any extensive way, aptly compared the two: "The Goodrich grab bag was chock-full of a great variety of the curious and the unusual; Abbott's neat information kit was packed with careful specifications and blueprints which described the way things worked—the common and ordinary as well as the more spectacular . . . Both were instrumental in widening the time and space world of

[14] Jacob Abbott, *Gentle Measures in the Management and Training of the Young* (New York, 1871), 297. Samuel Goodrich, *Peter Parley's Farewell* (1839; Philadelphia, 1841), ii. Goodrich, *Recollections of a Lifetime* (cit. n. 6), 2:331 (emphasis in original).

[15] On education in this period, see: Lawrence A. Cremin, *American Education: The National Experience, 1783–1876* (New York, 1980); Carl Kaestle, *Pillars of the Republic: Common Schools and American Society, 1780–1860* (New York, 1983); Barbara Beatty, *Preschool Education in America: The Culture of Young Children from the Colonial Era to the Present* (New Haven, Conn., 1995); Clarence Karier, *The Individual, Society, and Education: A History of American Educational Ideas*, 2nd ed. (Urbana, Ill., 1986); and Joseph F. Kett, *The Pursuit of Knowledge under Difficulties: From Self-Improvement to Adult Education in America, 1750–1990* (Stanford, Calif., 1994). Within the history of science discipline, greater attention to the history of science education is under way; as an introduction, see Kathryn M. Olesko, "Science Pedagogy as a Category of Historical Analysis: Past, Present, & Future," *Science & Education* 15 (2006): 863–80; and David Kaiser, ed., *Pedagogy and the Practice of Science: Historical and Contemporary Perspectives* (Cambridge, Mass., 2005).

[16] Jacob Abbott, *Rollo's Philosophy: Air* (Boston, 1842), 10–11. Samuel Goodrich, *The Tales of Peter Parley about America*, 3rd ed. (1827; Boston, 1830), 61.

the readers of this period."[17] Such a view, however, runs counter to that expressed by many commentators for whom children's literature is a fallow field unless it features Alice and her forerunners. A contemporary British author such as Charles Kingsley might mock Goodrich as "Cousin Cramchild" of Boston, U.S., in his children's fantasy *The Water-Babies,* and be pleased at calling forth a transatlantic Gradgrind as the dry-as-dust foil to his own cleverness; but there was more than a touch of chauvinism in this critique.[18] Later commentators from the British perspective have been similarly dismissive, as in Mary Jackson's evaluation of Goodrich's books as "mediocre and preachy . . . [and] sternly moralistic"—the antithesis of the "imaginative, light-hearted, humorous, and adventurous books" that had "gone underground" during the early nineteenth century when "the war in Lilliput" had temporarily been won by the forces of dullness who were fed by "an airless, inhumanly narrow view of the child's mind, capacities and needs."[19] Goodrich's and Abbott's books deserve to be read from within a more carefully considered historical context than such sweeping judgments suggest. Their work might indeed be modest, but it is, in fact, by means of the very modest compass of Goodrich's and Abbott's framework that the outlines of a convergence between American identity and scientific experience occurs: the promotion of a "plain style" scientific Americanism.

"PLAIN STYLE" SCIENTIFIC AMERICANISM: RAMIFICATIONS

In thinking about what values these science-themed works conveyed to their readers in this period of emergent debates on America's national identity, what is *not* to be found between the books' covers in terms of form and content plays as important a role as that which is there affirmatively. Declarations of American partisanship, for example, can be discerned, as when an author such as Goodrich urges his child readers to move beyond the books themselves and to collect native plants. Supplying instructions, he asks them to enlist the help of their parents in making a herbarium of dried specimens. In doing so, the children will be on their way to becoming botanists—and, as he confides, "I wish to have all the children who read this book become botanists." Not the least of the reasons for this, he states, is that "much more is known about plants in Europe than in America, because it is an older country. Therefore I wish to have the children here begin to learn about it, for the Americans ought not to be excelled by Europeans, since we have many more plants than they have."[20] But such sentiments rarely interrupt the flow of the business at hand, which is typically a straightforward set of descriptions of various topics on scientific themes or evocations of how one goes about seeking knowledge of the natural world.

Goodrich's and Abbott's books carried nationalistic overtones not just in content but in form and style as well. It is true that the authors did not strive after Anglo-European refinement and that their works in general were rough-hewn and lacked literary pretension: they were designed to fit current needs, the taking on of the fireside

---

[17] John Curtis Crandall, "Images and Ideals for Young Americans: A Study of American Juvenile Literature, 1825–1860" (PhD diss., Univ. of Rochester, 1957), 381.

[18] Charles Kingsley, *The Water-Babies: A Fairy-Tale for a Land-Baby* (1863; New York, 1891), 54, 67, 142.

[19] Mary V. Jackson, *Engines of Instruction, Mischief, and Magic: Children's Literature in England from Its Beginnings to 1839* (Lincoln, Neb., 1989), 189, 190.

[20] Samuel Goodrich, *The Child's Botany* (Boston, 1831), 17, 27.

schooling of a wide range of common people. As such, the authors drew on an American discursive tradition of plain talking that had characterized the Puritan preachers, who designed their sermons to be understood by "the common auditory." Such a style could not afford rhetorical embellishments that foregrounded an author's studied brilliance in a virtuosic turn; instead it needed to be clothed in "homely dress and a coarse habit because it came 'out of the wilderness.'" The plain style evolved over time, as Perry Miller characterized it, "through the prose of Benjamin Franklin, through the cultivated rusticities of the epistles of several Revolutionary 'farmers,' into the very language of the Declaration of Independence." Literary theorist Cristanne Miller notes that "by the mid-nineteenth century Puritan 'plain style' had become the language of self-expression, the trusted idiom in America, although—or perhaps because—it had lost its bolstering doctrinal and political contexts."[21] Timeliness and breadth of reach were of the essence to get through to the common auditory in the antebellum years; the goal was not to save souls but to save the nation through the creation of republicans.

Goodrich's and Abbott's adoptions of a plain style, then, can be understood as not merely an expedient choice, or the making of a virtue out of lack of skill, but rather as affirmative stances that sought to establish a partnership in learning with their far-flung audience of small scholars. In Goodrich's case, this bond was struck through the character of Peter Parley. A picture of Parley appeared opposite the first page, which opened with: "Here I am! My name is Peter Parley! I am an old man. I am very gray and lame. But I love to tell stories to children, and very often they come to my house, and they get around me, and I tell them stories of what I have seen and what I have heard."[22] Indeed, the numerous pictures of Parley that appeared invariably had a smiling group of children clustered around the old man, who seemed as an image to be an amalgam of a favorite elderly relative with all the time in the world to spare for young boys and girls, a Santa Claus who came bearing stories as gifts, and a historical figure, such as one who, when a child himself, might have sat on the knees of his Revolutionary forefathers. Parley often addressed his readers in first-person from the pages of his books. For example, at the end of *Tales about the Sun, Moon, and Stars*, Parley speaks to his audience directly, stating, "It would give me great pleasure to know that my little readers have been all of them pleased with these stories. To an old man that is now grey and lame, it would be a matter of delight, as he hobbles about the streets, to see in the bright faces of the little boys and girls, a smiling 'Thank you, Mr. Parley; thank you for your Stories!'"[23] Goodrich's works were written to be understood by the common auditory at its most foundational level, collapsing the distance between elite science and everyday experience.

Abbott's companion-like bond with his readers would come not from creating a fictional persona as the author but from creating child characters such as the young Rollo Holiday, whose actions and thoughts Abbott sketched with such a sympathetic pen that reviewers often remarked that he captured the way in which real children

[21] Perry Miller, "An American Language," in his *Nature's Nation* (Cambridge, Mass., 1967), 233. Cristanne Miller, *Emily Dickinson: A Poet's Grammar* (Cambridge, Mass., 1987), 143–44. For another use of this concept, see Brandon Brame Fortune, "Charles Willson Peale's Portrait Gallery: Persuasion and Plain Style," *Word and Image* 6 (1990): 308–24.

[22] Goodrich, *Tales of Peter Parley about America* (cit. n. 16), 9.

[23] Samuel Goodrich, *Peter Parley's Tales about the Sun, Moon, and Stars* (1830; Philadelphia, 1850), 115.

thought and acted. Abbott devised scenarios that brought to life scenes of everyday childhood with a kind of casual fidelity that allowed his readers to recognize reflections of their own experiences within the pages of his books, providing his audience with the sense that he was writing *for* them, not *at* them. This perspective was enhanced by the fact that Rollo was the star of a series of books, which followed his growth in a sequence, just as each member of Rollo's audience had done, and continued to do, themselves.

In British didactic fare of the previous generations, it is true that authors sought to produce texts that would appeal to children as children, and in this, Goodrich and Abbott were part of a broader movement. These earlier books especially favored the dialogue format, sometimes in a more formal question and answer mode, and sometimes in terms of the presentation of fictional conversations meant to simulate ordinary discourse. Aileen Fyfe, in an essay on "Young Readers and the Sciences" in the period from 1780–1820, takes note of a relevant example in Sarah Trimmer's *An Easy Introduction to the Knowledge of Nature and the Holy Scripture* (1780; 8th ed., London, 1793), which "appears to consist of conversations, but there is no real dialogue." In the passage she cites, two children are taking a walk with their mother, who directs their attention in the following manner: "Do you not smell something very sweet? Look about in the hedge, Henry, and try if you can discover what it is. See, Charlotte, what a fine parcel of woodbines he has got; they are quite delightful: but notice the woodbine is very different from the oak." As Fyfe observes, "[S]uperficially this conversation uses a child's curiosity to proceed to a lesson on botanical identification, but neither Henry nor Charlotte has a voice, and their mother dictates what they should investigate."[24] It is not that adults are never sources of information in Goodrich's and Abbott's works, for they are, with knowledge sometimes verging on the encyclopedic. But there are differences, both subtle and more overt, between their work and that of other authors that temper the status quo ante of the legacy of the didactic literature they were recasting.

For example, Abbott often shows us *how* Rollo thinks (rather than simply describing for us what Rollo should be learning, ventriloquizing his authorial voice through parental stand-ins), and it is Rollo's curiosity that often drives the narrative (with adults responding to him, rather than the other way around). Most striking of all are the lacunae and inconclusive tangents that appear in the course of Rollo's adventures, with proper pedagogical closure sometimes taking a back seat to the contingencies of the mundane world, with all its competing dynamics. This less confined approach is exemplified in an early passage in *Rollo's Museum*, in which Abbott seeks to capture how the curiosity and imagination of a child let loose on his own on a summer day would work. Having set Rollo on a perch above a running brook, Abbott releases the boy's running stream of thought in like manner:

> Rollo lay down upon the bridge, and looked into the water. There were some skippers and some whirlabouts upon the water. The skippers were long-legged insects, shaped somewhat like a cricket; and they stood tiptoe upon the surface of the water. Rollo wondered how they could keep up. Their feet did not sink into the water at all, and every now and then they would give a sort of leap, and away they would shoot over the surface, as if it had been ice. Rollo reached his hand down and tried to catch one, to examine his feet; but he could not succeed. They were too nimble for him. He thought that, if he could only

---

[24] Fyfe, "Young Readers and the Sciences" (cit. n. 5), 287–88, 288.

catch one, and have an opportunity to examine his feet, he could see how it was that he could stand so upon the water. Rollo was considering whether it was possible or not, that Jonas might make something, like the skippers' feet for *him*, to put upon his feet, so that *he* might walk on the water, when suddenly he heard a bubbling sound in the brook, near the shore. He looked there, and saw some bubbles of air coming up out of the bottom, and rising to the top of the water. He thought this was very singular. It was not strange that the air should come up through the water to the top, for air is much lighter than water; the wonder was, how the air could ever get down there.[25]

Nor is this all. Rollo begins to question where the water in the brook comes from and is surprised "that he had never wondered at it before . . . Where can all the water come from?" He then observes more bubbles coming up and thinks that finding a stick to poke about in the mud would be useful, for perhaps there is some kind of animal blowing the bubbles. Rollo's reflections are broken when Jonas, the young man who is his father's hired helper, comes by. Together the two try to reason through the mystery of the bubbles in the water, but this only leads to more questions. Jonas relates that he does not know "how the bubbles of air get down into the mud, at the bottom of the brook," but he tells Rollo he can think of another extraordinary phenomenon: the rain, which "is water coming down out of the air . . . the air gets down into the water, and you wonder how it can, when it is so much lighter than water. So water gets up into the air, and I wonder how it can, when it is so much heavier." It is a question that does not get resolved.[26] Whereas in the British books it is generally the case that for every question there is an answer, in Abbott's pages, it is instead the case that questions multiply and answers may or may not be forthcoming. The open-endedness of this inquiry not only reflects a characteristic of a child's curiosity rendered in naturalistic tones but also carries a message about democratic versus hierarchical polities.

The extent to which this American style could be seen as deviant is marked out by a British commentator in a lengthy essay in 1842 in the *Quarterly Review*.[27] Authors such as Goodrich and Abbott, the commentator claimed, had fallen "into the mistake of addressing [children] in print as they suppose them to talk to one another in every-day life"—this was "an empty simplicity" that encouraged "the love of too-easy reading in a child, like the taste for low company in an adult."[28] The writer found the concept of presenting scientific knowledge in such easygoing terms to be a categorical mistake—education being "a thing of seriousness and solemnity," and science an especially difficult subject area. As the author remarks, "[T]he difficulty of clothing the highest subjects in the meanest language is fortunately what most effectually unmasks the futility of this 'high life below stairs' kind of proceeding."[29] The class distinctions of referring to the American productions as "high life below stairs" was echoed in other descriptions that referred to the American products as "Transatlantic abominations" and "vulgar." Goodrich is scored for adapting material from his

[25] Jacob Abbott, *Rollo's Museum*, rev. ed. (1855; New York, 1867), 15–16 (emphasis in original).
[26] Ibid., 17, 18, 28, 29.
[27] The *Wellesley Index to Victorian Periodicals* indicates that the author was Elizabeth Eastlake, a journalist and commentator on art; for biographical information, see the entry in the *Oxford Dictionary of National Biography*.
[28] Anonymous, "Article II," *Quarterly Review* 71 (1842): 54–83, 59.
[29] Ibid., 57, 64.

[British] betters in the following manner: "Mr. Goodrich reminds us of those taste-less and irreverent workmen who, in the building of modern Rome, pounded the most beautiful antique marbles *to make mortar!*" At heart, as the author attests up front, "to combine instruction designedly with amusement is, we firmly believe, like uniting authority with familiarity, a sophistry which ends by equally destroying both."[30]

This fear of the ramifications of placing the authoritative in a dangerous proxim-ity to the familiar does not appear in American evaluations of Goodrich's or Abbott's work. A review in the *Christian Examiner* in 1831 states that:

> Peter Parley is, we believe, a great favorite with children, because in a simple way, which they like and understand, he has been telling them tales about almost every thing. This is an intimacy, which we have no intention or wish of disturbing. *The Tales about the Sun, Moon, and Stars*, will furnish young people with about as much astronomy as they can comfortably bear, in such a manner as to engage their pleased attention, and imprint the facts permanently on their memory. They very early desire to know something about those splendid lights and sparkles, the sun, moon, and stars; and they may be made to know much, if their capacities are consulted as they are by their friend Parley.[31]

An 1836 comment on *Parley's Cyclopaedia* in the *New Yorker* states that "to the general excellence of the 'Parley' works we have already borne testimony, while their peculiar adaptation to the wants and tastes of children is a matter of perfect notoriety." The *Maine Monthly Magazine* likewise lauds *The Animal Kingdom* with the follow-ing comments: "We have here another effort of Peter Parley to please and instruct the youthful mind. The old gentleman is really indefatigable in his exertions to benefit the rising generation. He has done more in the space of a few years to raise the stan-dard of juvenile literature, than had been accomplished in a half a century previous." Speaking on the issue of tone in Abbott's work in 1839, one reviewer stated with ap-probation that "we should recommend every one who wishes to learn how to address the young, how to talk to them and how to instruct as well as to entertain them, to take a seat . . . in Mr. Abbott's study." At midcentury the *Methodist Quarterly* argued that, in large part, Abbott's success was due to the fact that "simplicity rather than force distinguishes his style . . . while a pleasing naturalness gives a lasting charm to the whole." The writer further judged that "probably no other writer in this country has so many readers, or is doing so much to form the taste and character, as well as to in-form the intellect, of the rising generation."[32]

"A MUTUAL AND UNITED EFFORT": ABBOTT AND HIS "LITTLE PHILOSOPHERS"

Abbott's own presence in many of his works is that of a dramatist limning scenes that enact the process of knowledge acquisition as occurring within a community of learners in which there is no one individual who knows everything, and understand-ing how to handle disputes over knowledge claims or presumed authority is part of

---

[30] Ibid., 70, 79, 76 (emphasis in original), 57–58.

[31] "Children's Books," *Christian Examiner* 10 (May 1831): 215.

[32] Review of *Parley's Cyclopedia*, *New Yorker*, 22 Oct. 1836, 77; review of *Parley's Cyclopedia: The Animal Kingdom*, *Maine Monthly Magazine*, July 1836, 48; "Jonas's Stories and Rollo's Experi-ments," *Christian Register and Boston Observer*, 9 Nov. 1839, 178; "Jacob Abbott's Young Christian Series," *Methodist Quarterly*, October 1852, 609.

the educational task. This is one instance in which the study of nature is depicted as a natural training ground for future members of a democratic polity.

Even in the simplified form in which natural knowledge is presented in Abbott's books, the learning of facts is no simple matter of having them transferred from an authoritative source to the child mind. As Gillian Avery has noted, Abbott's aim was to produce "capable, self-reliant children who questioned, discovered, and thought for themselves," and a stable American character required the ability to experience facts firsthand in order to acquire the ability to judge authoritatively for oneself.[33] This is what Marco Paul learns in his visit to the Erie Canal, for although knowledge from books has value, "when we learn by observation, we go out and see for ourselves, instead of taking the statements or explanations of others."[34] Another example is drawn from Lucy and her friend Marielle's method of tide telling when at the seashore. Although they had an almanac that stated when high tide would come, "they did not like to trust the almanac entirely, especially as it was so easy to make a mark, and see for themselves." Their method of marking the water entailed placing a broken shell "at the highest place where the water came to as it rolled up the slope of sand," and then, after some time had passed, "observe whether the waves came up higher than their mark, or not so high; and thus they satisfied themselves whether the tide was rising or falling."[35]

Even when it is a matter of Rollo's parents' or others' being able to provide the necessary information, the story often indicates that it is preferable for the children to try to work things through on their own. In an instance of this point of view in *Rollo's Experiments*, Rollo and his cousin Lucy have decided to construct a sundial, and Abbott has Rollo inquire of "his mother if she would not be kind enough to help them fix their apparatus; but she said she would give them particular directions, though she should prefer letting them do the whole themselves, and then, if they met with any difficulties, they might come and report them to her, and she would tell them how to surmount them."[36] Indeed, much of the investigation into nature in the books is self-generated on the part of the children, and nature is not always a willing or transparent participant: Rollo catches a fish in a dipper to study it—only to lose it; a plan to trap bees under a flowerpot in an effort to create a working beehive goes amusingly awry; Rollo's belief that an umbrella will prove to be a suitable parachute for a jump off the shed proves incorrect. The natural world is shown as open to individual inquiry but only partially knowable or malleable.

Clearly, adults are authority figures in Abbott's fiction, but they are not necessarily therefore all knowing. For example, when the family is observing the properties of magnets in *Rollo's Experiments*, Rollo asks his father, upon being told that the magnet's attractive powers are contained at the ends of the bar, "Well, father, what is the reason?" Mr. Holiday answers that he does not know, and the following exchange

---

[33] This evaluation, with which I agree, is one that is at odds with some literature that discusses Abbott's advice regarding obedience to parental authority. Among relevant literature is Jani L. Berry, "Discipline and (Dis)order: Paternal Socialization in Jacob Abbott's Rollo Books," *Children's Literature Association Quarterly* 18 (1993): 100–105; and Mark I. West, "Guilt and Shame in Early American Children's Literature: A Comparison of John S. C. Abbott's *The Child at Home* and Jacob Abbott's Rollo Books," *University of Hartford Studies in Literature* 18 (1986): 1–7.

[34] Jacob Abbott, *Marco Paul's Voyages & Travels: Erie Canal* (New York, 1852), 16.

[35] Jacob Abbott, *Cousin Lucy on the Sea-Shore*, rev. ed. (1842; New York, 1863), 103, 104.

[36] Jacob Abbott, *Rollo's Experiments*, rev. ed. (1855; New York, 1867), 20.

occurs: "Don't you know, father?" said Rollo. "I thought you were going to tell us all about it." "No," said his father. "I only know a very little about it myself. I am going to explain to you some of the facts,—such as I happen to know."[37] In the *Fire* volume of *Rollo's Philosophy*, when the family is exploring what happens when iron filings are tossed into an oil lamp, Rollo begins to speculate about what might happen with something thicker, such as a tip broken off a knitting needle. Rollo's mother takes up his train of thought and asks if there were some way "of suspending a piece of iron as large as the end of a knitting-needle in the lamp; do you think it would take the fire?" Mr. Holiday again indicates ignorance: "I don't think it would be heated hot enough. For some reason or other, I don't understand exactly what, a large piece of iron cannot be heated very hot in a small fire, even if the fire entirely covers it."[38] A bit later there is more: 'Does inflammable mean,' continued Rollo's mother, 'that a thing takes fire easily, or that it burns with a great flame when it does take the fire?' 'I don't know,' said Mr. Holiday. 'I never thought of that distinction.'"[39]

And so it goes. A great deal of knowledge is imparted in the books, but the emphasis is on the *search* for that knowledge, and a necessary corollary to this search is that a sincere exploration will bring the community of inquiry to points at which the answers are unclear, and this is as it should be if the dice are not loaded beforehand by pretensions of omniscience.

Abbott made this point clear at the very beginning of his authorship, in *The Little Philosopher.* "A wish has been frequently expressed," he explained in the new prefatory notes he added to the 1833 edition, "that more of the questions had been answered in the book, as many teachers find that they are unable to answer all themselves." Abbott explains that there were, in fact, "two very good reasons" why he had not done so: first, because there were some questions that "he did not know how to answer himself"; and second, "because *he did not wish the teachers to be able to at once to answer all.*" Why shouldn't teachers be supplied with the means to display such omniscience? Abbott declares such a demonstration to be indefensible, whether for himself as book author or for any adult authority (and note that his explanation below specifically references *paternal* authority as well):

> The old idea of a teacher's trying to keep up before his pupils the character of infallibility, is now exploded. All good teachers, all wise parents, are willing freely to acknowledge their ignorance, and to engage with their pupils on the understanding, that *they are themselves learners, too,* though in a somewhat more advanced stage. When a child brings to its parent or teacher any difficult or perplexing question, "I don't know, but I will help you find out," is the best answer . . . They then, teacher and pupil, occupy common ground,—there is sympathy between them,—the child is encouraged by observing that his father is a learner, as well as himself.

That a parent or teacher should not be immediately all-seeing brings the question of intellectual authority into a communal rather than a hierarchical relationship between those designated as teachers and those designated as students, for all belonged to the larger category of learners. Abbott indicated that "a great many [of the questions] are

---

[37] Ibid., 140.
[38] Jacob Abbott, *Rollo's Philosophy: Fire* (1855; New York, 1868), 41.
[39] Ibid., 47.

intended to awaken the common curiosity and interest both of parent and child, and to engage them in that most useful and delightful employment, *a united and mutual effort*, in search of knowledge."[40]

In Abbott's knowledge community, both men and women can be authoritative sources, and girls and boys alike take on the roles of inquirers. This is true when adults are present, as when Rollo encounters his former grade-school teacher, and she becomes involved in his investigations (into optical illusions such as the rainbow, or in regard to water pressure in the digging of a small canal from the brook), or when his mother proposes some experiments with feathers and air temperature on a rainy day when he and his brother are out of sorts. It is also true in moments of recreation among the children themselves. In one episode in which Abbott describes Rollo at play with Lucy and his little brother, Nathan, when out-of-doors, here is Rollo's suggestion for how they will amuse themselves: "'O Lucy we will play go up the mountains. There is a hill for us. That shall be Chimborazo.' . . . They played that Rollo was the guide, and that Lucy was the philosopher. Nathan was the philosopher's servant."[41] In *Rollo's Museum*, it is Lucy who is elected by the children to be in charge of their search for curiosities for their natural history cabinet, and they are to follow her decisions as to any questions that arise among them.[42]

The appearance of female characters was no accident. Certainly, the idea that mothers had a special role to play in educating children's moral development and guiding their early years within the domestic sphere was an increasingly prominent theme in this era. The vogue for the pedagogical theories of Rousseau, Pestalozzi, and Fellenberg, in which children were to be given greater freedom to learn at their own pace, and to involve all their senses via object teaching, also was highly adaptable to the home setting. But there is more to this gender emphasis than simply an interest in maternal guidance for small children, for in the antebellum period the idea that the female intellect was the equal of males gained increasing force, with a complementary surge in opportunities for girls to receive the same advanced educational training as boys (of which Abbott himself was a part, in his years running the Mount Vernon School). Recent histories of women's education in this period have found that although "excluded from most colleges in institutions called seminaries, academies, and high schools, women received education very much like the education men received in the colleges, academies, and high schools they attended." Ideologies that professed no gender distinctions in mental power became widespread in the antebellum era, because "the realm of the intellect was regarded as being separate from other arenas of life"—that is, one need not believe in equality for women in the legal, political, or domestic sphere to affirm that girls and women not only *could* engage in mental self-improvement but that they *should*, and thus "few gender distinctions were made in regard to academic studies." This includes science instruction. Kim Tolley has shown that instruction on science in academies for girls was as likely as for boys, and in fact girls could end up having more exposure to science, because of the greater emphasis on instruction in Greek and Latin for boys—success at which

[40] Jacob Abbott, *The Little Philosopher, for Schools and Families: Designed to Teach Children to Think and to Reason about Common Things; and to Illustrate for Parents and Teachers Methods of Instructing and Interesting Children* (Boston, 1833), 12 (emphasis in original).

[41] Abbott, *Rollo's Philosophy: Air* (cit. n. 16), 12–13.

[42] Abbott, *Rollo's Museum* (cit. n. 25), 80–83.

was needed for entry into college. Outside of formal schooling, Deborah Jean Warner noted some time ago that "in the antebellum years, before it had become a recondite professional specialty, science played an important and wide-ranging role in American culture. As members of that culture women were encouraged to learn about science and to involve themselves in its pursuit."[43] Although the relatively equitable role that the female presence played within Abbott's depictions of engagement with science may surprise us today, it should not.

In addition to the relatively free equality that existed between the sexes in this regard, Abbott also makes clear that the ability to impart knowledge could come from sources other than parents or classroom teachers. Jonas, for example, is often portrayed as being able to teach Rollo about the natural world, despite the fact that he is a hired worker, aged perhaps thirteen or so, who has received little in the way of schooling. When he offers one day to give Rollo and his cousin James a lecture on the displacement of water, Rollo laughs and challenges him, arguing that Jonas could not give lectures because he is not a teacher. But Jonas insists that he is, and "at any rate if you will get James to come and help you make an audience, you may see if I can't lecture." And so he did. In another volume, he shows Rollo and his younger brother, Nathan, several experiments that can be performed with a bellows, instructing them that "I will be the professor and you two shall be my class in philosophy, and I will direct you how to make the experiments."[44] Jonas, in fact, plays a large part in the discussions about how nature works, with some of what he knows coming from books he has borrowed from Rollo's father in a kind of self home schooling, and much more from his personal experiences stemming directly from his own labor. A discussion about the nature of oxygen, for example, proceeds from work that Jonas is doing in clearing a nearby field by piling the debris into heaps that will be set on fire. Jonas asks Rollo what he supposes it might be that makes anything burn, and Rollo answers that it burns "itself." Jonas answers no, explaining that "the air makes it burn: it must have good air around it, or else it won't burn. There is something in the air which I call the life of it; this makes the fire burn. But when this is all gone, then that air will not make fire burn any longer." He then explains that the fire that Rollo is building inside a stump will need to have an opening from underneath the debris he has set inside if it is to burn successfully. Rollo then asks what it is in the air that causes the fire to burn, and Jonas answers that he has forgotten the name—"I knew once . . . but it was a hard word, and I have forgotten it"—but that "it is some part of the air, which goes into the fire, and is all consumed, and then the rest of the air is good for nothing."[45]

Jonas is a key pivot for the plain style of Abbott's scientific books. He is clearly intelligent and reflective and goes about his work with a spirit of inquiry equal to his practical competence and the reliability of his judgment. Although it is the character of Rollo that is most often remembered in later generations when Abbott's work

---

[43] Margaret A. Nash, *Women's Education in the United States, 1780–1840* (New York, 2005), 54, 1, 1. For further elaboration of women's pursuit of intellectual equality and the support for it, see Mary Kelley, *Learning to Stand and Speak: Women, Education, and Public Life in America's Republic* (Chapel Hill, N.C., 2006); Kim Tolley, *The Science Education of American Girls: A Historical Perspective* (New York, 2003); and Deborah Jean Warner, "Science Education for Women in Antebellum America," *Isis* 69 (1978): 58–67, 67. See also Kohlstedt, "Parlors, Primers, and Public Schooling" (cit. n. 1).

[44] Jacob Abbott, *Rollo's Philosophy: Water*, rev. ed. (1842; New York, 1855), 109–10; the lecture occurs later in chapter 11. Abbott, *Rollo's Philosophy: Air* (cit. n. 16), 51.

[45] Abbott, *Rollo's Philosophy: Air* (cit. n. 16), 127–28, 140.

is recalled, the character of Jonas made a notable impression on contemporaries as well. Looking back in 1866, one commentator fondly recalled Jonas and the other Jonas-like figures Abbott invented, summing up his characteristics thusly:

> Jonas is an admirable creation—the typical New England boy; such a boy as every one of us has been or has known. Steady, sensible, sagacious . . . Domestic and agricultural virtues adorn his sedate career. His little barn-chamber is always neat; his tools are always sharp; if he makes a box, it holds together; if he digs a ditch, there the water flows. He attends lyceum lectures, and experimentalizes on his slate at evening touching the abstruse properties of the number nine. Jonas is American Democracy in its teens; it is Jonas who has conducted our town-meetings, built our commonwealth, and fought our wars.[46]

Nothing could be further from the case of Jonas's role in the British books to which Abbott's Rollo series might be compared most profitably: that of Maria Edgeworth's *Harry and Lucy* stories, which also show the acquisition of scientific knowledge within a domestic setting. Members of an at least upper-middle-class family, Harry and Lucy are being tutored at home by their parents—in the end, Harry hopes to become a great scientist, and Lucy is learning to train her mind so as to be a good intellectual companion to one such as Harry. There is no comparable figure in these British stories to Jonas, in which a working-class individual carries a large responsibility for embodying the values of securing knowledge about the natural world to the young who look up to him as a guide despite his lack of scientific credentials.

Unlike Harry and Lucy, whose lives are filled with instruments such as barometers, hygrometers, and portable camera obscuras, Rollo and his family and playmates are not supplied with handy scientific instrumentation by Abbott in order to provide regular classroom-type lessons in the guise of ordinary domestic life. Neither are well-endowed deus ex machina figures employed to provide sophisticated instruments, as Edgeworth does for her young scholars, as when an uncle who was formerly a physician moves into the neighborhood, bringing with him "a microscope, an electrifying machine, an airpump, and a collection of fossils" and, as the children discover when they visit him, an industrial-strength barometer, a wooden orrery, and a pair of very large globes.[47] In the sequel, the children visit Sir Rupert—representing scientific nobility at its most impressive—who has an exquisitely outfitted castle laboratory. In the American books there are no optical instruments of any kind, let alone anything more ostentatious to be had: all that is at hand is the family, broadly construed, everyday items for ordinary use, and the natural world in which the family is embedded. The tools of their impromptu "philosophy classes" are of the most familiar sort, such as a bellows from Jonas's workshop. In fact, the magnet mentioned above was not a special instrument or toy, but an ordinary worn-down steel file that Jonas had magnetized to use in the workshop for picking up nails. Science is not the "star" of these science-themed books: it is, instead, the members of the community of inquiry themselves, engaged in their united and mutual effort to learn about common things.

---

[46] Thomas Wentworth Higginson, "Children's Books of the Year," *North American Review*, Jan. 1866, 236–49, 246.

[47] Maria Edgeworth, *Harry and Lucy, with Other Stories* (New York, 1836), 1:34, 37.

"KNOWLEDGE IS COMMON PROPERTY": GOODRICH ON KNOWING ONE'S PLACE

Goodrich does not have a set of players to stride across his pages, as does Abbott, enacting the creation of scientific Americans: instead, it is Parley who constitutes the figurehead around whom this virtual community is brought into existence through the steady stream of scientifically based material he pens, in a variety of formats suitable for nonreaders, young children, older youth, and family circles. It would be possible during the middle decades of the nineteenth century for a child successively to grow out of and grow into a Parley-produced scientific text at every stop along his or her developmental path—and then turn around and introduce the same pattern for his or her offspring, such was the prolific nature and popularity of Goodrich's works. Goodrich indeed had his entrepreneurial eye focused on the economic return of exploiting a multiplicity of market niches, but these niches also are consonant with the need to supply materials suitable for the developmental trajectory of a particular kind of young American: that embodied by Abbott's Jonas.

Goodrich also held to a conviction that Americans formed a community of learners, not the least in having to teach each other how to establish a new nation that would endure. This was not a matter for a nation's leaders alone; Goodrich was a staunch republican. For example, he agrees that "it is a matter of necessity that professional men should possess extensive erudition. But there is no reason that it should be restricted to them." Indeed, Goodrich is thoroughly unimpressed with the idea that a university degree raises one above the common citizenry. He points to the illegitimate birth of the university as an institution in support of this view:

> From the time that Europe began to emerge from the dark ages, it had been a matter of pride with many sovereigns to aid the revival of learning. But how was this done? Not by attempting to enlighten the whole community, but by the founding of colleges or universities, where a chosen few might be instructed in every branch of human knowledge. The idea was to establish institutions on a magnificent scale, endow them with ample funds, store them with rich libraries, collect into them every kind of philosophical apparatus, and place them under the guidance of men distinguished alike for learning and genius. Here the sons of the rich or the favorites of the powerful were to be assembled and instructed. Thus, while the people at large were to be left in darkness, a blaze of glorious light was to be collected into one focal point.[48]

Universities the United States might have, but it is the education of all that is of prime importance, and "the great discovery of our pilgrim fathers" was that "in opposition to the scheme of despotism, which would concentrate and confine knowledge in a university, they sought its diffusion over the people at large." To focus a society's energy on an educational elite is inefficient, undemocratic, and immoral. On the following point, Goodrich is absolutely certain: "The truth is, that knowledge is common property, and those who possess it are bound to distribute it for the benefit of others."[49] He insists that "the mechanic, the farmer and the tradesman may be benefited by knowledge, and may, without neglect of their proper vocation, cultivate a love of letters." To enlighten the community at large, Goodrich argues, one cannot escape the fact that "the plain truth is, that human improvement, like heat in water,

---

[48] Samuel Goodrich, *Fireside Education* (New York, 1838), 391, 17.
[49] Ibid., 19, 59.

works upward and not downward. If you would warm the whole mass, begin at the bottom."[50] For those Jonases to be as well educated as they deserved to be, some means toward this end needed to be effected.

Goodrich's approach is emphasized in the preface to *A Pictorial Natural History: Embracing a View of the Mineral, Vegetable, and Animal Kingdoms*. In it, he explains that the text could be mastered in a few months—adaptable to use in a common school, but also, he states so that "common readers" can have at hand an outline of the natural world, such that those who have neither the time nor means to currently explore further will have embedded in their minds a logical structure that would allow an individual "to retain his acquisitions, and indeed to make constant accessions to them in after life." Goodrich notes that "the largest portion of society begin and finish their technical education" in a primary seminary, and numerous readers may have themselves fallen into this category. This book, he offers, is but "a humble attempt to aid and encourage the inquirer in the outset of his search after a kind of knowledge hitherto inaccessible to many learners."[51] Why inaccessible? In his *Illustrated Natural History*, he explains that without works such as his, "the world at large" would be prevented from becoming "participators in the golden fruit of scientific research," for the communications of scientific men would otherwise "remain beyond the reach of the million, locked up in quartos, hidden in the libraries of the learned."[52] The general public could not depend upon aristocrats of learning to share what should be "common property." In writing about science for children and young people—and, no doubt, adult readers of these works as well—Goodrich saw himself as doing more than simply increasing the amount of content stored in his readers' minds. He was convinced that he was setting natural knowledge at liberty and creating freer minds in doing so.

In his scientific presentations Goodrich takes special care with the opening pages of his works to establish the idea within his readers that they are already equipped to think deeply and broadly about the nature of the world they live in. The *Pictorial Natural History* opens with a chapter on "The Material Universe" and an exercise in limbering the mind straight off: the reader is asked to begin where he is seated and to acknowledge that what we survey "is limited to a very small part of the whole system of nature. If we look beyond the house in which we live we see other houses, and also fields, and hills, and plains." He then takes the reader in his or her mind's eye out further and further in greater circles until the reader is contemplating the immensity of the earth, then its diminished status within our solar system, and then on to the immensity of space, where "every little star which is seen twinkling in the sky, is a sun like ours, supposed to be surrounded, too, with a similar troop of planets, which like our earth, are the residences of animated creatures." He takes his youngest readers on a similar journey in *Tales about the Sun, Moon, and Stars*, although he cleverly uses the captivating image of a hot-air balloon to have his readers begin their journey from

[50] Ibid., 391, 348.

[51] Samuel Goodrich, *A Pictorial Natural History: Embracing a View of the Mineral, Vegetable, and Animal Kingdoms* (1842; Boston, 1854), iv, iii.

[52] S. G. Goodrich, *Illustrated Natural History of the Animal Kingdom, Being a Systematic and Popular Description of the Habits, Structure, and Classification of Animals from the Highest to the Lowest Forms, with their Relations to Agriculture, Commerce, Manufactures, and the Arts*, 2 vols. (New York, 1859), 1:vi. For a brief discussion of this work and of Goodrich, see Welch, *Book of Nature* (cit. n. 2), 138–39.

home to outer space. The text starts with: "Here is a picture of a balloon! It is an immense bag of silk, as big as a small house. A net is thrown over it. To the bottom of this net a little car, like a boat, is attached." As the pictured balloon begins to ascend, he asks his readers to notice "the little flag which the man is waving in his hand," the net that can still be seen over the balloon, and the fact that one "can easily distinguish the countenances of the fearless men who are now going to take a ride into the regions of the clouds." He offers a further picture showing how small the house-size balloon now looks, showing it to be "scarcely bigger than a pin's head."[53]

In the *Pictorial Natural History*, a large image of the earth as viewed from space— nearly half the page, with just six lines of text below it—dominates the book's first page in the opening chapter. This image is one of Goodrich's most beloved devices, used early on and recurring frequently. It appears, in fact, in the 1827 volume *The Tales of Peter Parley about America*, his very first children's book on any topic. In this book, he interrupts his narrative about early America midway through, in order to first "make you understand some things" about the nature of the earth and moon. The moon, Goodrich relates, "looks small, because it is very far off; but it is really a great world, with mountains, and rivers, and seas upon it." Now this is interesting information in itself, but Goodrich goes on next to point out that "if we were on the moon, the earth we live upon would look small and round, like the moon." And Goodrich supplies a picture of the world as it would be seen from outer space and remarks that, "I suppose it looks on this picture as it does to the people in the moon." For those children who went on to read *Tales about the Sun, Moon, and Stars* in 1830, they would learn that "for those people who inhabit the stars, our earth is itself a star. Look up at one of the stars in the sky, and imagine yourself upon it. The world, dwindled by the distance, would appear to you a little glimmering light, so faint and far, as to be scarcely visible," and a newer, more evocative engraved image appears.[54] These two images from *Tales about America* and *Tales about the Moon, Sun, and Stars* will appear again in later volumes, including the *Pictorial Natural History* (1842), *A Glance at the Physical Sciences* (1844), and *The Wonders of Geology* (1844). It is a thrifty recycling of useful material, but I would argue that its presence across his varied editions also indicates that it represents a conceptual point of significance to Goodrich. In particular, by encouraging a perspectival shift in his readers, Goodrich assists them, in word pictures and in illustrations, to look on their familiar world with alien eyes to better understand it.

[53] Goodrich, *A Pictorial Natural History* (cit. n. 51); Goodrich, *Tales about the Sun, Moon, and Stars* (cit. n. 23), 9–13.

[54] Goodrich, *Tales of Peter Parley about America* (cit. n. 16), 60, 60–61, 61. Goodrich, *Tales about the Sun, Moon, and Stars* (cit. n. 23), 22. The more scholarly *A Glance at the Physical Sciences* from 1844 hedges on the question of the moon's inhabitation. The text reports that it does not appear that there are fluids on the moon's surface, and without air and water it could not be inhabited (or at least life would be different from that on our planet). A footnote, however, offers two authorities who hold that the moon "is inhabited by rational creatures" (23). On the question of planetary habitation in this era, see Michael Crowe, *The Extraterrestrial Life Debate, 1750–1900: The Idea of a Plurality of Worlds from Kant to Lowell* (Cambridge, UK, 1986). For an introductory commentary about visual images in popular culture, along with a discussion of a more sophisticated use of the view from another planet concept, see Bernard Lightman, "The Visual Theology of Victorian Popularizers of Science: From Reverent Eye to Chemical Retina," *Isis* 91 (2000): 651–80. Lightman highlights an image from Jefferys Taylor's *A Glance at the Globe and at the Worlds around Us* (1848), in which the earth is seen from the vantage point of the moon, and that also includes a rendering of the lunar landscape (651–52).

It is not only in astronomical passages that such perspectival shifts occur. In his discussions of foreign peoples, Goodrich makes frequent use of brief, shorthand stereotyping to sum up their "key" characteristics, confirming conventional prejudices in the process. And yet, there are times when Goodrich interrupts the text to cast doubt on the generalizations that have been presented, a tactic that Bruce Harvey has characterized as "the Peter Parley conscience, a reluctance to endorse unqualified claims about any particular culture's alleged inferiority."[55] The result is again to introduce frameworks at odds with the expected, which—briefly—displace readers from their own cultural locations and to perceive the statements from the point of view of the subjects under discussion themselves. (Goodrich, however, never goes so far as to have these others actually speak for themselves.) This is a tactic that makes descriptions of human nature more complicated and confused than they would be otherwise in a rationally ordered universe.

For example, Goodrich argues, distorted perception is responsible for what we know of Indians. It may not be surprising that he depicts the Spanish as "conquerors and spoilers of America [who] had strong motives for first hating, and then defaming, the Aborigines." To appropriate the riches of Mexico, Cortez and his men slaughtered millions and enslaved the rest. To "justify his conduct to his own conscience and the world at large" he had therefore to represent these peoples "in the most degrading and revolting colors." But Goodrich holds the English responsible as well. It was true that the colonists were "almost constantly in a state of active hostility with the savages. . . The savages were, therefore, enemies, and how hard is it to judge fairly of those we hate!" The English also, "to make their conduct stand fair before the world," adopted the strategy of "portray[ing] the Indians in the most unfavorable light" and "the misrepresentations, proceeding from the early settlers of America . . . constituted the main sources of history."[56]

But self-interested motives are not the only form of distortion. Goodrich also points out that "the natural disposition to interpret the bosoms of others by our own, has led historians and philosophers to estimate the Indians by transatlantic standards of thought, feeling and action." However natural the disposition, it is likely "to lead to false conclusions." Anyone who has struggled to master a new language knows what it is like to:

> become acquainted with a new, and before unknown, region of thought; how original then, and how different from our own, must be the mind, soul and character of a people, who have grown up by themselves, shaping out, in isolation of all the rest of the world, and in utter ignorance of all but themselves, their own manners, customs and institutions! In analyzing such a race we should study facts—abstain long from theory, and constantly be on our guard against bringing them up to be measured by the artificial rules established in our own minds.[57]

That Goodrich cannot live up to his own standard does not make it any less interesting that he struggles with it.

This is a refrain that occurs in other contexts as well. In *Peter Parley's Tales of Af-*

---

[55] Harvey, *American Geographics* (cit. n. 7), 45. Harvey adds that in this regard Goodrich's "works do become at times self-conscious about the grounds of their own authority" (45).

[56] Samuel Goodrich, *Lives of Celebrated American Indians/by the Author of Peter Parley's Tales* (Boston, 1843), 1, 2.

[57] Ibid., 3, 4, 5.

rica, for example, Goodrich informs his readers that "formerly, the accounts given us of the people of Africa, represented the negro races, as a stupid, debased portion of the human family, only fit to be the slaves and servants, of the rest of mankind. But modern travellers, more worthy of credit, give more favorable representations . . . the Negroes of Central Africa are more intelligent, and more civilized, than the world has been led to believe them." It is also the case that "the Caffrees and the Hottentots are now known to be superior in every respect, to what their Dutch neighbors, used to say they were."[58] This uneasiness concerning what had been taken to be fact appears even within the conservative confines of a family library, prompted by an encounter with human nature in a variant beyond the Anglo-European norm. As more facts are known, the old ones shift in relation to the new frame or are even discredited outright. What is taken to be natural is not necessarily stable knowledge, for a lack of perspective can compromise the basis for producing authoritative knowledge.

CONCLUSION

In the United States, the explosion of natural knowledge that occurred in the nineteenth century intersected with cultural patterns still very much in flux, and this fluidity possessed interesting implications for the organization of knowledge. It was by no means an assured fact that deferential patterns toward an intellectual elite would take hold, especially in a society that celebrated the "common man." It was a situation in which contemplating the nature of democracy could lead to imagining the democracy of nature, where scientific knowledge would be seen as a shared cultural possession, not simply the province of expert practitioners.

Goodrich and Abbott displayed a sense of mission embedded within the plain style imperative of their works, which was to reach as many young people as possible within the sphere of the common auditory, not simply to inculcate a rote morality or to impart facts for facts' sake, but to bring as many individuals as possible within a circle of lifelong learning and mutual improvement. They saw the burgeoning sphere of natural knowledge as belonging to everyone, and so they worked hard to put this knowledge into public circulation and into the hands of America's future citizens. In this sense, the larger community and the scientific community were coextensive, not disjunctive—as long as the circulation of knowledge flowed freely through republican channels. This was, in some ways, an innocuous educational goal: but to hold it as normative would be to place this goal at odds with the growing assertion of scientific leaders—even if mostly uttered *sotto voce* in these middle decades, but to become more dominant toward the century's end—that what occurred within the vernacular sphere was too superficial to count as meaningful or significant in terms of the constitution of a community of knowers.

Historians of science may still tend to evaluate past progress for a society by the numbers of real scientists produced, the pace of real research conducted, and the amount of real recognition bestowed upon professional aspirants by those of higher status, but these are not the only measuring sticks by which to assess the nature of a society's scientific character. In the mid-nineteenth century, Almira Hart Lincoln Phelps spoke for the standards applied by the plain style scientific Americans, at the end of her *Familiar Lectures in Botany*: "The spirit of our government is highly

[58] Samuel Goodrich, *The Tales of Peter Parley about Africa*, rev. ed. (Philadelphia, 1836), 123.

favorable to the promotion and dissemination of knowledge; and although Europe may boast of many stars which irradiate her firmament of letters, shining with brilliant lustre amidst the surrounding darkness of ignorance, may we not justly feel a national pride in that more *general diffusion of intellectual light, which is radiating from every part, and to every part of the American republic!*"[59] Phelps may have been overly enthusiastic in her estimation, but she describes well the goal toward which these midcentury educationists were striving.

Further exploration of the connections between childhood nature and the issue of scientific citizenship within American nature offers a key pathway to insights about the relationships between science and national identity. As the nation was on the threshold of evolving into a changed relationship with the Old World in the postbellum period, another commentator also focused on the image of the child, this time as reflected in the children's literature of the mid-nineteenth century, positing that it still had a role to play in the coming cultural transition. "Every American child, unless he has the misfortune to be transplanted across the Atlantic for schooling, is American in early associations; while every highly educated man among us has half his thoughts in Europe," this author noted. The opportunity to read through a wide array of children's books proved to be a pleasant "reversion" to "the vernacular" for adults who were "oversaturated with Transatlantic traditions," and the author commented that "it is singular how much more of the aroma of American nationality one can get from our children's books than from any others . . . External nature itself seems more sincere and genuine." Some aspects of American identity that had been nurtured in the national nursery seemed to be harder to cultivate in the later decades, and the writer imagined that "an epoch may yet come" where "a maturer civilization . . . shall grow from the common ground, and be as fresh and healthful as this childish society."[60] Peter Parley and Rollo would no doubt have agreed.

---

[59] Almira Hart Lincoln Phelps, *Familiar Lectures on Botany,* rev. ed. (Hartford, Conn., 1836), 235 (emphasis in original).
[60] Higginson, "Children's Books of the Year" (cit. n. 46), 249, 248, 249.

Points Critical:

Russia, Ireland, and Science at the Boundary

Michael D. Gordin*

ABSTRACT

This essay compares the way in which Russia and Ireland have defined themselves since the mid-nineteenth century as scientific nations (or not) by following the careers of D. I. Mendeleev (1834–1907) and Thomas Andrews (1813–85), both of whom were involved in the discovery of the "critical point" boundary between liquids and gases. Mendeleev and Andrews deployed their critical-point research in a similar fashion to integrate science into the national identity for their respective countries, a strategy that proved far more successful for Mendeleev than for Andrews.

INTRODUCTION

What could possibly be learned about science by discussing Russia and Ireland together? One could easily compile a list of dissimilarities between Europe's largest country and one of its smallest that would make the endeavor appear distinctly unworthwhile. I wish, however, to point to ways in which they can be usefully and fruitfully compared for a specific time period (the nineteenth century) and in a specific area of cultural development (science).[1] In that period, elite laboratory science had a small but noteworthy penetration into both Russia and Ireland, and in both places it was associated with a particular quasi-foreign stratum: either Russo-Germans or German-educated Russians, and the Anglo-Irish Protestant Ascendancy, respectively. By comparing the two, we can better understand the relationship of national identity to science in some of the "later emerging" nations of Europe—that is, those countries that developed nationalist movements after patterns of industrialization and national self-identification had already congealed into fairly standard forms in England, France, and the German states. In the particular case of Russian and Irish scientists, we can see how nearly identical professional strategies yielded

* Department of History, Princeton University, 136 Dickinson Hall, Princeton, NJ 08544; mgordin@princeton.edu.

This paper has benefited enormously from comments by participants at the conference Science, Technology, and National Identity, held at the University of South Carolina, September 20–22, 2007, especially the suggestions by Ann Johnson, Carol Harrison, and Alfred Nordmann, as well as the perceptive observations by the three *Osiris* reviewers. Peter Brown's comments on the Anglo-Irish have also proven most helpful.

[1] My approach here is inspired by the stimulating comparison of twentieth-century political culture in Italy and Japan offered in Richard J. Samuels, *Machiavelli's Children: Leaders and Their Legacies in Italy and Japan* (Ithaca, N.Y., 2003).

© 2009 by The History of Science Society. All rights reserved. 0369-7827/09/2009-0006$10.00

very different results. For instance, how did the career of a specific scientist come to be marked as positive or negative in specific national contexts? As it happened, in the Russian context, the career of the scientist was built up into a source of national pride and aspiration beginning in the late nineteenth century, while in Ireland the same period marked its decline in prestige. Comparing the parallel (and somewhat related) developments in both contexts will help illuminate how certain scientists became icons for national identity, while others were deliberately written out of the national pantheon, irrelevant in large part *because* their primary personal identities were as scientists.

One common strategy of considering a broader context for the scientific endeavor has been to situate historical narratives in a nation-state context. For some concerns—the formation of educational institutions, state-funding patterns, questions of citizenship—this is a very fruitful approach. For the question of "national identity," however, it proves more problematic.[2] This term bears two distinct but related meanings. First is the notion that individual scientists approach the study of nature in ways strongly correlated with a frame of mind set by their national origins. Taken to absurd extremes, the result replicates naïve (and chauvinist) cultural essentialism. Consider my favorite example:

> If the average quality of the German scientist is heavy thoroughness, that of the French-man clearness and lucidity of thought combined with an impulse to treat science as art, that of the British extraordinary positivity and that of the American an ability to combine specialization with mass production, then the distinctive character of the Russian may be seen in the restlessness of his spirit and the striving to embrace a wide field of knowledge, to find answers to questions which are ever present in his thoughts and which once raised may not be lightly put aside, but must be settled the one way or the other, if only for the satisfaction of his own soul.[3]

No historian of science today would endorse this view without reservation. But similar tropes find their way into a wide variety of mainstream accounts and remain popular among scientists. One way of getting away from this kind of reductionism is to approach national identity in its second sense: as the identity of a larger collective, a rhetorical device for public figures (including scientists) in specific nations to define the political and economic contours of their nation-state as being "scientific." That is, the nation as a whole comes to acquire the identity as being a "scientific nation," without our having to project that political strategy—for it *is* a political strategy—into the minds of its citizens.

Such professions of scientific national identity are historically situated in space and time, a point that can be illustrated most effectively through comparison.[4] The historical scholarship on Ireland—and even more so on science in Ireland—is riddled with brief as well as extended comparisons. The most common comparator is Canada,

[2] An exception, specific to the case of Irish mathematics in the early nineteenth century, is David Andrew Attis, "The Social Contexts of W. R. Hamilton's Prediction of Conical Refraction," in *Science and Society in Ireland: The Social Context of Science and Technology in Ireland, 1800–1950*, ed. Peter J. Bowler and Nicholas Whyte (Belfast, 1997), 19–35.

[3] Alexander Petrunkevitch, "Russia's Contribution to Science," *Transactions of the Connecticut Academy of Sciences* 23 (1920): 211–41, on 222.

[4] Maurice Crosland, "History of Science in a National Context," *British Journal for the History of Science* 10 (1977): 95–113, 111.

either in toto or just the individual case of Quebec,[5] followed by the obvious comparisons to other British colonies, such as India, New Zealand, Australia, and colonial America.[6] Historians interested in the economics of colonialism have considered Ireland in relation to the more obviously "colonial" cases of Africa and Asia.[7] A smaller subset compares Ireland with other "late-emerging" nations on the Continent (Italy, Spain, Portugal),[8] and one sociologist of science has noted some intriguing parallels with another recently independent, formerly colonized, northern European country, Finland.[9] Given all of this frantic comparison with the Irish case, it would be surprising if Russia were left out. In fact it has not been, although the comparisons have remained on the level of economics and rebellions. (Both countries experienced shocks to their peasant populations in midcentury, with the disastrous famine in Ireland in 1846 and the emancipation of the serfs in Russia in 1861, and subsequently suffered a wave of political terrorism by the Fenians and the Populists, respectively.)[10]

This essay explores both the fruitfulness and the limits of extending the Ireland-Russia comparison to the history of science. There are a large number of ways this

[5] John Wilson Foster, "Strains in Irish Intellectual Life," in *On Intellectuals and Intellectual Life in Ireland: International, Comparative, and Historical Contexts*, ed. Liam O'Dowd (Belfast, 1996), 71–97, 84; Richard A. Jarrell, "Colonialism and the Truncation of Science in Ireland and French Canada during the Nineteenth Century," *HSTC Bulletin* 5 (1981): 140–57; Jarrell, "Differential National Development and Science in the Nineteenth Century: The Problems of Quebec and Ireland," in *Scientific Colonialism: A Cross-Cultural Comparison*, ed. Nathan Reingold and Marc Rothenberg (Washington, D.C., 1987), 323–50.

[6] Richard A. Jarrell, "Technical Education and Colonialism in Ireland in the Nineteenth Century," in *Prometheus's Fire: A History of Scientific and Technological Education in Ireland*, ed. Norman McMillan (Dublin, 2000), 170–86; and Tony Ballantyne, "The Sinews of Empire: Ireland, India, and the Construction of British Colonial Knowledge," in *Was Ireland a Colony? Economics, Politics, and Culture in Nineteenth-Century Ireland*, ed. Terrence McDonough (Dublin, 2005), 145–61.

[7] Liam Kennedy, *Colonialism, Religion, and Nationalism in Ireland* (Belfast, 1996), 169–71. Kennedy concludes that Ireland should not be compared with other colonies but with western European countries. This, of course, merely continues the tradition of thinking of Ireland juxtaposed with another place. Opposed to Kennedy on the issue of "colonialism" as a category for Ireland are: Terry Eagleton, "Afterword: Ireland and Colonialism," in McDonough, *Was Ireland a Colony?* (cit. n. 6), 326–33; and the very influential Michael Hechter, *Internal Colonialism: The Celtic Fringe in British National Development* (1975; New Brunswick, N.J., 1999).

Much of the discussion of science in a colonial framework has centered on the classic model of "colonial science" as provided by George Basalla, "The Spread of Western Science," *Science*, N.S. 156, no. 3775, 5 May 1967, 611–22, which explicitly names Russia as effectively "colonial" with respect to science on 613. Basalla has been criticized from all sides for oversimplification and lack of a clear mechanism. For two examples, see Ian Inkster, "Scientific Enterprise and the Colonial 'Model': Observations on Australian Experience in Historical Context," *Social Studies of Science* 15 (1985): 677–704; and V. V. Krishna, "The Colonial 'Model' and the Emergence of National Science in India: 1876–1920," in *Science and Empires: Historical Studies about Scientific Development and European Expansion*, ed. Patrick Petitjean, Catherine Jami, and Anne Marie Moulin (Dordrecht, Netherlands, 1992), 57–72. This relates to whether one can truly consider Irish science to be "colonial." See, e.g., Steven Yearley, "Colonial Science and Dependent Development: The Case of the Irish Experience," *Sociological Review* 37 (1989): 308–31; Richard A. Jarrell, "The Department of Science and Art and Control of Irish Science, 1853–1905," *Irish Historical Studies* 23 (1983): 330–47; and Roy MacLeod, "On Science and Colonialism," in Bowler and Whyte, *Science and Society in Ireland* (cit. n. 2), 1–17.

[8] Dorinda Outram, "Negating the Natural: Or Why Historians Deny Irish Science," *Irish Review*, no. 1 (June 1986): 45–49, 48–49.

[9] Yearley, "Colonial Science and Dependent Development" (cit. n. 7), 327

[10] Eoin MacWhite, "Ireland in Russian Eyes under the Tsars," *Australian National University Historical Journal* 1 (1965–66): 5–13; and Derek Offord, "Political Terrorism in Russia in the 1880s: The Fenian Lesson," *Irish Slavonic Studies*, no. 5 (1984): 27–31.

could be done, and I have selected a very small example as my starting point: the priority dispute over the discovery of the liquid-gas critical point between Dmitrii Ivanovich Mendeleev (1834–1907) and Thomas Andrews (1813–85). This was a particularly one-sided priority dispute: in 1870 Mendeleev, having read Andrews's article on the critical point published in 1869, claimed to have discovered the phenomenon a decade earlier, a claim to which Andrews never responded. (This silence on Andrews's part is in itself significant.) The priority dispute is slightly technical, and I explore it in some detail because it teases apart a distinction between conflict among international chemists within their discipline (resolved largely in terms of the generations of chemists) and the presentation of a style of science to one's immediate national interlocutors (the nationalist importance of showing one's science as fundamentally *non*national). In addition, the specificity and clarity of the issues eliminate any consideration of radical incommensurability between these two rather different national contexts.

Then I show how that original context for both research projects (Heidelberg for Mendeleev, midcentury Belfast for Andrews) and the scientists' subsequent careers illuminate the role of the science of this period in the construction of national identities. I will develop the Irish case more extensively than the Russian one not just because the latter is better treated in recent historiography (especially in terms of the identification of the Soviet Union as a scientific state) but also because the Irish case represents a counterintuitive instance where scientific self-identification *failed* to take root in the national self-concept.[11] As a result, the circumstances by which Irish nationalism came to exclude Andrews from iconic status demand greater explication.

WHO DISCOVERED THE CRITICAL POINT, AND WHO CARES?

Mendeleev saw threats to his scientific status everywhere, and his career was studded with efforts to establish priority for his most noted work, the periodic system of chemical elements, against all claimants (but especially from German chemist Julius Lothar Meyer).[12] The first serious priority dispute he engaged in petered out quickly, yet in many ways it set the pattern for the more famous disputes that followed. Mendeleev in 1870 nicely presented his vision of the substance of the "critical point," and by working forward from Mendeleev to Andrews, we can see how the concept had been repackaged by Mendeleev to lend credence to his priority claim.

Mendeleev's 1870 article in *Poggendorff's Annalen* on the critical point highlights two tropes common to all of Mendeleev's priority claims (and not his alone): first, he claimed that he was not really interested in priority and recognized the value of the *data* being offered by his opponent, if not the opponent's *originality* or *interpretation*; and that the two of them were talking about the same effect—in this case, what Andrews called "the critical point" (Mendeleev cited it in English), and Mendeleev's own "absolute boiling point" (*absolute Siedetemperatur* [Ger.], *temperatura*

---

[11] See, e.g., the essays included in Michael D. Gordin, Karl Hall, and Alexei B. Kojevnikov, eds., *Intelligentsia Science, Osiris* 23 (2008), and references therein.

[12] The literature on this priority dispute is voluminous. For an introduction, see J. W. van Spronsen, *The Periodic System of Chemical Elements: A History of the First Hundred Years* (Amsterdam, 1969); and Eric R. Scerri, *The Periodic Table: Its Story and Its Significance* (Oxford, 2007).

absoliutnogo kipeniia [Rus.]).[13] What exactly is this phenomenon? As Mendeleev paraphrased Andrews, it is "a critical temperature [that] exists for all bodies, so that by a higher temperature vapour will not condense—and thus presents as a gas— while by a lower temperature the same would under a certain pressure become condensed, and thus be a true vapour."[14]

Mendeleev argued that Andrews drew the wrong conclusions from this effect, and for the wrong reasons, because Andrews insisted on confusing what Mendeleev saw as a clear line between liquids and gases:

> From the above [quotation], as well as in many other places in Andrews's article, the impression could easily be formed as if the transition of a gas into a liquid at a defined temperature were less clear and sharp than under usual conditions. The incorrectness of such a conclusion, so far as the absence of a clear difference between gases and liquids in Andrews's article, compels me to make some comments on this situation, all the more as similar investigations were undertaken by me about 10 years ago.[15]

Interestingly, it is precisely this blurring of the boundary that makes Andrews's work seem so original today and what makes the critical point of liquids so important.

The question then is why Mendeleev did not see this blurring as either correct or significant. He ignored the question because he believed the importance of the "absolute boiling point" was to generate natural "laws," thus introducing more bright lines in nature, not erasing those already recognized. This interest in finding laws stemmed directly from the origins of Mendeleev's research on this topic in organic chemistry: "So one would probably find for hydrocarbons C^nH^{2n+2} a regularity [*Gesetzmäßigkeit*] between the change in the absolute boiling point and its composition, in that these temperatures for the liquid homologues C^6H^{14}, C^7H^{16}, are either measured directly or are derived from their coefficients of cohesion."[16] At the time Mendeleev made his claim for priority, he was indeed shifting away from the study of periodicity and toward state-funded research—an innovation Mendeleev encouraged as part of a vision of Russia as a scientific nation—on the compressibility of gases, a project that would eventually end in shambles in 1880.[17] One function of Mendeleev's prior-

[13] D. Mendelejeff, "Bemerkungen zu den Untersuchungen von Andrews über die Compressibität [sic] der Kohlensäure," *Poggendorff's Annalen der Physik und Chemie* 141 (1870): 618–26, 622–23. Mendeleev would continue to insist on the identity of these concepts and his priority in them into the twentieth century. In the seventh edition of his textbook *The Principles of Chemistry*, published in 1903, Mendeleev included a long footnote that recounted many of the same arguments as this 1870 piece and concluded: "Obviously, it [the critical point] is identical with the absolute boiling temperature." Mendeleev, *Osnovy khimii*, 7th ed. (1903), reproduced in Mendeleev, *Sochineniia* (Leningrad, 1947), 5:291n29.

[14] Mendelejeff, "Bemerkungen zu den Untersuchungen von Andrews" (cit. n. 13), 618. On the predecessors that Andrews does cite, see Duane H. D. Rollder, "Thilorier and the First Solidification of a 'Permanent' Gas (1835)," *Isis* 43 (1952): 109–13; and Yorgos Goudaroulis, "Searching for a Name: The Development of the Concept of the Critical Point (1822–1869)," *Revue d'histoire des sciences* 47, nos. 3/4 (1994): 353–79.

[15] Mendelejeff, "Bemerkungen zu den Untersuchungen von Andrews" (cit. n. 13), 619.

[16] Ibid., 623. Today these formulas would be written with subscripts, but Mendeleev characteristically preferred superscripts (as did many of his contemporaries).

[17] This gas research is detailed in Michael D. Gordin, *A Well-Ordered Thing: Dmitrii Mendeleev and the Shadow of the Periodic Table* (New York, 2004), chap. 3. Mendeleev specifically invoked both Regnault and Mariotte's law of the temperature-volume relationship in his priority article: Mendelejeff, "Bemerkungen zu den Untersuchungen von Andrews" (cit. n. 13), 625.

ity claim, therefore, was to characterize his new path of research as no departure at all but rather continuity from his earlier work; another was to carve out a reputation for himself as a leading European chemist. Both of these conditions, as we shall see, were intimately linked to the context of St. Petersburg in the late 1860s.

As far as priority disputes go, this one had no legs. Andrews never bothered to respond publicly to Mendeleev's position. This has nothing to do with prejudice against the claims of a Petersburger. Early in his career, in the 1840s, Andrews did engage in a priority dispute with St. Petersburg academician Hermann Hess over thermochemistry.[18] Even more important than national identity for priority claims, in this case, was one's professional stage: earlier in one's career it is more vital to stake such claims to build a reputation—a strategy both Andrews and Mendeleev deployed at the equivalent point in their own careers. Andrews did not see the criticisms as serious enough to answer and likely interpreted the Russian's experiments as being fundamentally about a different question. Nobody else seemed to take Mendeleev's claim seriously, either. It has very rarely been addressed in the chemical or historical literature, and the few glancing references to it leave no doubt as to where the priority belongs: "to Andrews and to him alone."[19]

The question is less why Andrews did not respond—such claims of priority were ubiquitous in this period and were rarely answered unless taken up by a third party—than why Mendeleev chose to make the claim when this research had lain fallow for a decade. Why did he initiate the dispute at all? First, because he was at an early stage of his career. Second, because he had much to gain in confronting an established (and, even better in the Russian context, *foreign*) scientist and little to lose, even in the case of complete silence on his opponent's part (which was in fact the case). But this was not simply a case of jockeying for prestige. For Mendeleev, there was a deep philosophical interest in the effect of the absolute boiling temperature, an interest he felt Andrews had ignored in his articles. The Petersburger was interested in the critical point as a locus for finding natural laws between organic homologues. This interest emerged directly out of his research on the absolute boiling temperature, conducted in his apartment laboratory while on a two-year state-funded postgraduate fellowship during 1859–61 spent mostly in Heidelberg. This research on organic liquids occupied essentially the entirety of his time abroad, and he published on it widely and in several languages (although not English).[20] In these delicate experiments, Mendeleev took an array of very pure liquid hydrocarbon homologues and studied their capil-

[18] Thomas Andrews, "On the Thermal Changes Accompanying Basic Substitutions," *Philosophical Transactions of the Royal Society of London*, pt. 1 (1844), 21, reproduced in Andrews, *The Scientific Papers of the Late Thomas Andrews*, ed. P. G. Tait and A. Crum Brown (London, 1889), 107.

[19] A. C[rum] B[rown], "Obituary Notices of Fellows Deceased," *Proceedings of the Royal Society of London* 41 (1886): i–xv, on xv.

[20] These are gathered, in Russian translation, in Mendeleev, *Sochineniia* (cit. n. 13). The original French and German articles are: "Notiz über die Ausdehnung homologer Flüssigkeiten," *Annalen der Chemie und Pharmacie* 114, no. 2 (1860): 165–9; "Sur la cohésion de quelques liquides et sur le rôle de la cohésion moléculaire dans les réactions chimiques des corps," *Comptes Rendus* 51 (1860): 97–99; "Sur la cohésion moléculaire de quelques liquides organiques," *Comptes Rendus* 50 (1860): 52–54; "Ueber die Molecularcohäsion einiger organischen Flüssigkeiten," *Zeitschrift für Chemie und Pharmacie* 3 (1860): 49–52; "Ueber die Ausdehnung homologer Flüssigkeiten," ibid., 397; "Ueber die Cohäsion einiger Flüssigkeiten, und über die Rolle, welche die Molecularcohäsion bei den chemischen Reactionen der Körper spielt," ibid., 481–84; and "Ueber die Ausdehnung und das specifische Gewicht der Flüssigkeiten beim Erwärmen über ihren Siedepunkt," *Zeitschrift für Chemie und Pharmacie* 4 (1861): 33–37.

lary rise in narrow tubes. Capillarity was his proxy for measuring the "cohesiveness" of the various fluids. The original idea, therefore, had absolutely nothing to do with the transition between gas and liquid; it was concerned exclusively with a property belonging to liquids.[21] (Nonetheless, capillarity is a very good way to find the critical point of various liquids.)[22]

Mendeleev's goal here was always to find microphysical laws concerning the forces that held liquids together and thus the forces that drove chemical reactions. Of course, he did not receive perfect results: "This law, however, is not completely exact, but only approximate—as are all laws known to date in physical chemistry."[23] Yet Mendeleev was convinced that eventually he would find the kind of regularity that would provide a connection between the microphysical (cohesion) and the macrophysical (capillarity):

> Selecting cohesion of liquids as the subject of my researches, I indeed supposed that it would be subject to such a law. Continuing the investigations of this subject, I have in mind above all a collection of data. The measure of the cohesion of bodies, doubtless, is a property more characteristic than, for example, the boiling point, and we have to date very little data about it. As a consequence, probably, one will discover the dependence between cohesion and many other physical properties, such as specific weight, expansion, heat capacity, latent heat, etc. With the development of molecular mechanics the measure of cohesion should enter as a necessary data for the resolution of a majority of questions.[24]

From his position in Belfast, Thomas Andrews had long established a reputation in thermochemistry, but his most famous work, that of the critical point, was an original departure for him in the 1860s, rather late in his career.[25] After a period of some silence in the 1860s, Andrews chose to deliver his findings in the distinguished Bakerian Lecture of 1869. With a decided emphasis on the transition between states and not either gases or liquids, he titled it "On the Continuity of the Gaseous and Liquid

[21] Mendeleev, "Chastichnoe stseplenie nekotorykh zhidkikh organicheskikh soedinenii," *Khimicheskii zhurnal N. Sokolova i A. Engel'gardta* 3 (1860): 81–97, reproduced in Mendeleev, *Sochineniia* (cit. n. 13), 12. Mendeleev was also interested in thermodynamic questions such as latent heat, but he made it clear that these were secondary issues: Mendeleev, "Sur la cohésion de quelques liquides et sur le rôle de la cohésion" (cit. 20), 99; Mendeleev, "Ueber die Ausdehnung und das specifische Gewicht der Flüssigkeiten" (cit. n. 20), 33.

[22] J. Livingston, R. Morgan, and Reston Stevenson, "The Weight of a Falling Drop and the Laws of Tate: The Determination of the Molecular Weights and Critical Temperatures of Liquids by the Aid of Drop Weights," *Journal of the American Chemical Society* 30 (1908): 360–76; Eric Higgins, "The Temperature Coefficient of the Weight of a Falling Drop as a Means of Estimating the Molecular Weight and the Critical Temperature of a Liquid" (PhD diss., Columbia Univ., 1908); and Eldred H. Chimowitz, *Introduction to Critical Phenomena in Fluids* (New York, 2005), 290–92.

[23] Mendeleev, "Chastichnoe stseplenie nekotorykh zhidkikh organicheskikh soedinenii [II]," *Khimicheskii zhurnal N. Sokolova i A. Engel'gardta*, 1 (1860): 145–70, reproduced in Mendeleev, *Sochineniia* (cit. n. 13), 30.

[24] Ibid., 32. This obsession with finding laws of nature runs like a scarlet thread throughout Mendeleev's career; the periodic system case is only the most obvious, prolonged, and successful example. See Gordin, *A Well-Ordered Thing* (cit. n. 17), chap. 7.

[25] For secondary studies of Andrews's work on the critical point, see: J. S. Rowlinson, "The Work of Thomas Andrews and James Thomson on the Liquefaction of Gases," *Notes and Records of the Royal Society of London* 57 (2003): 143–59; Rowlinson, "Thomas Andrews and the Critical Point," *Nature* 224, 8 Nov. 1969, 541–43; Allan A. Mills, "The Critical Transition between the Liquid and Gaseous Conditions of Matter," *Endeavour* 19, no. 2 (1995): 69–75; H. Mackle, "Thomas Andrews, Calorimetrist," *Nature* 224, no. 5219, 8 Nov. 1969, 543–44; and Cyril Domb, *The Critical Point: A Historical Introduction to the Modern Theory of Critical Phenomena* (London, 1996).

States of Matter." This lecture may never have been delivered, and if it was, very few people attended, as most of Andrews's more illustrious peers were in London that day attending the first Faraday Lecture at the Royal Institution (Andrews's close friend Michael Faraday had died two years earlier), delivered by the French chemist Jean-Baptiste Dumas and chaired by English chemist Alexander Williamson. Andrews, typically, declined to leave Belfast.[26] It is clear from the discussion in the published article based on the ostensible lecture, however, that he approached the phenomenon from the gaseous side and then worked his way back to the liquid, increasing the pressure on a sample in a thin capillary tube. In the paper, Andrews strongly empha-sized the *physical* (rather than chemical) properties of the phase transition, mostly measured on carbonic acid (carbon dioxide). Even "transition" puts it a bit strongly because Andrews contended there was no real boundary at all:

> The answer to the foregoing question, according to what appears to me to be the true in-terpretation of the experiments already described, is to be found in the close and intimate relations which subsist between the gaseous and liquid states of matter. The ordinary gaseous and ordinary liquid states are, in short, only widely separated forms of the same condition of matter, and may be made to pass into one another by a series of gradations so gentle that the passage shall nowhere present any interruption or breach of continuity. From carbonic acid as a perfect gas to carbonic acid as a perfect liquid, the transition we have seen may be accomplished by a continuous process, and the gas and liquid are only distant stages of a long series of continuous physical changes.[27]

In fact, he explicitly excluded the very microphysical speculations Mendeleev was so keen on illuminating with his organic homologues, although he did recognize that his findings implied strong intermolecular forces.[28] So Andrews defined the "criti-cal point" as the temperature above which it was impossible to condense a gas into a liquid, no matter how high the pressure; in other words, the critical point marked an *absolute* phase transition. Below that point, any substance could exist in either or both states simultaneously. The critical point of carbon dioxide was 31°C, of ether 200°C.

Much as in the case of Antoine Lavoisier's "oxygen" and Joseph Priestley's "de-phlogisticated air," it should be clear that in this case the two scientists were not looking at (or for) "the same" effect: Andrews sought clarity of concepts at the ex-pense of blurriness in the phenomena before him; Mendeleev wanted rigorous laws of natural phenomena but remained vague about categories such as "cohesion." An-drews worked with gases, whereas Mendeleev's interest lay only in the liquids. And, finally, Mendeleev cared about the microphysical features of the actual substance, while Andrews was interested in macroscopic, indeed thermodynamic, qualities. In what follows, I explore how these divergent approaches to what we now consider to be a single phenomenon were deeply embedded in professional choices made by the two scientists, and how those choices then shaped the readiness with which later Rus-

---

[26] Rowlinson, "Work of Thomas Andrews and James Thomson" (cit. n. 25), 146.

[27] Thomas Andrews, "On the Continuity of the Gaseous and Liquid States of Matter: The Bakerian Lecture," *Philosophical Transactions*, pt. 2 (1869), 575, reproduced in Andrews, *Scientific Papers of the Late Thomas Andrews* (cit. n. 18), 315. As he further articulated it, the transition was not only continuous but also symmetrical with respect to direction. See Andrews, "On the Gaseous and Liquid States of Matter," *Royal Institution of Great Britain*, 2 June 1871, reproduced in Andrews, *Scientific Papers of the Late Thomas Andrews*, 343.

[28] Andrews, "On the Continuity of the Gaseous and Liquid States of Matter" (cit. n. 27), 315.

sians could unflinchingly adopt Mendeleev the scientist as a "Russian" icon, while the parallel choices by Andrews led to his being excluded from claims to "Irishness" in the twentieth century.

Andrews and Mendeleev came from and lived in two very different contexts, but they employed almost identical strategies and drew on very similar resources to establish their critical point/absolute boiling temperature research in their respective sites. Mendeleev situated this research from the start of his career as part of an ambitious program of pure science designed to secure him a position in the sparse academic environment of Petersburg. Andrews deployed similar moves to validate a position as a public intellectual in a troubled emerging nation. For Mendeleev, the purpose was both professionally and politically to point himself toward the future; for Andrews, it was to salvage some continuity with the past. By looking at these strategies as parallel yet divergent, I argue that we can see much of what gets attributed to "national character" in the sciences as the effects of a rather standard strategy in different contexts.

Mendeleev's case was doubly unusual: He was a scientist from a scientifically underrepresented country, and he performed his research not from his home base of Petersburg but in Heidelberg.[29] Mendeleev ended up in Heidelberg after the end of a long train of events initiated by Russia's loss of the Crimean War (1854–56). After the defeat, the government of the new tsar, Alexander II, began a series of reforms designed to bolster the fiscal and military stability of the Russian Empire. Although today these so-called Great Reforms are often (retrospectively) viewed as liberalizing measures—especially the 1861 emancipation of the serfs—they are more accurately seen as conservative efforts to avert unrest.[30] The development of the industrial sector of the economy was one of the prime targets of these reforms. To facilitate this in the late 1850s, a series of talented young Russian specialists in fields such as law, medicine, and the sciences were sent on state-subsidized two- or three-year trips (*komandirovki*) to Europe (and primarily to the German universities) to learn new techniques, acquire new research materials, and in general be trained in the structures and practices of a modern research university. The architects of the program, which lasted for about a decade, hoped to reimport these students so that technical education in the empire could be modernized.[31] Science was part of Russia's reorientation for the future.

Mendeleev was among the first cohort of such students who had just finished their Russian "candidate" or "master's" degrees, and he elected, along with a sizable cadre

[29] See M. D. Mendeleeva, "Novye materialy o zhizni i tvorchestve D. I. Mendeleeva v nachale 60-kh godov," *Nauchnoe Nasledstvo* 2 (1951): 85–94. On his capillarity work while there, see V. P. Veinberg, "Raboty D. I. Mendeleeva po kapilliarnosti i temperature absoliutnago kipeniia," in *Trudy Pervago Mendeleevskogo s"ezda po obshchei i prikladnoi khimii, sostoiavshagosia v S.-Peterburge 20-go po 30-go dekabria 1907 g.*, ed. V. E. Tishchenko (St. Petersburg, 1909), 89–106.

[30] W. Bruce Lincoln, *The Great Reforms: Autocracy, Bureaucracy, and the Politics of Change in Imperial Russia* (DeKalb, Ill., 1990); and Alfred J. Rieber, "Alexander II: A Revisionist View," *Journal of Modern History* 43 (1971): 42–58.

[31] Michael D. Gordin, "The Heidelberg Circle: German Inflections on the Professionalization of Russian Chemistry in the 1860s," in Gordin, Hall, and Kojevnikov, *Intelligentsia Science* (cit. n. 11), 23–49.

of Russian chemists, to spend his time in Heidelberg. Most of these Russians hoped initially to work with Robert Wilhelm Bunsen but eventually congregated instead in the organic chemistry laboratory of privatdozent Emil Erlenmeyer.[32] Mendeleev was not one of the Russians in Erlenmeyer's laboratory; he set up his own in his apartment where he conducted all of his capillarity research. He had been sent abroad in early 1859 for a two-year fellowship from the physical-mathematical faculty of St. Petersburg University (where he had begun his graduate work), with the possibility of a renewal for a third year. He attempted to parlay his findings on "absolute boiling temperature" into an extension for a third year.

Mendeleev's claim began as such things often did in Imperial Russia: with a petition. In a draft of his request to the physical-mathematical faculty for a further year abroad, composed on December 18, 1860, Mendeleev extolled his work on capillarity:

> But the largest part of my time in my stay abroad was dedicated to studies on that special field which connects chemistry with physics and mechanics. Convinced of the identity of the forces of chemical affinity and the force of cohesion and confident that a possibly full solution of questions on the causes of chemical reactions could not be done without knowledge of the magnitude of molecular cohesion, I chose as my specialty this very poorly worked out area.[33]

Note that he made no reference to the absolute boiling point here (or to the rather obvious practical implications for the liquefaction of gases, to which Andrews's work was almost instantly applied). Instead, he focused on highly ambitious potential theoretical insights. Or, as he wrote in a separate request to the petitioner of the St. Petersburg educational region, his hope was to partially efface the boundary between chemistry and physics: "The brilliance of purely chemical discoveries made contemporary chemistry a completely specialized science, tearing it away from physics and mechanics, but doubtless there should come a time when chemical affinity will be seen as a mechanical phenomenon, similar to how it has already come time for us to consider light and heat to be such phenomena."[34]

Mendeleev's move reflected a combination of factors: his own considerable ambition; his conviction that chemistry should grow closer to physics; and his sense of what would *work*, what would lead to an extension of his stay. Mendeleev believed—based on the surrounding environment in Heidelberg and in western European chemistry more generally, as well as on his sense of the aspirations of the Russian chemical community back home—that careful laboratory studies at the forefront of risky and speculative areas of physical chemistry would be precisely the kind of work the

---

[32] On why Heidelberg and why Erlenmeyer, see ibid. (Coincidentally, Andrews also spent the summer of 1854 in Heidelberg, working with Bunsen. P. G. Tait and A. Crum Brown, "Memoir," in Andrews, *Scientific Papers of the Late Thomas Andrews* (cit. n. 18): ix–lxii, xxvii.) The Russian higher educational system had three tiers: the "candidate," which roughly corresponds to an advanced undergraduate degree; the "master's," roughly equivalent to the degree of the same name in the West and which enabled a scholar to teach at the university level for a fixed period of time; and the "doctoral," roughly equivalent to a German *Habilitation*. Mendeleev earned his master's in 1858 but received his doctorate in 1864, after he returned from Heidelberg.

[33] Reproduced in M. N. Mladentsev and V. E. Tishchenko, *Dmitrii Ivanovich Mendeleev: Ego zhizn' i deiatel'nost'* (Moscow, 1938), 1:223.

[34] Reproduced in ibid., 226. The implication with respect to light as mechanical concerned the luminiferous ether.

modernizing state would encourage. Science, in this framing, was not yet a feature of *national* identity; it was intrinsically cosmopolitan. Since the established social position of the scientist was still rare in Russia, Mendeleev cast himself as international, thus drawing on other features of Russia's rather inclusive political culture. That is, he thought he should make a case to the Russians that he was a European chemist, reflecting their own European identifications, and that they should bankroll his further development along these lines.

They apparently did not agree. Mendeleev's friend and fellow chemist N. P. Il'in, who spied for him on the academic politics back home, wrote on February 22, 1860 (upon publication of Mendeleev's first major article on capillarity):

> At the faculty meeting or the university council [dean of the physical-mathematical faculty Heinrich Friedrich Emil] Lenz, on the communication to him of this work or he himself read it in the C.R. [*Comptes Rendus*], said, that for this, in order to do this, what you are doing now, it was not especially necessary to travel abroad, one could do that here as well; where this wind blows from, I do not know, but obviously your work is not understood as having the significance you would like them to ascribe to it.[35]

Mendeleev's mentor Aleksandr Voskresenskii seconded this in a letter on March 19, 1860: "[I]t wouldn't hurt to present something else, something purely chemical."[36] Mendeleev was required to return to Petersburg in February 1861.

In short, he had miscalculated: he had assumed that what Petersburg really wanted was evidence that what was happening with their scientists was no different from what was happening with their counterparts in the West. Instead, Lenz and his colleagues wanted evidence that the students were taking unique advantage of their stay abroad to do work not possible back home. Mendeleev's strategy had been to show that work back home was of the same kind (although of a different degree) as that in Heidelberg; his superiors disagreed. And, indeed, in the 1850s and early 1860s, Russian chemists were in general underfunded, overworked, and poorly organized.[37] There was no reason for locals to think that they were doing the same kind of work as their foreign peers—yet. The identity of "scientific" was not yet inherently included within "Russia." There are two points to take away from this: First, that Mendeleev's priority claim in 1870 was not only an attempt to set up his new gas research but also a retrospective justification of his Heidelberg research and a rebuke to his local colleagues. Second, that a plea for universal and pure science was not what his local context demanded in 1860. By the time of his priority claim a decade later, however, when Mendeleev reiterated the importance of his capillarity work, the situation was rather different. The Russian Chemical Society was founded in November 1868, providing a professional structure for chemists in the imperial capital, and Mendeleev had already begun to receive some attention for his periodic system of chemical elements—another research project that was cosmopolitan, not specifically Russian. With a professional community and local academic power

---

[35] Reproduced in ibid., 237.

[36] Reproduced in ibid.

[37] On the parlous state of Russian chemistry in this period, see Nathan M. Brooks, "The Formation of a Community of Chemists in Russia: 1700–1870" (PhD diss., Columbia Univ., 1989); V. V. Kozlov, *Vsesoiuznoe khimicheskoe obshchestvo imeni D. I. Mendeleeva, 1868–1968* (Moscow, 1971); and Iu. I. Solov'ev, *Istoriia khimii v Rossii: Nauchnye tsentry i osnovnye napravleniia issledovanii* (Moscow, 1985).

(he became professor of general chemistry at St. Petersburg University in 1867) to back him, he could successfully make the claim that Russian science competed on an international level for fundamental laws of nature.[38] He was well on the way to becoming an icon for Russian science precisely *because* his science was not specifically local.

For his part, Thomas Andrews deployed moves similar to those of Mendeleev and achieved completely opposite results—in both the short and long terms. That is, in the 1860s he turned claims of pure science into local credibility; and after his death, these very claims *excluded* him from the status of national icon for Ireland. Andrews belonged to the generation of chemists immediately preceding Mendeleev's, and he spent almost the entirety of his career happily ensconced in the relative backwater of Belfast. He had been born at 3 Donegall Square in Belfast on December 19, 1813, the eldest of six children of a noted linen merchant. Like many aspiring Belfast Protestants (he was Church of England and Ireland, not Presbyterian), he studied at the only local establishment to provide advanced education, the Belfast Academical Institution, but he left it in 1828 to study chemistry at the University of Glasgow with Thomas Thomson. By age fifteen, he had published his first scientific paper, "On the Action of the Blowpipe on Flame." Yet he remained on the move, studying in Paris at the laboratory of J.-B. Dumas, which he enjoyed enormously before returning to the British Isles to complete his education as a physician, earning a diploma from the Royal College of Surgeons of Edinburgh on April 25, 1835, and on August 1 that same year received his MD from Edinburgh. Somewhat unusual for his generation and origin, he had no deep interest in medicine but had undertaken a medical education since it was the best way for an Irishman to learn something about chemistry. In 1835, he received several jobs offers to teach chemistry in London and Dublin, but he turned them all down to practice medicine in Belfast. In 1842, he married a Scotswoman, Jane Hardie, and in 1845 began his career at the newly founded Queen's College in Belfast, where he remained until his death.

Andrews led a disciplined life. He was a man of Spartan qualities: he had an early breakfast and ate nothing until his late dinner after a full day of work. He lectured each day from Monday to Friday at 3 pm and had an exam each week; he taught practical chemistry each day to medical students for one term; and he presided over experimental work in his own lab, which was open from 9 am to 3 pm every day (and on Saturdays 9–12).[39] His laboratory research before discovering the critical point consisted of several peaks: he performed important work on the blood of cholera patients, showing a depletion of water and therefore that death was caused by dehydration; and he determined the composition of ozone (that it consisted solely of oxygen). Over the course of this provincial career, he attained a series of high honors: fellow of the Royal Society in 1849, president of the Chemistry Section of the British Association in 1852 and 1871 (declined to serve again in 1880 due to poor health), president of the British Association itself in 1876, and offered a knighthood in 1880

---

[38] See Gordin, *A Well-Ordered Thing* (cit. n. 17), chaps. 2 and 7, for the growth of Mendeleev's ambitions with respect to laws of nature.

[39] Students who performed exceptionally well on a special exam were allowed to work in the laboratory free of charge; others had to pay a fee of ten pounds for six months or thirteen pounds for the college session. T. W. Moody and J. C. Beckett, *Queen's, Belfast, 1845–1949: The History of a University*, 2 vols. (New York, 1959), 1:162.

(again declined because of illness).[40] Through his entire later career, he rarely budged from Belfast.

Belfast proved to be a rather interesting context. Although it only officially became a city in 1888 (after Andrews's death), it was clearly one of the more dynamic urban areas in mostly rural Ireland throughout the nineteenth century. A provincial town of 13,000 people in 1782, it experienced explosive growth in the following decades, with a population 50,000 by 1831 and 350,000 by 1901. This growth was fueled by linen, the trade of the Andrews family. By 1870, 99 percent of Irish linen was shipped from the port of Belfast, and 21,000 of the 50,000 industrial and commercial workers in the city labored in some aspect of the textiles trade, which experienced extraordinary growth during the "cotton famine" caused by the American Civil War.[41] Along with this demographic and economic growth, Belfast began to develop more of an autonomous cultural life, independent of the massive influence of Dublin, becoming (according to some lights) a "Northern Athens." Thus, despite disadvantages such as the lack of a local scientific instruments trade, Andrews was able to develop a formidable chemical laboratory at Queen's College, Belfast.[42]

The college and Andrews grew in prominence together and fed off each other. In 1845, just before the Great Potato Famine reconfigured Irish politics, the island was in the throes of the repeal movement headed by Daniel O'Connell. The goal of the movement was to repeal the 1801 Act of Union that had disbanded Irish self-governance and fully incorporated it into Britain. One contributing issue, with increasing traction in the regions outside Dublin, was the very limited state of higher education at home in Ireland. For the most part, only two serious institutions existed: Trinity College Dublin, the flagship of Irish education, which was closed to Catholics until Catholic Emancipation; and Maynooth College, a state-subsidized Catholic college to train priests, thus preventing an outflow of seminarians to France during decades of turbulent politics across the English Channel. In partial response to worries that Ireland was losing all of its best intellectual talent to Britain or to the Continent—and not least to dampen Protestant support for repeal—the Queen's College

[40] These biographical details are culled from: Tait and Crum Brown, "Memoir" (cit. n. 32); Henry Riddell, "Dr. Thomas Andrews: The Great Chemist and Physicist," *Proceedings of Belfast Natural History and Philosophical Society* (1920–21), 107–35; Crum Brown, "Obituary Notices of Fellows Deceased" (cit. n. 19), xi–xv; "Thomas Andrews," *Journal of the Chemical Society* 49 (1886): 342–44; and "Thomas Andrews, F.R.S.," *Nature* 33, 17 Dec. 1885, 157–59.

[41] Emily Boyle, "'Linenopolis': The Rise of the Textile Industry," in *Belfast: The Making of the City, 1800–1914*, ed. J. C. Beckett et al. (Belfast, 1983), 41–55; W. A. Maguire, *Belfast* (Staffordshire, UK, 1993), 59; and W. H. Crawford, *The Impact of the Domestic Linen Industry in Ulster* (Belfast, 2005). On the substitution of other fabrics for American cotton, see Sven Beckert, "Emancipation and Empire: Reconstructing the Worldwide Web of Cotton Production in the Age of the American Civil War," *American Historical Review* 109 (2004): 1405–38. On the growth and development of the city's infrastructure as the Ulster linen trade shifted from its rural origins to a seat in Belfast, see: Philip Ollerenshaw, "Industry, 1820–1914," in *An Economic History of Ulster, 1820–1940*, ed. Liam Kennedy and Philip Ollerenshaw (Manchester, UK, 1985), 62–108, 66; Helena C. G. Chesney, "Enlightenment and Education," in *Nature in Ireland: A Scientific and Cultural History*, ed. John Wilson Foster (Dublin, 1997), 367–86, 376; Raymond Gillespie, *Early Belfast: The Origins and Growth of an Ulster Town to 1750* (Belfast, 2007); Stephen Royle, "The Growth and Decline of an Industrial City: Belfast from 1750," in *Irish Cities*, ed. Howard B. Clarke (Dublin, 1995), 28–40; and Gerard James Slater, "Belfast Politics, 1798–1868" (PhD diss., New Univ. of Ulster, 1982).

[42] John Hewitt, "'The Northern Athens' and After," in Beckett et al., *Belfast* (cit. n. 41), 71–82; Peter Brooke, "Religion and Secular Thought, 1800–1875," in ibid., 111–28, 123; and J. E. Burnett and A. D. Morrison-Low, "Irish Provincial Instrument Making," in *"Vulgar and Mechanick": The Scientific Instrument Trade in Ireland, 1650–1921* (Dublin, 1989), 71–88.

system was established in 1845 to create three explicitly nondenominational colleges in Belfast, Cork, and Galway, each headed by a president and vice president to coordinate the development before they opened officially to students in 1849.[43] The vice president of Queen's College Belfast (QCB) was Thomas Andrews.

The fear of students emigrating was real. At the very moment that the Russian Empire was initiating its own effort to encourage study abroad for the present in order to minimize its necessity in the future, Ireland was facing a real hemorrhage of its young scholars. Before midcentury in Ireland, a significant number of students would travel abroad for at least part of their education, especially if they were Catholic (and thus had limited options for higher education at home). In these cases, the students would seek out universities in Rome, Paris, and other Continental Catholic cities. This experience on the Continent directly shaped the patterns of Irish science in terms of topics studied, structure of scholarship, and so on, upon their return.[44] Although not comprehensive, demographic data on Irish scientists in the nineteenth century is highly suggestive. Among scientists working in Ireland, 25 percent were born in Britain, 8 percent on the Continent, and 3 percent elsewhere. Of those born in Ireland, or to Irish parents temporarily abroad at the moment of birth, 36 percent were wholly or partially educated in Britain. Only 12 percent of them were educated on the Continent—although it has to be said that for many of these latter the decision to study abroad depended more upon whether one was a chemist than whether one was Catholic. Chemists experienced a tremendous pull to go to the Continent—and here Andrews was no exception.[45] Yet he himself insisted that even this limited leakage of students abroad had to stop: "The country which depends unduly on the stranger for the education of its skilled men, or neglects in its highest places this primary duty, may expect to find the demand for such skill gradually to pass away and along with it the industry for which it was wanted."[46] The Russians would have agreed.

Andrews's 1867 pamphlet on the Queen's College system, *Studium Generale: A Chapter of Contemporary History*, one of his two major publications on public policy issues,[47] was a clearheaded and vigorous defense of the nondenominational, decen-

---

[43] Kevin B. Nowlan, *The Politics of Repeal: A Study in the Relations between Great Britain and Ireland, 1841–50* (London, 1965), 174. On the general history of Irish higher education in this period, see T. W. Moody, "The Irish University Question of the Nineteenth Century," *History* 43 (1958): 90–109; W. G. Scaife, "Technical Education and the Application of Technology in Ireland, 1800–1950," in Bowler and Whyte, *Science and Society in Ireland* (cit. n. 2), 85–100; Richard A. Jarrell, "Some Aspects of the Evolution of Agricultural and Technical Education in Nineteenth-Century Ireland," in ibid., 101–17. On the repeal movement's relationship with revolutionary Fenianism after the Potato Famine, and the role of Irish nationalism more broadly, see Brian Jenkins, *Irish Nationalism and the British State: From Repeal to Revolutionary Nationalism* (Montreal, 2006).

[44] Chesney, "Enlightenment and Education" (cit. n. 41), 371, 383.

[45] James Bennett, "Science and Science Policy in Ireland in the Mid-Nineteenth Century," in Bowler and Whyte, *Science and Society in Ireland* (cit. n. 2), 37–47, 37–38. Justus von Liebig, in particular, had a tremendous influence on science pedagogy in nineteenth-century Ireland, as discussed in Enda Leaney, "Science and Conflict in Nineteenth-Century Ireland," in *Culture, Place and Identity*, ed. Neil Garnham and Keith Jeffery (Dublin, 2005), 66–77, 67.

[46] Andrews, "Presidential Address," delivered at the Glasgow meeting of the BAAS, 6 Sept. 1876, reproduced in Andrews, *Scientific Papers of the Late Thomas Andrews* (cit. n. 18), 414.

[47] I exclude for the purposes of this essay Andrews's brief venture into alcohol policy. In 1867, he was president of the Education Section of the Social Science Association, which had a meeting in Belfast that year. He presented a paper called "Suggestions for Checking the Hurtful Use of Alcoholic Beverages by the Working Classes," displaying an interest in alcohol policy that eerily shadows Mendeleev's. (See Gordin, *A Well-Ordered Thing* [cit. n. 17], 165.) Andrews wanted to regulate the sale of hard liquor, advocating a system in which publicans had to offer food and that no pub could serve

tralized structure of the Queen's University system against the University of London, Maynooth, and the newly established Catholic University in Dublin.[48] (The Queen's University administered the degrees, but the examinations were given at each of the three colleges.) In many ways, QCB was the most successful of the Queen's Colleges; it was the only one not incorporated into the National Universities of Ireland (NUI) in 1908, and it exists to this day.[49] Unabashedly pro-Irish, Andrews insisted that differences in performance were not due to any particular defect with the Irish students. Test results were "at least strongly in favour of the truth of my position, that the mental culture of the graduates of the universities of Ireland is, on the whole, not inferior to the mental culture of the graduates of the University of London."[50] Part of the success, in retrospect, can be seen to be Andrews's vigorous advocacy of a medical school for QCB (not intended in the original 1845 plans) and his effective administration in building up a strong science curriculum.[51]

But the real target of Andrews's pamphlet was not the chauvinists in England who would decry the quality of the Irish schools but the Catholics at home who blasted the Queen's University system as the "Godless Colleges" and tried to dissuade Catholics from attending them. This, he felt, was unworthy of the scientific tradition of great Catholics such as Blaise Pascal, Alessandro Volta, and even Galileo Galilei.[52] Catholics could also be scientific, and so could a Catholic Ireland, maintained Andrews. Yet, at the same time that they tried to prevent Catholics from studying alongside Protestants, Irish prelates made an effort to control the content of the curriculum in the recently established Catholic University to the point of harming free inquiry, in Andrews's presentation, and certainly had the effect of diminishing Catholic representation in science (and thus reinforcing the Protestant Ascendancy).[53] QCB was, in fact, a success in being nondenominational, despite the fact that its first two presi-

---

beverages stronger than 17 percent alcohol, the strength of sherry, eventually creeping down to the burgundy standard of 12 percent. Tait and Crum Brown, "Memoir" (cit. n. 32), xliv–xlv.

[48] Thomas Andrews, *Studium Generale: A Chapter of Contemporary History* (London, 1867), 29, 33–34.

[49] On QCB in post–World War II Northern Ireland, see L. A. Clarkson, *A University in Troubled Times: Queen's Belfast, 1945–2000* (Dublin, 2004). On Queen's College, Cork, see John A. Murphy, *The College: A History of Queen's/University College Cork, 1845–1995* (Cork, 1995).

[50] Andrews, *Studium Generale* (cit. n. 48), 21.

[51] On medicine, see Arthur Deane, ed., *The Belfast Natural History and Philosophical Society: Centenary Volume, 1821–1921* (Belfast, 1924), 63. On medical education in Andrews's Ireland, see Peter Froggatt, "Competing Philosophies: The 'Preparatory' Medical Schools of the Royal Belfast Academical Institution and the Catholic University of Ireland, 1835–1909," in *Medicine, Disease and the State in Ireland, 1650–1940*, ed. Greta Jones and Elizabeth Malcolm (Cork, Ireland, 1999), 59–84. The faculty insisted on the development of the medical school, which built on the lately established (1835) medical school at the Belfast Academical Institution and its central urban location. Moody and Beckett, *Queen's, Belfast* (cit. n. 39), 1:xlvii, 44, 86. On chemistry, see: Cecil L. Wilson, "Schools of Chemistry in Great Britain and Ireland—XXIX: The Queen's University of Belfast," *Journal of the Royal Institute of Chemistry* 81 (1957): 16–29.

[52] Andrews, *Studium Generale* (cit. n. 48), 63, 84–85.

[53] G. T. Wrixon, "Irish Science and Technology: The Changing Role of the Universities," *Irish Review*, nos. 17–18 (Winter 1995): 118–26, 119. For more on the church's objections, see Moody and Beckett, *Queen's, Belfast* (cit. n. 39), 1:277. On the identification of science with the Protestant Ascendancy, see David Andrew Attis, "The Ascendancy of Mathematics: Mathematics and Irish Society from Cromwell to the Celtic Tiger" (PhD diss., Princeton Univ., 2000); Gordon L. Herries Davies, "Irish Thought in Science," in *The Irish Mind: Exploring Intellectual Traditions*, ed. Richard Kearney (Dublin, 1985), 294–310, 305; James A. Bennett, *Church, State, and Astronomy in Ireland: 200 Years of Armagh Observatory* (Belfast, 1990), 56–57; and Terry Eagleton, *Scholars and Rebels in Nineteenth-Century Ireland* (Oxford, 1999).

dents, Pooley Shuldam Henry and Josias Leslie Porter, were Presbyterian ministers, and Andrews, as a long-serving vice president, was a devout Church of Ireland man.[54] For Andrews, being nondenominational was more than a convenient political pose. In a stance analogous to Mendeleev's in his very different context, Andrews insisted on a position of pure science—religious tests were irrelevant to the pursuit of pure knowledge—as a way of minimizing otherwise inevitable strife.[55]

Andrews's conciliatory defense of a scientific Ireland that links its past with its present was even more evident in his *Second Chapter of Contemporary History,* intervening in a much more explosive issue of public policy: the disestablishment of the minority Church of England and Ireland, undoing an aspect of the Act of Union that had grown increasingly unpopular on the island due to the tithe paying required of Catholic farmers. Andrews, unlike many of his ascendancy peers, favored disestablishment. He offered his value-neutral scientific pose here to present an account of Irish ecclesiastical history that demanded removal of the Church of Ireland's privileges, which "must, in his opinion, be eventually settled, if discontent and turbulence are to be banished from the soil of Ireland, and the inhabitants of the British Islands knit into a compact and United people."[56]

The main target of this pamphlet was again the clergy, the same group Andrews had lambasted in the *Studium Generale* for warping the charter of the Catholic University. The clergy was the opposite of what Andrews thought an elite should be to shepherd its people into peaceful coexistence; he offered, implicitly, a vision of a scientific priesthood that through its very nondenominational impartiality would be better suited to govern the hodgepodge of faiths on the island.[57] This was so because the Catholic priesthood fundamentally misunderstood Ireland; it was simply not a Catholic country:

> The people of Ireland is an expression frequently employed to describe the Roman Catholics alone, and Ireland is often spoken of as a Catholic country. In number, the Catholics greatly exceed all the other inhabitants, but their preponderance is only numerical; and to ignore the powerful Protestant minority, which forms the great majority among the classes of higher intelligence, and has succeeded in planting on Irish soil the flourishing industry of Ulster, is altogether inexcusable. No greater mistake can be made than to con-

[54] Brooke, "Religion and Secular Thought" (cit. n. 42), 113; and Moody, "Irish University Question" (cit. n. 43), 99.

[55] This notion that value blindness in science would covertly work in favor of nonsectarianism and thus union was common in many contexts in nineteenth-century Ireland. See the interesting analysis of roving lecturers in Enda Leaney, "Missionaries of Science: Provincial Lectures in Nineteenth-Century Ireland," *Irish Historical Studies* 34 (May 2005): 266–88.

[56] Thomas Andrews, *The Church in Ireland: A Second Chapter of Contemporary History* (London, 1869), 2. For a recent discussion of the surrounding issues, see Oliver P. Rafferty, *The Catholic Church and the Protestant State: Nineteenth-Century Irish Realities* (Dublin, 2008). On the creation of the Church of Ireland, see Alan Ford, "Dependent or Independent? The Church of Ireland and Its Colonial Context, 1536–1649," *Seventeenth Century* 10 (1995): 163–87.

[57] Edward Brynn, *The Church of Ireland in the Age of Catholic Emancipation* (New York, 1982). The famine had a disproportionate impact on Catholics, however, and so the numbers became (slightly) less lopsided for the rest of the century. In 1861, the number of Catholics had fallen by 30 percent since 1834, while those of Anglicans and Presbyterians had only fallen by 19 percent each. This meant Ireland's population was 77.7 percent Catholic, with Anglicans at 12 percent and Presbyterians at 9 percent. By 1901, the figures were Catholics at 74.2 percent, Anglicans at 13 percent, and Presbyterians at 9.9 percent. Sean Connolly, *Religion and Society in Nineteenth-Century Ireland* (Dublin, 1985).

sider Ireland to be a Catholic country, in the same sense in which France is a Catholic, or England a Protestant country.[58]

The solution was not to privilege Catholics over Protestants but to avoid privileging any religion. He advocated disestablishment of the Church of England and Ireland and the allocation of its endowment and tithes among the various churches proportionately by population, to be revised periodically to account for demographic fluxes (say, every twenty-five to fifty years).[59] As a scientist, he took a fundamentally conservative stance, in the sense of only undertaking radical change in times of crisis and in general advocating small and incremental changes. In this case, the same applied: "To overturn a great national institution is always an operation of danger, and should never be attempted, except on the strongest grounds of necessity."[60] As an impartial scientist, he was also the right person to be able to tell when that moment of crisis had come.

Andrews was on the side of the victors here, although it is uncertain how much impact his second pamphlet had. The Church of Ireland was disestablished in 1869. Ironically, although Andrews may have supported the policy, the consequences worked against all of his arguments in its favor: it did not serve to promote union; in fact, it even weakened the support for science that the Protestants had provided. The union weakened in part because disestablishment meant ipso facto the relative strengthening of the Catholic Church, which after the famine had become a rallying point for national identity, coupled with a general intensification of the bureaucratic and authoritarian tendencies in the church of this period.[61] Retrospectively speaking, this was Andrews's final victory, for the rest of his project—a unified notion of Irishness that was nonsectarian and embedded as an equal partner in the British Isles—was obliterated with the coming of independence for the Republic of Ireland in 1921. What followed was not just the obliteration of Andrews's vision for a nonsectarian Ireland but the erasure of Andrews as an "Irish scientist" in a "scientific Ireland."

CONCLUSION: THE MAKING AND UNMAKING OF ICONS

The divergence between Mendeleev's and Andrews's views of the critical point did not prevent the possibility of mutual intelligibility, and their strategies of defining themselves as both national and cosmopolitan scientists proved to some extent fruitful for both in their lifetimes. After their deaths, however, their fates in national memory took radically different paths. Mendeleev came to be seen as quintessentially Russian, in a sense somewhat unmoored from his own self-identification as a Russian *scientist*; while Andrews was written out of Ireland's identity altogether.

Andrews's nineteenth-century peers widely considered him to be an *Irish* scientist, and he considered himself this way as well, a situation not atypical for the sizable minority of non-Presbyterian Protestants who would by the 1890s be known as the

---

[58] Andrews, *Church in Ireland* (cit. n. 56), 31.

[59] Ibid., 56.

[60] Ibid., 38. On the connection between this Burkean conservatism and science, see Gordin, *A Well-Ordered Thing* (cit. n. 17), chap. 1.

[61] Thomas A. Boylan and Terrence McDonough, "Dependency and Modernization: Perspectives from the Irish Nineteenth Century," in *Ideology and Ireland in the Nineteenth Century*, ed. Tadhg Foley and Seán Ryder (Dublin, 1998), 113–29, 125.

Anglo-Irish or (due to their predominant position in almost every sphere of public life) the Protestant Ascendancy.[62] There is today a growing historiography on the position of science in Irish history, which is mostly excellent (outside of a few instances of special pleading or tokenism), but it tends to focus almost exclusively on developments in Dublin—the cultural as well as the administrative capital of nineteenth-century Ireland—and more generally in places that ended up in the Irish Free State after 1922.[63] That is to say, almost all of the scholarship on science in Ireland excludes Belfast, and it is Belfast where Andrews staked his claim.[64] It was relatively unproblematic in the nineteenth century to be a scientist and Irish *and* British. After 1922, the triangulation became much more difficult.

To some extent, the eclipse of Andrews had a great deal to do with who he was—an Anglo-Irish Protestant—but perhaps even more so it had to do with his profession as a scientist. A historical puzzle remains as to why Irish nationalism, when it emerged, identified itself specifically with an ultramontane Catholicism and not with the state-building potential of science and technology—as, for example, anticolonialism and nationalism in India did.[65] The fear of an antiscientific free Ireland pushed several leading Irish scientists to advocate retaining the union for as long as possible—not least because a rupture with England might damage scientific ties to their peers in

---

[62] On this point, see the classic study by J. C. Beckett, *The Anglo-Irish Tradition* (London, 1976), 96. Andrews is identified as specifically "Irish" in almost all of the obituaries, but see, especially, Tait and Crum Brown's preface in Andrews, *Scientific Papers of the Late Thomas Andrews* (cit. n. 18), iii. On the conceptual ambiguity of who counts as an "Irish scientist," see Frank A. J. L. James, "George Gabriel Stokes and William Thomson: Biographical Attitudes towards Their Irish Origins," in *Science in Ireland, 1800–1930: Tradition and Reform*, ed. John R. Nudds, Norman McMillan, Denis L. Weaire, and Susan M. P. McKenna Lawlor (Dublin, 1988), 75–82, 76.

[63] See, e.g.: Attis, "Ascendancy of Mathematics (cit. n. 53); David Berman, "Enlightenment and Counter-Enlightenment in Irish Philosophy," *Archiv für Geschichte der Philosophie* 64 (1982): 148–65; Outram, "Negating the Natural" (cit. n. 8); Seán Mac Cartáin, "Technical Education in Ireland 1870–1899," in McMillan, *Prometheus's Fire* (cit. n. 6), 188–210; McMillan, "Ireland and the Reform of the Politics and Government of British Science and Education," in ibid., 481–524; McMillan, "Mathematical, Scientific and Engineering Reform before the Twentieth Century in Dublin University," in ibid., 138–69; McMillan, "The Transmogrification of the Colonial Tradition of Mathematics, Science and Engineering," in ibid., 74–105; McMillan, "Organisation and Achievements of Irish Astronomy in the Nineteenth Century—Evidence for a 'Network,'" *Irish Astronomical Journal* 19 (March and Sept. 1990): 101–18; Sean Lysaght, "Themes in the Irish History of Science," *Irish Review*, no. 19 (Spring/Summer 1996): 87–97; A. D. Morrison-Low, "The Trade in Scientific Instruments in Dublin, 1830–1921," in Burnett and Morrison-Low, *"Vulgar and Mechanick"* (cit. n. 42), 39–69; Herries Davies, "Irish Thought in Science" (cit. n. 53), 294–310; John Wilson Foster, "Natural Science and Irish Culture," *Éire-Ireland* 26 (1991): 92–103; Foster, *Nature in Ireland* (cit. n. 41); Foster, "Out of Ireland: Naturalists Abroad," in ibid., 308–65; Desmond Clarke, "An Outline of the History of Science in Ireland," *Studies* 62 (1973): 287–302; Ronald Cox, ed., *Engineering Ireland* (Cork, 2006); Nicholas Whyte, "'Lords of Ether and of Light': The Irish Astronomical Tradition of the Nineteenth Century," *Irish Review*, nos. 17–18 (Winter 1995): 127–41; Whyte, *Science, Colonialism, and Ireland* (Cork, 1999); Roy Johnston, "Science and Technology in Irish National Culture," *Crane Bag* 7, no. 2 (1983): 58–63; James Bennett, "Why the History of Science Matters in Ireland," in *Science and Irish Culture*, ed. David Andrew Attis (Dublin, 2004), 1:1–14; Brian B. Kelham, "The Royal College of Science for Ireland (1867–1926)," *Studies* (Autumn 1967): 297–309; and Enda Leaney, "Phrenology in Nineteenth-Century Ireland," *New Hibernia Review* 10 (2006): 24–42.

[64] Noteworthy exceptions to the silence on what would later become Northern Ireland include: Deane, *Belfast Natural History and Philosophical Society* (cit. n. 51); John Wilson Foster, "Natural History, Science, and Irish Culture," *Irish Review*, no. 9 (Autumn 1990): 61–69; and Ruth Bayles, "Understanding Local Science: The Belfast Natural History Society in the Mid-Nineteenth Century," in Attis, *Science and Irish Culture* (cit. n. 63), 139–69.

[65] Gyan Prakash, *Another Reason: Science and the Imagination of Modern India* (Princeton, N.J., 1999); and Deepak Kumar, *Science and the Raj: 1857–1905* (New York, 1995).

Britain.[66] After the disestablishment for which Andrews had argued so forcefully, his fellow Protestants were disempowered both institutionally and symbolically. Once the union had eliminated Ireland's parliament, the Church of Ireland became the most important symbol of corporate identity for Irish Anglicans. With the church's status diminished, many influential Protestants shifted toward the Gaelic-centered cultural revival, which vaunted the folklore and superstition that rationalists such as Andrews had sought to minimize. Cultural dominance came to replace ecclesiastical dominance.[67] In this frame of reference, after disestablishment, it was difficult in some circles (especially Ascendancy ones) to be both Irish and a scientist—science being tagged as *English*. And after 1922, Andrews's heirs could not even count him as a scientist who worked in Ireland, for the very simple reason that the place he had lived and worked was no longer *defined* as Ireland: it was British Northern Ireland and part of the United Kingdom. For this reason most post–Free State commentators on Andrews find themselves in the unenviable position of engaging in sometimes torturous pleading for his inclusion as an Irish icon.[68]

The contrast with Mendeleev could hardly be starker. Even during his lifetime Mendeleev came to represent the incarnation of "the scientist" as public intellectual—with his opinions on capital punishment, local politics, and the state of the world taken quite seriously. Much of this was the result of his careful self-fashioning, but it was also an indication that the program Mendeleev had pushed for—to have scientists be considered as exemplars of Russianness, not as foreign to it (and not as transplanted Europeans)—was quite successful.[69] To some extent, this success relied on developments beyond Mendeleev's direct control: the establishment of multiple professional communities of scientists following the foundation of the Russian Chemical Society in 1868; the industrialization of the Russian economy demanding more up-to-date (and simply *more*) technical experts on a par with those in central and western Europe; and the extensive travel by Russians across the Continent, which tended to give Russians a more European frame of reference for their local icons. But the manner in which Mendeleev as an *individual* was appropriated as a tool for Russian self-identification was also intimately connected with his rejection by the St.

[66] Greta Jones, "Scientists against Home Rule," in *Defenders of the Union: A Survey of British and Irish Unionism since 1801*, ed. D. George Boyce and Alan O'Day (London, 2001), 188–208.

[67] Beckett, *Anglo-Irish Tradition* (cit. n. 62), 104; and James H. Murphy, *Ireland: A Social, Cultural, and Literary History, 1791–1891* (Dublin, 2003), 69. On the Cultural Revival as reducing the influence of scientific thinking, see: Foster, "Natural Science and Irish Culture" (cit. n. 63), 95; Foster, "Natural History in Modern Irish Culture," in Bowler and Whyte, *Science and Society in Ireland* (cit. n. 2), 119–33, 127; Patrick Carroll, *Science, Culture, and Modern State Formation* (Berkeley, Calif., 2006), 169; Thomas Duddy, *A History of Irish Thought* (London, 2002), xiv; Sean Lysaght, "Science and the Cultural Revival: 1863–1916," in Bowler and Whyte, *Science and Society in Ireland* (cit. n. 2), 153–65; Sinéad Garrigan Mattar, *Primitivism, Science, and the Irish Revival* (Oxford, 2004); Oliver Macdonagh, *States of Mind: A Study of Anglo-Irish Conflict, 1780–1980* (London, 1983), 105; and Timothy G. McMahon, *Grand Opportunity: The Gaelic Revival and Irish Society, 1893–1910* (Syracuse, N.Y., 2008). For arguments against this position, see Whyte, *Science, Colonialism, and Ireland* (cit. n. 63), 169–70.

[68] See, e.g.: William J. Davis, "Thomas Andrews," in *Some People and Places in Irish Science and Technology*, ed. Charles Mollan, William J. Davis, and Brendan Finucane (Dublin, 1985), 46–47; Davis, "In Praise of Irish Chemists: Some Notable Nineteenth-Century Chemists," *Proceedings of the Royal Irish Academy* 77B, no. 18 (1977): 309–16; D. Thornburn Burns, "Thomas Andrews, 1813–1885," in *Physicists of Ireland: Passion and Precision*, ed. Mark McCartney and Andrew Whitaker (Bristol, 2003), 77–84; and Wilbert Garvin and Des O'Rawe, *Northern Ireland Scientists and Inventors* (Belfast, 1993), 14–17.

[69] Gordin, *A Well-Ordered Thing* (cit. n. 17), chap. 9.

Petersburg Academy of Sciences for the chair of technology in November 1880.[70] Blaming this incident on the so-called German party in the academy (many of whom, despite the name, were Russian-born, and some of whom were ethnically Russian) promoted Mendeleev as quintessentially *Russian*, despite the fact that foreign scientists were the ones buttressing his credentials and his own career began with capillarity research conducted in the German town of Heidelberg! In a similar fashion, Ivan Pavlov survived his transit across the Russian Revolution to achieve new heights of fame in Stalinist Russia precisely because the Soviet state wanted to define itself as "scientific," and Pavlov was a living icon who could serve for this end.[71] No matter that Pavlov and Mendeleev were both conservative and decidedly anti-Marxist; their pasts could be airbrushed when politically convenient. It was just this kind of inclusive impulse that was lacking on the part of Irish public intellectuals with respect to Andrews's legacy.

We have come a long way from an 1870 article claiming priority in an obscure dispute over credit for discovery of the critical point. I now must return to answer the question implied by the posing of this essay: What do we learn from the juxtaposition of Andrews and Mendeleev—and Ireland and Russia—that we would not have learned by looking at them separately? The first answer is that for all of their divergences (in age, nationality, language, specialization, reputation, and so on), Andrews and Mendeleev employed very similar strategies in arguing for a place for elite laboratory science in a modernizing, late-emerging nation-state. For both of them, the scientist's position was to advocate conservative reforms focused on stability and to bolster those positions precisely by insisting on the neutrality and universality of their science.[72] Because of the specific circumstances of Ireland's and Russia's histories, the effects of almost identical research and professional strategies was, in the case of Andrews, to minimize the importance of national identity—with lasting consequences for his status as an Irish scientist and for Ireland as a scientific nation—and, for Mendeleev, to intensify it.

The parallels between them extend further and tell us something about the common origins of the rampant national identification of scientists across Europe in the second half of the nineteenth century. The most common biographical experience shared by the two men was extensive education in Continental chemistry laboratories. The difference, however, was where they went: Germany or France. Perhaps a further investigation of the two models of Continental education and how they spread to different contexts can deepen our understanding of how very similar patterns seem to appear in such different peripheries of Europe (and beyond).

Finally, there is an interesting contrast between the positions of national identity in science in both countries that indicates a deeper anxiety about the rise or decline of science in fin-de-siècle Europe. Both Ireland and Russia present themselves in today's imaginary as folksy, peasant, traditional cultures, saturated with alcohol, music, and artistic creativity. But both also produced first-rate science, a fact somehow eclipsed in present discourse. In the case of Russia, the eclipse is a partial result of the sheer incandescence of the artistic developments of the late nineteenth century,

---

70 Ibid., 138–41.

71 Daniel P. Todes, "Pavlov and the Bolsheviks," *Studies in History and Philosophy of the Life Sciences* 17 (1995): 379–418.

72 I draw here on the classic discussion in Joseph Ben-David, *The Scientist's Role in Society: A Comparative Study* (1971; repr., Chicago, 1984), 125.

as well as the tremendous identification of science and technology with the Soviet regime that succeeded that period. For Ireland, the question of how and why (and whether) science declined as home rule agitation increased has been a topic of extensive historical inquiry.[73]

The point is not so much about why we *now* might perceive those national identities as "scientific" or not but about what historical actors thought *then*. Mendeleev was fairly confident that Russian science would continue to reach higher and higher triumphs, while Andrews had to contend with the possibility that the great age of British (and Irish) science was approaching an end.[74] This indicates one of the central contrasts between the two men: Andrews directed his attention toward the past (the system of education, the established church), defended through historical inquiry; Mendeleev's public efforts (tariff reform, educational transformation, military modernization) were pointed to the future, defended through predictions based on economics. Mendeleev spoke to the future, and the future was kinder to him. Andrews remains rooted in the past, lauded as the historically significant discoverer of the critical point, yet forsaken as a national icon.

[73] Herries Davies, "Irish Thought in Science" (cit. n. 53), 306–9; Greta Jones, "Catholicism, Nationalism and Science," *Irish Review*, no. 20 (June 1997): 47–61; Whyte, "Lords of Ether and of Light" (cit n. 63), 136; and Morrison-Low, "Trade in Scientific Instruments in Dublin" (cit. n. 63), 63.

[74] Andrews, for what it's worth, did not believe that science was declining in the British Isles generally, or Ireland specifically. Andrews, "Presidential Address" (cit. n. 46), 409.

Germans in Russia:
Cold War, Technology Transfer, and National Identity

By Asif A. Siddiqi*

ABSTRACT

In the aftermath of World War II, the Soviet government captured and put to work thousands of German scientists and engineers in support of domestic military projects. Many were assigned to aid in the development of ballistic missiles but were repatriated back to the German Democratic Republic in the early 1950s. This much-invoked but poorly understood chapter in the history of Soviet science brought into relief a larger set of issues on the constitutive role of science and technology in the articulation of a Soviet national identity during the early cold war. These factors, which included a resurgence of postwar nationalism and the culture of extreme secrecy, forced an unlikely outcome to the question of how best to make use of German expertise: reframe the Germans as "less useful" and send them home. Here, the intersection of cold war imperatives, technology transfer, and national identity produced a condition in which the Germans' ultimate fate had less to do with their expertise (which was quite impressive) than the perception of their expertise. The latter was easier to manipulate and eventually overshadowed the former.

INTRODUCTION

In October 1951, Soviet minister of armaments Dmitrii Ustinov prepared a short report summarizing the contributions of several hundred German scientists and engineers working under him. These men and women had been forcibly relocated from Soviet-occupied Germany in 1946. For five years, they had worked on the Soviet ballistic missile program at a remote and isolated island in Lake Seliger, about 300 kilometers northwest of Moscow. After a long debate among industrial leaders during the summer, the government had finally decided to send the Germans home. In his summary, Ustinov noted that "the Germans working in the area of reactive technology rendered significant help to re-create and reconstruct German designs, especially in the first period of time." However, he concluded, "[o]wing to the long isolation from modern science and technology, the work of the German specialists has become less effective, and at the present time when principally new and more

*Dept. of History, Fordham University, 441 E. Fordham Road, Bronx, NY 10458; siddiqi@fordham.edu.
The author would like to thank all the participants in the Science, Technology, and National Identity workshop at the University of South Carolina in September 20–22, 2007.

© 2009 by The History of Science Society. All rights reserved. 0369-7827/09/2009-0007$10.00

modern models of [rockets] are being created, they would not be able to provide significant help."[1] Thus was justified the end of one of the most infamous episodes in the history of the Soviet space program: the kidnapping and exploitation of German rocket scientists after World War II.

Ustinov's report, and Stalin's decision to repatriate almost all of the German scientists back to the German Democratic Republic, represented the end result of a set of political, scientific, and rhetorical conflicts over the "proper" role of German scientific and technical expertise within the Soviet context. On one level, these deliberations, involving Soviet engineers, security service chiefs, and influential architects of the Soviet military-industrial complex, were about the "usefulness" (or as Ustinov noted, the "uses"—*ispol'zovanii*) of the Germans. However, as I will show here, in the evaluation of the role of Germans in the USSR, "usefulness" was only one criterion among many, and ultimately not the most important one. Beyond the efficacy of the Germans, these discussions implicitly and often explicitly touched on a larger set of issues rooted in the constitutive role of Russian science and technology in the articulation of a cold war national identity in the Soviet Union. These factors, which included a resurgence of postwar nationalism and the culture of extreme secrecy that characterized Soviet science during the cold war, forced an unlikely outcome to the question of how best to make use of German expertise: reframe the Germans as "less useful" and send them home. Here, the intersection of cold war imperatives, technology transfer, and national identity produced a case in which "science" proved to be more portable than "scientists."[2]

BACKGROUND

The transfer of German technology and expertise after the war to the Western Allied powers remains one of the most well-known episodes of "intellectual plunder" in recent history. Thousands of (principally) American engineers, managers, and industrial representatives scoured through the detritus of German industry for documents, hardware, and expertise. Under programs with enigmatic names such as *Overcast* and *Paperclip*, a vast amount of matériel and personnel were secretly and sometimes illegally transported to the United States.[3] Undoubtedly the most enduring legacy of this episode was the contribution of Wernher von Braun's "rocket team" to the U.S. space program. Von Braun had gained notoriety as the chief designer of the infamous German V-2 (or A4), the world's first long-range ballistic missile, with which the Nazis had terrorized civilian populations in Europe in the final years of the war. Despite concerns about his sympathies to the Nazi cause, von Braun rose quickly in the U.S. Army's ballistic missile program. Based in Huntsville, Alabama from the 1950s, von Braun's team was eventually absorbed by the newly formed National Aeronautics

---

[1] D. Ustinov, "Spravka ob ispol'zovanii nemetskikh spetsialistov v NII-88 ministerstva vooruzheniia," Oct. 1951, fond [f.] 8157, opis' [op.] 1, delo [d.] 1454, listov [ll.] 118–21, Russian State Archive of the Economy, Moscow (hereafter cited as RGAE).

[2] I use the term "science" here in the generic way in which it was often used in the Soviet national context to include a variety of practices and epistemologies, which might be more finely disaggregated into subcategories such as science, applied science, and technology.

[3] John Gimbel, *Science, Technology, and Reparation: Exploitation and Plunder in Postwar Germany* (Stanford, Calif., 1990); Linda Hunt, *Secret Agenda: The United States Government, Nazi Scientists, and Project Paperclip, 1945 to 1990* (New York, 1991); C. G. Lasby, *Project Paperclip: German Scientists and the Cold War* (New York, 1971).

and Space Administration (NASA), in 1960. This engineering team produced the rockets that launched the first American satellite, the first American astronaut, and eventually the first humans to set foot on the Moon.[4] Historians continue to debate the team's effective legacy, with many now arguing that lavish publicity surrounding von Braun's efforts overshadowed a vibrant and indigenous strand of rocketry research in the United States that contributed equally if not much more to the U.S. space program.[5]

Unlike in the case of von Braun's rocket team, the discussion on Soviet plunder of German knowledge was always defined by and limited to a single dominant question: How much did the Germans contribute to Soviet scientific and technological prowess? In other words, how much of Soviet science was really "Soviet"? Westerners were aware of this movement of matériel and expertise soon after the war, but the episode acquired a particular urgency after the launch of the first *Sputnik* satellite in 1957. In a perhaps apocryphal story passed around in the aftermath of *Sputnik*, one highly placed American general complained that they "had captured the wrong Germans."[6] At the time, leading U.S. officials, including President Dwight Eisenhower and his secretary of state John Foster Dulles, suggested that the Soviets owed the success of *Sputnik* to German scientists and technology.[7] Within a decade, however, that notion was no longer the prevailing consensus: most Western analysts and historians accepted the perspective—reinforced by a steam of official Soviet pronouncements on the history of their space program that denuded it of any German role—that the dramatic successes of the Soviet space program were largely due to indigenous efforts.

This perspective was muddied by revelations from several different types of sources in the era of *glasnost'* and the post–cold war period. They included the memoirs of Soviet veterans and Germans who had returned to the West in the 1950s but had remained silent through much of the cold war. American intelligence agencies also declassified several reports from the early cold war—based on extensive interviews with the returned Germans—that contributed to a reassessment of the German episode.[8] From these sources, it seemed likely that the Germans *had* made significant contributions at the birth of the Soviet ballistic missile program but that their con-

[4] Michael J. Neufeld, *Von Braun: Dreamer of Space, Engineer of War* (New York, 2007).

[5] The new literature challenges the so-called Huntsville school of space history, disseminated by the custodians of von Braun's legacy, which reiterates the notion that von Braun and his team were central to the rise of the U.S. space program. See Roger D. Launius, "The Historical Dimension of Space Exploration: Reflections and Possibilities," *Space Policy* 16 (2000): 23–38.

[6] Albert Parry, *Russia's Rockets and Missiles* (Garden City, N.Y., 1960), 111.

[7] "Transcript of the President's News Conference on Foreign and Domestic Matters," *New York Times*, 10 Oct. 1957; "Draft Statements on Soviet Satellite," 5 Oct. 1957, John Foster Dulles Papers, Dwight D. Eisenhower Library, Abilene, Kansas.

[8] For the most famous Russian memoirs, see Boris Chertok's *Rockets and People,* vol. 1, ed. Asif A. Siddiqi (Washington, D.C., 2005), and *Rockets and People,* vol. 2, *Creating a Rocket Industry*, ed. Asif A. Siddiqi (Washington, D.C., 2006). For German testimonies, see Kurt Magnus, *Raketensklaven: Deutsche Forscher hinter rotem Stacheldraht* (Stuttgart, 1993); Werner Albring, *Gorodomlia: Deutsche Raketenforscher in Russland* (Hamburg, 1991). For a representative intelligence document, see CIA Office of Scientific Intelligence, *Scientific Research Institute and Experimental Factory 88 for Guided Missile Development, Moskva/Kaliningrad*, Scientific Intelligence Research Aid, 4 March 1960, OSI-C-RA/60-2. See also James Harford, "What the Russians Learned from German V-2 Technology," in *History of Rocketry and Astronautics,* ed. Donald C. Elder and Christophe Rothmund (San Diego, 2001), 23:401–24.

tributions (and utility) diminished in the late Stalin years. When Soviet bureaucrats no longer perceived an urgent need for their presence, the Germans were sent home. Scholars remained markedly silent on possible reasons for *why* or *how* the Germans became less "useful."[9]

Much of the discussion on the German-Soviet interaction has been circumscribed by a few intellectual and methodological limitations. First, the analysis of German contributions to the Soviet missile program has conflated two related issues: the Soviet exploitation of German technological matériel and the Soviet use of German human expertise. There is no doubt that the Soviet missile industry developed from the former, but its debt to the latter remains clouded in simplifications. Second, these works have focused on narrowly defined issues that see the German contribution as disembodied and existing without context—the broader concerns of postwar Soviet society, culture, and politics in the late Stalin era are invisible in these narratives. Third, cold war–era ideological anxieties colored much of the scholarship, conveying a reflexive and simplistic black-and-white view of technology transfer that may be characterized as "the Soviets always lagged in industrial technology because of their incapacity to innovate, then they stole everything from the West, yet their products remained inferior." These works assumed a reductive relationship between nation and technology, as if the margins of both concepts were fixed in time and immutable.[10] Finally, restrictions on archival research meant much of the results were based on conjecture, rumor, or at best, memory. Two important recent works that benefit from archival documents, while vital in ferreting out salient aspects of the transfer of German technology to the Soviet Union, favor detail over analysis.[11]

In this essay, I use recently accessible Russian archival sources to reconstruct the experience of German rocket scientists forcibly relocated to the Soviet Union during the early cold war. My goal here is not to measure the "proper" contribution of the Germans to the Soviet missile project but to bring into relief the collision of two very powerful and countervailing forces: an urgent national need to make use of German expertise, and a concomitant necessity to disavow it. By reconstructing how the main actors in this story—Soviet industrial managers, Soviet scientists, and German engineers—viewed the episode at the time, and by situating those perspectives into the broader cultural context, I hope to offer some insight into the process by which scientific achievement and national identity enjoyed a complicated, often contradictory, but reciprocal relationship in the final years of the Stalinist era.

[9] A few influential German historians have also made the counterclaim that the Germans made *crucial* contributions to the Soviet space program that are evident well into the 1960s, in some sense returning the discussion full circle to the original reactions circa post-*Sputnik*. See Olaf H. Przybilski, "The Germans and the Development of Rocket Engines in the USSR," *Journal of the British Interplanetary Society* 55 (2002): 404–27; Jürgen Michels, *Peenemünde und seine Erben in Ost und West* (Bonn, 1997).

[10] The most famous in this category were A. C. Sutton's influential *Western Technology and Soviet Economic Development* (Stanford, Calif., 1965–1971), published in three volumes covering 1917–30, 1930–45, and 1945–65. In a similar vein, see Ulrich Albrecht, *The Soviet Armaments Industry* (Langhorne, Pa., 1993).

[11] Cristoph Mick, *Forschen für Stalin: Deutsche Fachleute in der sowjetischen Rüstungsindustrie 1945–1958* (Munich, 2000); Matthias Uhl, *Stalins V-2: Der Technolgietransfer der deutschen Fernlenkwaffentechnik in die UdSSR und der Aufbau der sowjetischen Raketenindustrie 1945 bis 1959* (Bonn, 2001).

RUSSIANS IN GERMANY

The German journey to the depths of Lake Seliger began with the Soviet demand for reparations as the war ground to a halt in Europe. Based on its own evaluation of the cost of material losses during the war, the Soviet government drew up plans to take advantage of areas under its control. In the postwar years, Soviet officials consistently quoted a figure of $10 billion worth of reparations owed by Germany.[12] Through 1945 and 1946, survey teams known as "trophy brigades"—composed of scientists, soldiers, industrial managers, and party representatives—scoured Europe and Asia, cataloging and collecting a huge amount of material for delivery back to the Soviet Union. The USSR, like its Western allies, engaged in a systematic "exploitation and plunder" of Axis resources unrivaled in history.[13] A vast quantity of this "reparation" was scientific and military in nature. Over a period of two years, Soviet teams collected and transported equipment from 4,786 German and Japanese enterprises, 655 of which were of an explicitly military nature.[14] Norman Naimark speculates that by 1948, the Soviets had transported about $2.68 billion worth of equipment back to the USSR, a volume equivalent to one-third of the Soviet-occupied zone's productive industrial capacity in 1945.[15]

Rocket technology was not a large portion of this take, but it was undoubtedly one of the most important. Beginning in May 1945, Soviet teams representing the aviation and artillery industries capably spread out through Soviet-occupied Germany, evaluated the state of German rocket weapons ("reactive armaments" in Soviet parlance), and began the intensive process of sending what was valuable to the Soviet Union. Qualified experts, both military and civilian, gleaned an enormous amount of information about German achievements in the field despite the lack of technical documentation. By the end of August 1945, the official military printing press was able to issue a secret seventy-page monograph on the V-2 rocket, detailing every major system of the missile and its capabilities and operational procedures.[16]

The Soviets had made modest advances in rocket technology in the 1930s, but their development had been marred first, by internecine conflicts over proper technological paths and second, by World War II, which had redirected funding away from innovations with uncertain prospects to weapons of immediate utility.[17] The Germans, benefiting from persistent institutional and financial support, a powerful industrial base, and highly qualified engineers, had produced a series of increasingly

---

[12] According to the Yalta agreements in February 1945, the Soviets asked for 50 percent of $22 billion. See *Foreign Relations of the United States: The Conferences at Malta and Yalta, 1945* (Washington, D.C., 1955).

[13] The phrase is from Gimbel, *Science, Technology, and Reparation* (cit. n. 3). See also Matthias Judt and Burghard Ciesla, *Technology Transfer out of Germany after 1945* (Amsterdam, 1996).

[14] For overall figures, see f. 5446, op. 52, d. 2, ll. 45–116, State Archive of the Russian Federation, Moscow (hereafter cited as GARF).

[15] Norman M. Naimark, *Russians in Germany: A History of the Soviet Zone of Occupation, 1945–1949* (Cambridge, Mass., 1995), 169.

[16] A. A. Vekser, *Raketa dal'nego deistviia Fau-2 (po trofeinym materialam)* [The V-2 Long-Range Rocket (Based on Trophy Materials)] (Moscow, 1945). The monograph is stored in f. 4372, op. 94, d. 314, ll. 90–57 (pages numbered in reverse order), RGAE. For a detailed account of the Soviet capture and collection of German spoils in 1945–46, see Asif A. Siddiqi, "Russians in Germany: Founding the Postwar Missile Programme," *Europe-Asia Studies* 56, no. 8 (2004): 1131–56.

[17] Asif A. Siddiqi, "The Rockets' Red Glare: Technology, Conflict, and Terror in the Soviet Union," *Technology and Culture* 44 (2003): 470–501.

sophisticated missile weapons culminating in the V-2.[18] Boris Chertok, one of the young Soviet designers who flooded Germany after the war in search of rockets, later wrote extensively about his experiences. He recalled that "the actual scope of work on rocket technology in Germany was far superior to what we had imagined." Looking at abandoned equipment in German laboratories, he lamented, "Not one of our institutes, factories, or laboratories could even imagine such abundance."[19]

By the fall of 1945, rumors had spread throughout local areas about the generous Soviet treatment of German skilled labor, the good housing and working conditions, and perhaps most important, the Soviet decision to forego denazification procedures for valuable scientists. These policies worked effectively. In the fall of 1945, the Soviets recruited several accomplished German academics and industrial scientists. Although none had worked with von Braun's Peenemünde team, most had higher academic degrees and experience in Germany industry. They included Dr. Kurt Magnus, a specialist in theoretical mechanics, and Dr. Johannes Hoch, an expert in automatic control from Göttingen University. Dr. Manfred Blasing, a former employee of Askania, helped the Soviets in organizing a laboratory to test control surface actuators. The institute also recruited Professor Waldemar Wolff, the former chief ballistics expert for Krupp, and Dr. Werner Albring, an aerodynamics expert from Dresden.[20] These men played crucial roles in helping the Soviets fill gaps in both understanding the operation of German missiles and setting up experiments to improve subsystems.

The Soviets managed to obtain at least one fairly important V-2 engineer, Helmut Gröttrup, who had served as von Braun's deputy for guidance, control, and telemetry at Peenemünde. Although approached by the Americans, Gröttrup communicated to the Soviets through his wife, Irmgard, that he would work for the Soviets if offered complete freedom.[21] Unlike most of the Peenemünde team, Gröttrup did not like the idea of relocating to the United States and hoped that working for the Soviets would offer a modicum of stability for his family. In late September 1945, along with his wife and two children, he moved from the American to the Soviet zone. Gröttrup did not disappoint. As head of the Gröttrup Bureau, a Soviet-created team in occupied Germany, he wrote a "report in the middle of 1946 that was the most complete and objective history of Peenemünde and of the technical problems that were solved during the development of [the V-2]."[22] Gröttrup helped in two more ways: by telling the Soviets about other skilled Germans who might be useful to the Soviets, and by directing the Russians to find more V-2 equipment. Chertok noted later, "I would say that we were right on the money with Gröttrup."[23] The Soviets compensated him accordingly, providing him with a mansion for his family (owned by a rich German merchant who was thrown out), a salary of 5,000 marks per month (an average German income), the highest among the Germans and higher than the

[18] For German rocket work, see Michael J. Neufeld, *The Rocket and the Reich: Peenemünde and the Coming of the Ballistic Missile Era* (New York, 1995).

[19] Chertok, *Rockets and People* (cit. n. 8), 1:236, 242.

[20] "Spisok nemetskikh spetsialistov dlia ministerstva vooruzheniia sssr po institutu 'nordkhauzen' (ob'ekt bleikherode)," f. 397, op. 1, d. 3, ll. 36–39, RGAE; Chertok, *Rockets and People* (cit. n. 8), 1:299–300.

[21] "OMGUS Special Intelligence," 20 March 1947, RG 260, box 65, National Archives, Washington, D.C. (hereafter cited as NA).

[22] Chertok, *Rockets and People* (cit. n. 8), 1:307.

[23] Ibid., 302.

salary of almost every Soviet engineer working in Germany. His wife got her own sports car and a riding horse.[24]

By the end of 1945, through enticements of wages and food or sometimes by coercion, a total of 1,200 skilled Germans spread through the Thuringia region labored to help the Soviets found their postwar rocket program. An additional 1,800 unskilled German laborers worked to disassemble the facilities at the famous Mittelwerk factory, most of them young men from families living in and around the city of Nordhausen, Thuringia. They worked under the guidance of 248 Soviet engineering and military personnel.[25]

By April 1946, the feared Soviet security services chief, Lavrentii Beriia, could report confidently to Stalin on the valuable work in Germany during the previous twelve months. In that time, Soviet personnel had collected and translated into Russian a vast amount of German documentation on missiles; created a special rocket institute at Nordhausen as a nerve center for its operations; established five technological and design bureaus focusing on various areas of missile development; reconstructed the Mittelwerk factory adequately enough to resume production of the V-2; refurbished various testing laboratories; and put together seven V-2 rockets using German components, of which four were ready for firing.[26] As operations sped up, by the end of 1946, Soviet teams in occupied Germany had assembled forty V-2 rockets—using parts discovered and parts manufactured in recent months.[27] Although some rockets were still missing key components (such as graphite radars), the first order of business was now to test fire these rockets and simultaneously establish a production line back in the Soviet Union to acquire the capacity to produce the V-2 in a *domestic* setting.

OPERATION *OSOAVIAKHIM*

Soviet industrial managers and security service chiefs had no plans to indefinitely continue the valuable work being done in occupied Germany. By February 1946, Soviet and German workers combined had expended about 80,000 person hours of labor on the disassembly of rocket facilities. From Mittelwerk—the heart of the former Nazi rocket production program— they had loaded 717 wagons to carry 5,647 tons of equipment back to the Soviet Union.[28] Three months later, in May, Stalin signed a top-level government decree establishing a skeletal network of research, development, and production facilities in and around Moscow to support work on ballistic missiles. Stalin's decree gave the missile a priority second only to that of the atomic bomb project as "one of the most important government tasks." The May 13 decree also determined the fate of the Germans then working in occupied Germany; they

[24] Frederick I. Ordway III and Mitchell R. Sharpe, *The Rocket Team: From the V-2 to the Saturn Moon Rocket—The Inside Story of How a Small Group of Engineers Changed World History* (New York, 1979), 320.

[25] "Spisok rabotnikov i spetsialistov organizatsii upolnomochennogo spetskomiteta v germanii," f. 397, op. 1, d. 3, ll. 112–20, RGAE.

[26] Beriia et al. to Stalin, 17 April 1946, f. 3, op. 47, d. 179, ll. 28–31, Archive of the President of the Russian Federation, Moscow (hereafter cited as APRF); Vannikov to Beriia and Malenkov, 22 Jan. 1946, f. 8495, op. 1, d. 71, l. 115, RGAE.

[27] Malenkov et al. to Stalin, 31 Dec. 1946, f. 3, op. 47, d. 183, ll. 1–6, APRF.

[28] F. 8044, op. 1, d. 6313, 154–55, RGAE.

would all be "transfer[ed] to the USSR by the end of 1946," where "they would be paid a higher salary" if they were working on missiles. For their accommodations, the Soviet government would provide them with "150 prefabricated sectional Finnish-style houses and 40 eight-apartment log houses."[29] Few of the Soviet engineers and none of the Germans had any idea of these decisions made back in Moscow—the Germans seemed to sincerely believe that their work would be continued in occupied Germany, a belief reinforced by their Soviet counterparts, who repeatedly reassured the Germans that they would never be taken from their homeland.

In the early fall of 1946, the Soviet Ministry of Internal Affairs (MVD) drew up detailed plans for the forced relocation of German scientists, engineers, and technicians from Soviet-occupied Germany. One of Beriia's deputies, MVD chairman Ivan Serov, personally supervised the project; he visited Thuringia to ask the head of the Soviet rocket effort in Germany, Lev Gaidukov, for a list of Germans who might be essential to continue rocket development back in the Soviet Union. Serov's instructions, as remembered by Chertok, were simple: "We will allow the Germans to take all their things with them . . . If the head of the family demands that [the whole family goes], we will take them. No action is required of you except a farewell banquet. Get them good and drunk—it will be easier to endure the trauma."[30] The operation, code-named Osoaviakhim (the name of an interwar mass paramilitary organization), would be supported by trains, food, and supplies from the Soviet Military Administration in Germany and troops from the MVD.

The actual deportation proved to be a total surprise to the Germans. On October 21, Gröttrup and several of the leading Germans attended a meeting in Nordhausen on possible improvements to the V-2. As planned, afterward there was a party at a local restaurant for the attendees that ran into the wee hours of the night. At 3:00 am, after most had gone home inebriated and were in bed, army trucks drove up to each selected German's place of living. An interpreter read out a grim order for the transfer while a counterintelligence officer and several soldiers stood guard:

> According to the arrangement of the [Soviet Military Administration in Germany], the organization where you have worked so far is being transferred with its technical personnel for further work to the Soviet Union. Along with the organization, you are likewise mobilized for work in the Soviet Union . . . Further information about the contract of your work in the Soviet Union will be communicated to you after passage into the USSR . . . you will work in the Soviet Union for 5 years. The living conditions and wages will be the same as that of Soviet specialists.[31]

According to MVD records, 90 percent of the Germans resisted the order. Even those who agreed to go pleaded with the Soviets to be taken only by themselves, that is, without their families, an appeal that was largely ignored. Amid pandemonium and panic, 2,552 German "specialists" boarded the trains on the morning of October 22, 1946; with their families, a total of 6,560 Germans left their home country. Just over half the Germans went directly to work for the Soviet aviation industry. A

[29] "Voprosy reaktivnogo vooruzheniia," in *Khronika osnovnykh sobytii istorii raketnykh voisk stra-tegicheskogo naznacheniia*, ed. I. D. Sergeev (Moscow, 1994), 227–34.
[30] Chertok, *Rockets and People* (cit. n. 8), 1:365.
[31] Uhl, *Stalins V-2* (cit. n. 11), 132.

smaller number, 302 German men and women, were deported for work on the missile program.[32] The remainder was split between work in the nuclear program, electrical industry, optics sector, radio engineering, and chemical engineering. The caravan of trains made their way directly to Moscow efficiently and with few obstacles—logistical or otherwise. Within two weeks, the more than 6,000 Germans were spread among thirty-one different industrial institutions in the European landmass of the Soviet Union.

GERMANS IN RUSSIA

The nerve center of the Soviet missile program was located northeast of Moscow in the suburb of Kaliningrad. Here, Soviet administrators and engineers were struggling to set up a rocket research institute at an old factory—built originally by the German firm Rheinmetall (later Rheinmetall-Borsig) in the 1920s. Both Germans and the returning Soviets were shocked by the abysmal state of affairs at the factory—the facility lacked storage areas for rocket parts, workspaces for technicians, and even the most elementary tools. Work conditions were appalling, especially in the winter when it was often colder inside than outside.

The institute, officially known as NII-88 (or "nee-88" from the Russian abbreviation for Scientific Research Institute 88), was organized around a structure that replicated the Soviet-German operation in occupied Germany. The institute had three main structural divisions—a design bureau to develop missiles, an applied science branch, and a factory to produce test models of missiles. The design bureau itself was divided into a number of units—department no. 3, headed by the charismatic chief designer Sergei Korolev, was responsible for fully reproducing the V-2 missile using homemade materials. Korolev, later to become the tragic hero of the Soviet space program, was already a headstrong figure in the Soviet missile industry. Only forty years old at the time, he had an uncanny eye for devoting resources to the most effective technical path, a skill that he combined with a penchant for negotiating the often harrowing labyrinths of Soviet industrial culture. His key liability was the six years spent in various gulag camps as a result of trumped-up charges of sabotage and "wrecking," accusations that remained officially on his record.

Korolev was embedded in a Byzantine hierarchy that stretched up to Stalin. His immediate superiors included institute director Lev Gonor, a war-tested veteran who had guided the mass production of artillery weapons at the famous Barrikady artillery in Stalingrad during the siege. Gonor was also one of the few Jewish managers in the weapons industry, and he was active in the Soviet Jewish Anti-Fascist Committee. Given the growing anti-Semitism in the late Stalin years, his appointment was an anomaly, but Gonor had a strong patron behind him, Minister of Armaments Dmitrii Ustinov, appointed by Stalin to manage the high-priority missile program. A loyal Stalinist, Ustinov had joined the Bolshevik steel factory in Leningrad in 1937 as a junior engineer, but owing to the massive arrests during the Great Terror and the patronage of Leningrad party secretary Andrei Zhdanov, he became the youngest de-

---

[32] These numbers vary depending on the source. The overall numbers are from a memo dated 2 April 1949 summarizing the work of German specialists. See f. 5451, op. 43, d. 767, l. 196, GARF. According to data from February 1947, 378 Germans were slated for work in the missile industry, but after further clarification of duties, a total of 302 ended up working on missiles. See Kirpichnikov to Malenkov, 16 Feb. 1947, f. 4372, op. 94, d. 1838, ll. 39–35 (pages in reverse order), RGAE.

fense manager in the Soviet war effort, in control of some of the biggest and most famous Soviet factories. Through the war, his commissariat supplied artillery cannons,
machine guns, rifles, revolvers, and mortars for the Soviet armed forces.[33]

In Kaliningrad, Ustinov had at his disposal 302 Germans who were considered the
cream of the crop of the deported Germans. According to Soviet classifications—
based more often than not on the verbal claims of the Germans rather than any documentation—they included 5 "doctor-professors," 54 "doctors," 30 engineering graduates, and 123 "engineers and engineer-designers."[34] A relatively small number of
them (about 20 or so) actually worked on the German rocket program during the war;
the rest were from academia or German industry. The Soviets came to value some of
them greatly. Chertok calls Magnus, a former student of Ludwig Prandtl's, a "first-
class theoretician and engineer in the field of gyroscopes and theoretical mechanics,"
and Hoch, a scientist from Hettingen University, a "brilliant experimenter in automatic control."[35] Their qualifications would seem to contradict many later claims that
the German experts in the Soviet Union were low-level people with hardly any scientific or engineering skills. Only a small minority lacked formal educational training,
that is, were workshop foremen or secretarial help. The distribution of those with and
without formal education, in fact, was quite similar to that of the von Braun rocket
team who went to the United States.

The Germans' pay scales were adjusted according to their educational distinctions
and compared favorably with the scales of the highest ranking Soviet designers they
worked with (see table below). By comparison, at the time, the average Soviet engineer received a monthly salary of a thousand rubles.

Most of the leading Germans (ninety-nine specialists with their families) shared
houses and dachas along the road from Moscow to Kaliningrad. As was common in
the postwar period, whole families typically shared one- or two-room apartments.
University graduates were more fortunate. Gröttrup and his wife and two children, for
example, lived in the six-room villa belonging to a former Soviet minister, complete

Monthly Salary (in rubles)	German designers	Soviet designers
6,000	Kurt Magnus Theodore Schmidt Joachim Umpfenbach	Sergei Korolev (chief designer)
5,500	Heinz Zeise	
5,000		Iurii Pobedonostsev (chief engineer)
4,500	Herman Gröttrup Willi Schwarz	
4,000	W. Kessel	
3,000		Boris Chertok (deputy chief engineer)
2,500		Vasilii Mishin (deputy chief designer)

[33] D. F. Ustinov, *Vo imia pobedy: Zapiski narkoma vooruzheniia* (Moscow, 1988); Iu. S. Ustinov,
ed., *Narkom ministr marshal* (Moscow, 2002).
[34] Kirpichnikov to Malenkov, 16 Feb. 1947, f. 4372, op. 94, d. 1838, l. 37, RGAE.
[35] Chertok, *Rockets and People* (cit. n. 8), 1:299–300.

with all necessary amenities, including Mrs. Gröttrup's chauffeur-driven BMW and her transplanted horse.

COLD WAR, SECRECY, AND AMBIVALENCE

Stalin had given Minister Ustinov and NII-88 director Gonor one of the most important national security assignments in the early cold war. Ballistic missiles represented one of three high-priority cold war military projects with origins in wartime scientific work. For all three of them—the atomic bomb, ballistic missiles, and radar—Stalin had sanctioned the creation of "special committees," interagency management bodies whose existence was never announced and whose needs had priority over those of any other industrial cause. Both Ustinov and Gonor reported to the Special Committee for missiles and often visited the Kremlin to report directly to Stalin on the status of the missile project.[36] Ustinov had taken on the project on his own initiative knowing full well the risks involved. Both during and after the war, Stalin, Beriia, and their security police surrogates considered ministers and commissars expendable. A perhaps apocryphal recollection hints at Ustinov's nervousness: after Stalin assigned Ustinov to manage the missile program, Ustinov told one of his aides that upon seeing Beriia's cold stare that night, he knew that the responsibility was "not to be taken lightly," that this was "a very, very serious matter," and that "there would be no excuses from me."[37]

A culture of secrecy was an unavoidable aspect of the ballistic missile program. For close to forty years, the Soviet media never announced the names of institutions, designers, or missiles. Only rarely did the media officially acknowledge the program's existence. The level of secrecy was particularly airtight during the late Stalin years, exacerbated both by cold war tensions and a domestic security service whose leaders cultivated and thrived on a culture of secrecy. Documents related to the missile program were all classified top secret—even documents that touched on the most mundane aspects of the program, such as opening a canteen for workers. Engineers were not allowed to tell their families where they worked or what they did beyond a generic description that it was part of the "machine-building" industry. Institutions were known openly (in order to send mail) not by their names but by post office box numbers. Information was also compartmentalized within organizations. Lower rank-and-file workers at the NII-88 institute, for example, had only the dimmest idea of the institutional subordination of the missile program in the government. Subcontractors were often not told the actual names of missiles but rather coded designations. A number of institute luminaries—such as Korolev—fully embraced this culture, having worked as scientists within the gulag system, which begat many traits of the Soviet defense industry's institutional culture. Korolev liked to joke that the same guards who used to protect him when he was in the gulag were still protecting him at NII-88; in his former life, they were there to keep him in—now they were there to keep others out.

Given the secrecy requirements, the use of German expertise to help design and build the Soviet Union's first ballistic missile posed a number of institutional, techno-

---

[36] Ustinov visited with Stalin in the Kremlin at least seven times in 1947.

[37] Mikhail Pervov, *Zenitnoe raketnoe oruzhie protivovozdushnoi oborony strany* (Moscow, 2001), 32.

logical, and procedural problems. For the Germans to be fully effective, they needed to be well integrated into the Soviet program and have access to information, ongoing work, and programmatic decisions. Yet, the Germans—only recently, the sworn fascist enemies of the Soviet Union—were working on one of the most top-secret weapons program in Stalinist Russia. The early solution to this problem was to divide the Germans into two groups. Of the total 175 Germans "employed" at the NII-88 institute, 99 were assigned to work at the main institute campus at the decrepit factory in Kaliningrad. The remainder were sent off to Gorodomlya, a small isolated island in the middle of Lake Seliger, about 300 kilometers northeast of Moscow.[38] The deserted island, surrounded by a beautiful lake, was virtually uninhabited until the twentieth century; in the 1920s, the Soviet government had established a biological research laboratory there to study diseases, but the program was abandoned with the onset of World War II. Gorodomlya was virtually cut off from Soviet society and culture—the ideal working environment for the Germans, or so the Soviets believed.

According to Chertok, the living and working conditions at Gorodomlya were comfortable:

> When I arrived on the island [to visit the Germans], I could only envy the way they lived, because in Moscow my family and I lived in a communal four-room apartment, in which we occupied two rooms. . . . Many of our [own] specialists . . . still lived in barracks, where they did not have the most elementary conveniences. . . . On weekends and holidays [the Germans] were permitted to make excursions to the regional center of [the nearest town] Ostashkov and to Moscow to go to shops, markets, theaters, and museums. Therefore, life on the island surrounded by barbed wire could not in any way be considered comparable to the status of prisoners of war.[39]

The later recollections of the Germans seem to suggest exactly otherwise. Magnus rejected Chertok's claim as a "fantasy," asserting that at most small German groups were allowed to visit Moscow perhaps only ten times in seven years.[40] Like the gunmen surrounding the entire compound, a shortage of food and supplies was a ubiquitous part of daily life. When the Germans first arrived in Gorodomlya, all seventy-six of them (with their families) were housed in six two-floor apartment buildings with no running water, no plumbing, and no sewage system. An unintended consequence of the variegated salary system was that many "lower-ranked" Germans began to resent the ones who had higher salaries, thus sowing the seeds of dissension that would grow over the years.

The Germans' movements were strictly monitored and circumscribed. Recreational trips to Moscow were, in fact, extremely rare due to concerns over Germans' coming into contact with the "civilian" Soviet citizens—their MVD handlers were not only afraid of information leaks but also wanted to avoid potential tensions with Soviet citizens, still seething from the carnage inflicted upon them by the Nazis. Except for service personnel who staffed canteens and shops, few Soviet citizens lived on the island. Waldemar Wolff, a ballistics scientist from the famous Krupp firm, and

---

[38] S. Vetoshkin, "Poiasnitel'naia zapiska k spisku inostrannykh spetsialistov, rabotayushchikh v NII-88 ministerstva vooruzheniia sssr," f. 4372, op. 94, d. 1838, l. 67, RGAE.

[39] Chertok, *Rockets and People* (cit. n. 8), 2:44–45.

[40] Magnus quoted in James Harford, *Korolev: How One Man Masterminded the Soviet Drive to Beat Americans to the Moon* (New York, 1997), 79.

Josef Blass, a design engineer, shared directorship duties among the Gorodomlya Germans.

Initial assignments were handed out to the Germans as soon as they arrived in the Soviet Union. The primary goal was to prepare the first V-2 launches under Soviet command, an objective made more urgent by news that both the British and the Americans had already conducted similar tests. The Germans were also to assist their Soviet colleagues in developing a production infrastructure to manufacture V-2 missiles at the factory in Kaliningrad. In addition to these priority assignments, the Germans were given a variety of tasks related to technical aspects of missile design (static test stands, rudders, guidance systems, and telemetry systems) and applied scientific problems (design, ballistics, aerodynamics, thermodynamics, and chemistry).[41]

During the Germans' first years in the Soviet Union, their status within the institute fluctuated, undoubtedly reflecting anxiety over limiting access to secret information. For the first few months after their arrival, they remained underutilized. In contrast to the work situation in occupied Germany, where the Germans and the Soviets were integrated into a single organizational structure, in Kaliningrad, the two groups worked in total isolation from each other. Not surprisingly, they got little done. At a meeting of the Special Committee in March 1947, members expressed their displeasure at the fact that "the Germans were not being fully used," as if the outcome had somehow been out of the committee's control. Committee members decided to take deliberate measures to bring the Germans into the mainstream Soviet program, including making individual Germans accountable to the Soviets for particular areas of research and development. The Germans were also henceforth collectively renamed Department G and enjoyed the same rights as all other (Soviet) departments in the NII-88 institute. The department had its own subdivisions for ballistics, aerodynamics, engines, control systems, and missile testing, and a design bureau. The Special Committee specifically wanted to "ensure that all working groups of German specialists have the necessary technical materials and literature, and also to create conditions for their work in the laboratories and design bureaus."[42]

During their first year, those Germans who worked near Moscow enjoyed a great deal of support in the form of resources and equipment. Beyond helping the Soviets in preparing V-2 rockets for test launches, they played a critical role in certifying parts for tests, setting up testing equipment at the institute, and helping to organize the process of systems integration for the complex rocket. Access to information remained a barrier for fully effective work, however. When ministerial bureaucrats failed to outline a consistent strategy for integrating the Germans into the mainstream program, Soviet engineers on the ground took the initiative to implement their own measures: they made sure that all work that the Germans did was accessible to them but denied the Germans information about anything that they themselves were doing. Magnus later recalled:

> Only much later did we realize what tricky means the Soviets applied so that we would not learn what they were really doing: the Russian top experts, organized in work teams like us, worked in the same factory, but carefully separated from us. They worked on the

[41] Vetoshkin, "Poiasnitel'naia zapiska k spisku inostrannykh spetsialistov (cit. n. 38), ll. 67–66.
[42] "Predlozheniia po punktu 2 povestki zasedaniia spetsial'nogo komiteta reaktivnoi tekhniki," 7 March 1947, f. 8044, op. 1, d. 1609, ll. 216–18, RGAE.

same subjects as we did, and they studied our reports with greatest care. The contact between them and us was maintained by young engineers.[43]

In effect, the Germans worked within a parallel structure that served as a comparative model for the Soviets' own work. Not surprisingly, the German community gradually began to feel cut off, not only from the Soviets but also from the broader community of scientists and engineers.

The high point for the Germans was undoubtedly their participation in the first successful V-2 firings from Soviet soil. Soviet and German engineers jointly prepared twenty V-2 rockets, half assembled back in Germany ("Series N") and the rest at the NII-88 factory in Kaliningrad ("Series T"). All were built from equipment the Soviets discovered in Germany after the Nazi surrender. In September 1947, hundreds of administrators, engineers, and technicians set out from Moscow on trains for Kapustin Iar, an old village in a sparsely inhabited area between the Black and Caspian seas where the military had built a makeshift firing ground for long-range missiles. Accompanying the Soviet contingent were twenty-one German scientists and engineers—the leading lights of Department G—headed up by Gröttrup and Lehman.[44] On hand also was Ivan Serov, the deputy minister at the MVD, almost as feared as his boss, Lavrentii Beriia. Serov's presence added a dead air of anxiety to the proceedings. The Germans were especially nervous, knowing full well Serov's historic disdain for non-Russians.

The Germans fully expected not to be involved in the launches—but they were in for a surprise. During the first V-2 launch, on October 18, 1947, the rocket lifted off and flew 207 kilometers although it landed 30 kilometers from its target. A second launch, two days later, fared much worse; the rocket hit ground a dismal 181 kilometers from its target.[45] Three days later, after a third rocket performed poorly, Serov intervened in the process, issuing veiled threats and hinting of sabotage by the designers. Both Soviet and German engineers, fearing for their jobs, kept quiet. Ustinov, technically in charge of the whole affair, took a big risk and decided to ask not the Soviet engineers but the Germans for advice, telling Gröttrup and his men, "This is your rocket and your instruments, go figure it out."[46] Hoch and Magnus, surprised at the public invitation—which visibly annoyed the entire Soviet team—quickly determined the problem and proposed a solution, which to their joy solved the problem. Chertok later recalled that "[w]e celebrated the [next] successful launch together. The authority of the German specialists, whom up until then only the [technical people] had respected, immediately rose in the eyes of the State Commission."[47] Strikingly, in the official report on the launches to Stalin, the State Commission singled out "Doctor Hoch" for mention as having solved a critical problem during the testing.[48] A few weeks after the German team's return from Kapustin Iar, just before the New Year, Stalin personally signed an order giving "the foreign specialists" who had displayed

[43] Magnus, *Raketensklaven* (cit. n. 8), 107–8.

[44] V. Ivkin and A. Dolinin, "Oni byli pervymi," *Krasnaia zvezda*, 17 Jan. 1998.

[45] "Kratkii tekhnicheskii otchet o provedenii opytnykh puskov raket dal'nego deistviia A-4 (Fau-2) na gosudarstvennom tsentral'nom poligon MVS v oktiabre-noiabre 1947 goda (izvlechenie)," 28 Nov. 1947, f. 3, op. 47, d. 185, ll. 14–21, 42–48, 231–38, APRF.

[46] Chertok, *Rockets and People* (cit. n. 8), 2:37.

[47] Ibid., 38.

[48] N. Iakovlev et al. to Stalin, 28 Nov. 1947, f. 3, op. 47, d. 185, ll. 1–13, APRF.

"excellence during the test launches of the [V-2] rocket a one-time bonus equal to three months' salary for each person." He also sanctioned bonuses for the Germans to be given out every other month for "successful solution" of research and development work.[49]

PERSONALITY CONFLICTS

Within six months of these triumphant launches—the apotheosis of the German stay in Russia—a set of interrelated and evolving factors, including friction with Soviet engineers and the changes in the Soviet cultural landscape, significantly affected the fortunes of the Germans. Problems first arose over defining the most immediate goal following the V-2 launches. Already, five months before these launches, on May 7, 1947, Stalin had signed a secret decree authorizing the development of a "domestic" copy of the V-2, known as the R-1.[50] This decision, reinforced by several more in the following year, was a point of deep contention for many Soviet engineers. In particular, Korolev and his senior staff believed that once they had launched and tested the German V-2, they should be allowed to move directly to a new rocket with a range of 600 kilometers, more than twice that of the V-2. Korolev's engineers had begun work on such a missile practically from the moment that they had arrived in occupied Germany, making significant advances by early 1947.[51] Now, Korolev feared that the time-consuming creation of an exact Soviet duplicate of the German V-2 would delay the more ambitious project indefinitely. According to Chertok:

> [I]t offended him to . . . design a domestic R-1, which by government decree was an exact copy of the [German V-2]. Being by nature a commanding and ambitious person who was easily hurt, he could not conceal his feelings when they hinted to him that "you're not making your own rocket, you're reproducing a German one." On this topic, Minister Ustinov, who initiated the program for the exact reproduction of the German [V-2] rocket as practice for the production process, had serious conflicts with Korolev on more than one occasion.[52]

The headstrong Korolev also clashed with Ustinov over interactions with the Germans. Despite Ustinov's express orders, "Korolev simply and demonstratively, ignored everything that had to do with the work the German collective was doing on Lake Seliger. Not once did he visit Gorodomlya Island, nor did he associate with Gröttrup or with the other leading German specialists."[53] Chertok explains Korolev's animus as an outgrowth of his bitterness: having been one of the founders of Soviet rocket technology in the interwar years, he had spent his best years in the gulag while the Germans raced on ahead; now he was being forced to copy them to catch up.

---

[49] "Rasporiazhenie soveta ministrov sssr no. 19317-rs," 29 Dec. 1947, f. 3, op. 47, d. 185, l. 242, APRF.

[50] Kirpichnikov to Molotov, April 1948, f. 4372, op. 95, d. 437, ll. 19–15 (pages in reverse order), RGAE.

[51] Korolev defended the "draft plan" of this missile, known as the R-2, at a plenary session of the institute in April 1947. See S. P. Korolev, "Doklad na pervom plenarom zasedanii NTS NII-88 po rakete R-2," in *S. P. Korolev i ego delo: Svet i teni v istorii kosmonavtiki: izbrannyye trudy i dokumenty*, ed. G. S. Vetrov and B. V. Raushenbakh (Moscow, 1998), 120–22.

[52] Chertok, *Rockets and People* (cit. n. 8), 2:48.

[53] Ibid., 49.

Pride aside, there was also a moral dimension to Korolev's reluctance to interact with the Germans. As with many of his compatriots, wartime experience had deeply colored Korolev's impressions of Germans in general; for Korolev and others, it was not easy to separate "being German" from "being Nazi," especially in light of the horrific atrocities committed by the Nazis on the eastern front.

Friction between the Soviet and German engineers heated up once the ministry leadership gave each group the assignment to develop ideas for a new advanced rocket superseding the Nazi V-2. In mid-1947, the Germans proposed a new ballistic missile known as the G-1, one that would be about the same dimensions as the V-2 but be capable of flying more than double the distance, that is, about 600 kilometers.[54] Simultaneously and independently, Korolev's engineers worked to develop their missile with similar capabilities. In the conceptual design phase, it was possible to devote resources to both teams, but it became increasingly clear that once the "paper stage" was over, the ministry would have to decide how to allocate funding to implement the project. Chertok summarized the intractable problem facing the ministry:

> Out of political and security considerations, no one would allow us to create a mixed Soviet-German collective at NII-88 like the one we had in Germany. But even if they did give us permission, whose design would be developed there and who would be the chief designer? That Korolev would work under Gröttrup was absolutely out of the question. And if Gröttrup worked under Korolev? This too was unrealistic, because Korolev would immediately announce, "Why? We can handle it ourselves."[55]

The only realistic option was to set up two independent and parallel organizations to design and build ballistic missiles in the Soviet Union. But this was clearly beyond the means of the ministry and its dozens of subcontractors.

SCIENCE, *ZHDANOVSHCHINA*, AND THE GERMANS

Uncertainty about the proper use of Germans was severely exacerbated by the broader cultural climate in postwar Soviet Union, particularly the Communist Party leadership's drive to foster a national identity rooted in rehabilitating individuals and symbols from the Russian national past and reconciling them with more recent Marxist-Leninist statist ideology. Glimpses of this seemingly paradoxical campaign of rhetoric had appeared in the late 1930s; wartime experience added a powerful new dimension to the notion of a particularly Soviet (as opposed to Russian) national identity. Besides a rejection of non-Russian claims to Soviet identity and a valorization of the uniqueness of the Soviet wartime experience, a fundamental constituent of this ideological campaign was a complete repudiation of foreign influences in Soviet culture.[56] The campaign began in 1946 as the first cold war tensions were mounting,

---

[54] Vetoshkin to Kirpichnikov, 10 March 1948, f. 4372, op. 95, d. 431, ll. 43–40 (pages in reverse order), RGAE.

[55] Chertok, *Rockets and People* (cit. n. 8), 2:57.

[56] David Brandenberger, *National Bolshevism: Stalinist Mass Culture and the Formation of Modern Russian National Identity, 1931–1956* (Cambridge, Mass., 2002), 183–225. For other recent works that have explored similar themes in the construction of postwar Soviet identity, see Ronald Grigor Suny and Terry Martin, eds., *A State of Nations: Empire and Nation-Making in the Age of Lenin and Stalin* (Oxford, 2001); Yitzhak M. Brudny, *Reinventing Russia: Russian Nationalism and the Soviet State, 1953–1991* (Cambridge, Mass., 1998); Robert C. Williams, *Russia Imagined: Art, Culture, and*

when Communist Party secretary and former Leningrad Party boss Andrei Zhdanov demanded stricter ideological regulation of two publications that had dared to include works by writers considered "apolitical," "bourgeois," and "individualistic." Eventually, Zhdanov's ideas, which distilled cultural production down to two camps, imperialistic and democratic, were applied to all forms of intellectual activity, including literature, art, music, and finally, science. Actions considered bourgeois or projecting a recognizable servility to Western values were severely penalized, often with tragic outcomes. The campaign, known popularly as *Zhdanovshchina* (loosely translated as "the time of Zhdanov"), became a surrogate for rooting out any and all vestiges of "Western" influence in Soviet culture; it eventually acquired a distinctly anti-Semitic character in its critique of "cosmopolitanism."[57]

The Zhdanovshchina campaign affected several Soviet scientific and social scientific disciplines, including physics, biology, cybernetics, linguistics, psychology, and philosophy. In many cases, scientific communities took the initiative to reform their particular disciplines, often with chaotic or contradictory results that did not always conform to the party's relatively simplistic demarcations of what was acceptable (class-based science) and what was not (science without context). Scientists frequently used the campaign to bolster their own professional positions, helping to produce more discord. Most famously, party intervention, combined with internal professional conflicts, nearly ended the practice of at least two disciplines, Mendelian genetics—of Lysenko fame—and cybernetics.[58]

Zhdanovshchina explicitly linked science with national identity by deifying dead and living Russian intellectuals precisely for their Russian heritage. These claims were generated largely as a response to increasing cold war tensions and the rise of the United States (which represented a generic "West") as a global superpower. Soviet writers declared that Russians invented the steam engine, bicycle, airplane, electric light bulb, radio, insulin, vitamins, synthetic rubber, and even the game of baseball. The campaign was orchestrated and sanctioned at the highest levels; in a secret letter to the Communist Party Central Committee, both Zhdanov and Stalin provided examples of Russian inventions and discoveries now "misappropriated" by foreigners.[59] In 1949, the Academy of Sciences hosted a conference on the history of Soviet science, the goal of which was to eliminate a "contemptuous attitude" toward national achievement in science. The following year, the academy issued a volume, *Iz istorii otechestvennoi tekhniki* (From the History of Native Technology), whose

National Identity, 1840–1995 (New York, 1997); and Helena Goscilo and Andrea Lanoux, eds., *Gender and National Identity in Twentieth-Century Russian Culture* (DeKalb, Ill., 2006).

[57] Kees Boterbloem, *The Life and Times of Andrei Zhdanov, 1896–1948* (Montreal, 2004); Werner G. Hahn, *Postwar Soviet Politics: The Fall of Zhdanov and the Defeat of Moderation, 1946–53* (Ithaca, N.Y., 1982); Laurent Rucker, "La 'Jdanovshchina': Une campagne antisemite (1946–1949)?" *Bulletin de l'Institut d'Histoire du Temps Present* 35 (1996): 83–94.

[58] Kirill Rossianov, "Editing Nature: Joseph Stalin and the 'New' Soviet Biology," *Isis* 84 (1993): 728–45; Peter Kneen, "Physics, Genetics and the *Zhdanovshchina*," *Europe-Asia Studies* 50 (1998): 1183–202; Alexei Kojevnikov, "Rituals of Stalinist Culture at Work: Science and the Games of Intraparty Democracy circa 1948," *Russian Review* 57 (1998): 25–52; Slava Gerovitch, *From Newspeak to Cyberspeak: A History of Soviet Cybernetics* (Cambridge, Mass., 2002); Ethan Pollock, *Stalin and the Soviet Science Wars* (Princeton, N.J., 2006).

[59] Nikolai Krementsov, *Stalinist Science* (Princeton, N.J., 1997), 139.

explicit goal was to highlight "the grandeur and diversity of the fatherland's scientific thought."[60]

The Zhdanovshchina campaign put obvious conflicting pressures on industrial managers such as Dmitrii Ustinov, caught between fulfilling Stalin's mandate and responding to Zhdanov's call. Some Soviet engineers believed that the outcome of this tension was preordained. Iurii Pobedonostsev, the chief engineer at the NII-88 institute, told Chertok at the time that "no matter what [the Germans] come up with, it won't be in step with our current trend in ideology, which dictates that everything created recently or previously in science and technology be done without any foreign influence."[61]

On a personal level, Zhdanovshchina's pressures gave Korolev a practical rationale for avoiding contact with the Germans. Although Korolev was technically a "free" man, his conviction as an enemy of the state still remained on the books—formal "rehabilitation" would not come until the Khrushchev Thaw. His extant conviction made it very difficult—almost impossible—for him to join the Communist Party since the party accepted only individuals with clean records. His lack of party credentials, in turn, made his position as chief designer tenuous, especially in light of the Zhdanovshchina campaign, which necessitated "cleanings" and "purges" of "deviances" from party orthodoxy within all Soviet social and cultural institutions, including, especially, those deeply embedded in the Soviet military-industrial complex. Korolev was particularly vulnerable in 1947–48 as the ripples from Zhdanovshchina began to penetrate the NII-88 institute. In July 1947, at a party meeting at the institute, attendees for the first time discussed the Kliueva-Roskin affair, a case involving two Soviet scientists in Moscow who had welcomed the U.S ambassador and as a result were viciously attacked by Zhdanov and other party ideologues for "servility to the West."[62]

After Trofim Lysenko's triumph over genetics in August 1948, Korolev found himself under more scrutiny. Already, one of his deputies, who like him had been a gulag prisoner but now designed missiles, had been fired from the institute.[63] In December of that year, at a party conference, institute director Lev Gonor noted that "the great work on cleansing the institute from unreliable [elements] is not instilling confidence; this work needs to be continued"—words directly aimed at Korolev, disliked by many for his gruff style of work. More specifically, Korolev came under rebuke for not admitting his shortcomings—his acerbic character, his *bezpartiinost'* ("the absence of Party mindedness"), and his gulag sentence. Admitting fault in public was a ubiquitous aspect of party culture that, under pressure from the Zhdanovshchina campaign, had now spread into the scientific and technological communities. The most important of these party rituals involved *kritika* and *samokritika* ("criticism" and "self-criticism"), the practice of a high official accused of wrongdoing admit-

[60] V. V. Danilevskii, ed., *Iz istorii otechestvennoi tekhniki: Issledovaniia i materialy* (Leningrad, 1950), 3; Alexander Vucinich, *Empire of Knowledge: The Academy of Sciences of the USSR (1917–1970)* (Berkeley, Calif., 1984), 230.

[61] Chertok, *Rockets and People* (cit. n. 8), 2:46–47.

[62] F. 2416, op. 1, d. 37, l. 28 (30 July 1947), Russian State Archive of Socio-Political History, Moscow (hereafter cited as RGASPI).

[63] Korolev's deputy K. I. Trunov had been fired after a party committee meeting on his "suspect" past. See f. 2416, op. 1, d. 38, l. 43 (10 March 1947), RGASPI.

ting his or her mistakes at a meeting of superiors and subordinates. There were many functions of the rituals, including "initiating and socializing the speakers, their internalization of group norms and subordination of personal views to the collective, [and] stimulating scientific debate," but one of the most important was simply to purge scientists from the profession.[64] At the December meeting, Kaliningrad city's party committee noted:

> We know about the fact that in other fields, such as the biological front, there are lessons which can be a lesson to us . . . that in the field of scientific and technical progress we must have this *kritika* and *samokritika* . . . Symptoms of the absence of *kritika* and *samokritika* are found in [Korolev's department]. . . . Here, it is necessary to look to the Party committee and call Korolev to order and bring to order those comrades who in this respect are taking incorrect positions. I think that we need to keep a vigilant watch over this [situation].[65]

Given Zhdanovshchina's various dimensions, Korolev was forced to take great care in his actions. In addition to regulating his behavior to conform to prevailing party dictates, he had to steer clear of the Germans since they represented a foreign influence; yet, he had to account for the possibility that their help was essential to the success of his work, as mandated by Stalin.

Negotiating all of these concerns required a delicate dance from all three constituencies—the bureaucrats, the Soviet engineers, and the German specialists. The bureaucrats (Ustinov, Gonor, and others) needed to satisfy Stalin's whims to build long-range ballistic missiles, a goal that would fail, they believed, without the help of the Germans; they sought to give the Germans the resources they needed but recognized that parallel and independent work by Germans and Soviets was financially untenable. The designers (Korolev, Mishin, and others) needed to avoid the kind of behavior that would get them fired, purged, or worse, especially given the pressures to reinforce a new nationalist tenor in Soviet science in the early cold war years; they did not want to be working for the Germans or having the Germans work for them. Finally, the Germans (Gröttrup, Hoch, and the others), although forced to relocate to the USSR, were eager to move ahead with their own designs; they wished to be integrated into the mainstream Soviet ballistic missile program. All these forces existed in a setting in which the culture of secrecy frustrated easy solutions.

MAKING THE GERMANS LESS USEFUL

A solution to this Gordian knot was found through complex gymnastics that left the one constituency who had little or no power, the Germans, out in the cold. Taking advantage of the vigilant need for secrecy, industrial managers such as Ustinov and Gonor effectively slowed down German work on the G-1 missile until the Soviets matched the German quality of work. Once the Soviet side had eclipsed the Germans, the perceived utility of the latter plummeted.

This process took over a year. In September 1947, Gröttrup and the other leading Germans had enthusiastically "defended" the design (much like a thesis defense)

---

[64] J. Arch Getty, "*Samokritika* Rituals in the Stalinist Central Committee, 1933–38," *Russ. Rev.* 58 (1999): 49–70.
[65] F. 2416, op. 1, d. 71, ll. 15, 34, 39, 341 (3 Dec. 1948), RGASPI.

of their new G-1 ballistic missile at a session of the institute's "scientific and technical council." A qualitative advance over the now increasingly obsolete V-2, the G-1 incorporated a number of striking innovations; the missile was essentially the same dimension as the V-2 but was capable of flying much further, was significantly lighter, and was much more accurate.[66] By and large, the Soviet review panel found Gröttrup's missile very promising; they recommended that before the Germans finalize a workable design, they try out various elements of the missile on test rigs.[67] The Germans were elated with the decision as it seemed to underscore faith in their capabilities.

Within six months of this decision, as the Zhdanovshchina campaign reached its zenith, a series of blows seemed to cripple the German team. First, despite the earlier positive assessment of the G-1 rocket, the NII-88 institute leadership allocated few resources to the Germans to test equipment for their project. When Chertok went to see Gröttrup early in 1948, Gröttrup complained that "in spite of the [institute's] favorable decision regarding his design, he could not meet a single request on that document" as there was little equipment on Gordomlya Island to facilitate ground testing.[68] In the meantime, the limited access to the Soviet program that the Germans had previously been allowed became even more circumscribed. In a significant setback to the Germans, in June 1948, Gonor abruptly transferred *all* the Germans—including the ones who lived and worked at the institute in Kaliningrad—to Gorodomlya's remote location.[69] Furthermore, industrial managers began complaining about the "inadequate knowledge and poor qualifications" of the Germans. In July 1948, Ustinov transferred sixty-eight Germans from the Ministry of Armaments to the care of the Ministry of Internal Affairs for "civilian production" because they had displayed very little skill or qualifications in handling missile technology.[70] A senior government planning official also noted in a memo to the Special Committee that secrecy and compartmentalization, necessary to ensure that the Germans not have access to state secrets, had been deeply damaging to the future prospects of the Germans. He wrote that:

> during the period [of their] work in the Soviet Union, because of . . . secrecy, the German specialists were not connected to domestic developments, both in . . . models of reactive armaments and to developments in related fields of technology, and therefore even the most qualified German specialists have gradually fallen behind in technical discourse and at the present they remain at much the same level as German technology from 1944–1945.[71]

Despite these setbacks, the Germans persevered in their G-1 design. Against all odds, they managed to improve the paper design of the missile to make it more powerful and efficient—they increased the projected range from 600 to 810 kilometers. In December 1948, a number of the leading Germans were invited back to the NII-88

---

[66] Chertok, *Rockets and People* (cit. n. 8), 2:50–56; Vetoshkin to Kirpichnikov, 10 March 1948 (cit. n. 54).

[67] Chertok, *Rockets and People* (cit. n. 8), 2:55–56.

[68] Ibid., 58–59.

[69] Ustinov to Saburov, 31 July 1948, f. 8157, op. 1, d. 1206, l. 146, RGAE.

[70] Ustinov to Bulganin, 29 July 1948, f. 8157, op. 1, d. 1206, ll. 137–40, RGAE.

[71] Kirpichnikov to Bulganin, April 1948, f. 4372, op. 95, d. 431, ll. 59–56 (pages in reverse order), RGAE.

institute to defend their reenvisioned rocket. The response was the same: the insti-tute's senior engineers expressed generally "positive and amiable" opinions, and the institute asked the Germans to conduct experimental work to verify their theories, experimentation which proved impossible to conduct because the institute provided no resources to do so.[72]

Through 1949, the Germans continued work on the G-1, albeit with what little they had on their isolated island. By this time, Korolev's parallel team, unbeknown to the Germans, had achieved some significant successes. In October and November 1948, under Korolev's command, the NII-88 institute launched the first series of nine R-1 missiles, the domestic manufactured versions of the German V-2.[73] Although engineers faced substantial problems in launching these weapons, these experiments were significant as they confirmed that Soviet industry could marshal a scattered net-work of institutions to produce and test a relatively sophisticated piece of machinery. In copying the German missile, Korolev's team had altered the vehicle modestly to conform to Soviet industrial practices and available resources.

The Soviets independently moved very quickly past the R-1; less than a year later, in September and October 1949, they began testing an experimental version of *their* 600-kilometer rocket. This missile, the R-2E, was markedly superior to the German V-2—incorporating a set of innovations in line with Soviet industrial practices. It was not as technically elegant as the advanced German proposal, but it got the job done. The success of this missile finally and irrevocably eclipsed any hope that the compa-rable German G-1 project would ever come to fruition. After 1948, Soviet engineers continued to assign tasks to the Germans; each assignment—usually a demand to conceive an entirely new missile with ambitious capabilities—raised the Germans' hopes that *this* time the Soviets would recognize the value of German work. All these expectations were dashed.

The Germans grew increasingly desperate and listless; morale became a serious problem among the group, exacerbated by bitter infighting, alcoholism, and marital infidelity.[74] Gröttrup's wife, Irmgard, succinctly summarized the ennui and the disaf-fection of the Germans when she wrote in her diary on the first day of the year 1950: "Once again, we are on the threshold of a new year. It is futile to keep on wondering what lies in store for us: exactly nothing."[75] Only on rare occasions were the Ger-mans allowed to leave the island to visit Moscow. Under these conditions, the group worked on a series of missile proposals more advanced than the G-1, all of which momentarily diverted their attentions from their earthly problems.[76] The Germans dutifully fulfilled these requests, sent the reports to Kaliningrad, and never heard from the Soviets again, barring occasional requests communicated via intermediaries to clarify one point or another.

---

[72] Chertok, *Rockets and People* (cit. n. 8), 2:60–65; Vetrov and Raushenbakh, *S. P. Korolev i ego delo* (cit. n. 51), 655.

[73] For a summary of the launches and their results, see Pashkov to Kirpichnikov, 5 Nov. 1948, f. 4372, op. 95, d. 437, ll. 127–26 (pages in reverse order), RGAE.

[74] Ordway and Sharpe, *Rocket Team* (cit. n. 24), 336, 340.

[75] Irmgard Gröttrup, *Rocket Wife* (London, 1959), 134.

[76] These "projects" included the G-1M (range of 1,100 kilometers), the G-2 (2,500 kilometers), the G-3 (8,000 to 10,000 kilometers), the G-4 (3,000 kilometers), and the G-5 (a cruise missile). For a summary, see CIA Office of Scientific Intelligence, *Scientific Research Institute and Experimental Factory 88* (cit. n. 8). The Soviets never built any of these missiles.

THE END OF THE ROAD

In February 1950, Ustinov outlined the status of the 495 Germans (a total of 1,241 with their families) working in his ministry, distributed between the missile and optics industries. He argued that because the Germans had been isolated for so long from the "source stream of knowledge," they were no longer as valuable as they had been five years before. As such their wages should be renegotiated to lower levels. Furthermore, he recommended that all Germans be transferred out of his Ministry of Armaments since his ministry was wholly engaged in secret work. Finally, he noted that the work motivations of a "significant portion" of the Germans had sharply decreased because they no longer wanted to be in the Soviet Union.[77]

Ustinov consulted with the Ministry of State Security (MGB), the Soviet counterintelligence agency, on possible courses of action. After some deliberation, it was decided to repatriate at least some of the Germans back to the recently established German Democratic Republic (GDR), a decision approved by Stalin in August 1950. The logistics of facilitating the transfer took longer than expected because of concerns that the Germans would carry sensitive secrets about the Soviet missile program back to the West; the possibility of monitoring and limiting the repatriated Germans' movements in the GDR appears to have partly assuaged these concerns. In March 1951, ministry officials divided the Germans into two categories, representing those who could be allowed to go home and those who had done useful work in the missile program, that is, those who represented a higher security risk.[78] The MGB advised delaying the departure of the latter for a year and a half after they had stopped working on sensitive work so as to ensure that their knowledge would be obsolete by the time that they returned to Germany.[79] As a result, ministry bureaucrats devised an elaborate plan to assign the Germans a host of civilian tasks in the interim but without telling them that their work profile had changed.[80] In July 1951, Ustinov, his superiors, and the Soviet security services finally signed off on allowing 345 Germans (plus their family members) to return to East Germany.[81]

The Germans departed from Ostashkov (the closest town to Gorodomlya Island) for their homeland by train in three waves, in December 1951, June 1952, and November 1953. Families were allowed to exchange their savings in rubles for deutsche marks. The Soviets paid for the trip home and drew up an agreement whereby the East German government had to provide the returnees with accommodations and employment.[82] The last eight German scientists, which included Gröttrup, were given permission to leave the USSR on November 22, 1953. Within a week they were all gone, ending the seven-year existence of the NII-88's "Department G." The few Germans who voluntarily remained were allowed to move to Moscow, given good salaries, and five-year contracts in industries unrelated to missile development.

[77] D. Ustinov, "Spravka o nemetskikh spetsialistiakh, rabotaiushchikh v sisteme ministerstva vooruzheniia," 6 Feb. 1950, f. 8157, op. 1, d. 1379, ll. 26–28, RGAE.

[78] Sukhomlinov to Zubovich, 26 March 1951, f. 397, op. 1, d. 105, l. 98, RGAE.

[79] Pitovranov to Zubovich, July 1951, f. 397, op. 1, d. 105, l. 163, RGAE.

[80] Kurganov to Rudnev, 28 July 1951, f. 397, op. 1, d. 105, l. 183, RGAE.

[81] Ustinov to Bulganin, 18 July 1951, f. 8157, op. 1, d. 1454, l. 24, RGAE.

[82] For an overview of the fate of repatriated Germans in East Germany, see André Steiner, "The Return of German 'Specialists' from the Soviet Union to the German Democratic Republic: Integration and Impact," in Judt and Ciesla, *Technology Transfer out of Germany* (cit. n. 13), 119–30.

The fate of the Germans after their residency in the USSR was varied. Almost all were extensively interrogated by U.S. intelligence services, providing a wealth of information about the Soviet ballistic missile program, especially about the period from 1945 to 1947. Their information about the later years was frequently confused and incorrect, but for many decades their claims were the only information historians had on the early years of the Soviet missile program.[83]

CONCLUSIONS

On one level, the success of the Soviet ballistic missile program could be measured by the effectiveness of the missiles it produced. Here, some believed that German expertise was invaluable. Yet, the project's ultimate value was also inseparable from its standing as a Soviet effort directed by Soviet actors. Here, given the pressures generated by a new campaign to articulate postwar Soviet national identity, German expertise had to be repudiated. For several years after the kidnapping of German scientists, these two impulses existed in an uneasy balance until German expertise was rendered useless and thus offered a solution acceptable to all parties: send the German scientists back home. In some sense, both the arrival and the departure of the Germans were driven by cold war concerns. The Germans had been forcibly relocated to the Soviet Union to aid the Soviets in developing an important strategic capability, second only to the development of the atomic bomb. They were returned to Germany when cold war tensions—as manifested in domestic proscriptions—made their work in the Soviet Union untenable. In the end, the relationship between national identity and the practice of science was characterized not only by the former informing the latter but also by a deep-seated anxiety that the latter would deleteriously affect the former.

The most striking aspect of this episode is the extent to which Soviet actors, principally industrial administrators and engineers, intervened in the process of Soviet-German cooperation to render the Germans *less* useful. When the German work in the Soviet program became more difficult to integrate with concomitant Soviet work due to the needs of secrecy, personality conflicts, and the nationalist Zhdanovshchina campaign, the Soviets slowed down German activities until it could be conclusively shown that the fruits of their work had little utility. In other words, the Germans' ultimate fate had less to do with their expertise (which was quite impressive) than with the perception of their expertise. The latter was easier to manipulate and eventually overshadowed the former.

In revisiting the German contribution to the Soviet rocket program, care should be taken not to conflate two related but rhetorically different phenomena: the transfer of matériel, and the relocation of personnel. There is no doubt that technology transfer in the form of the V-2 played a fundamental part in the birth of the Soviet missile program (as it did in the United States). Soviet industry's plunder of rocket detritus from occupied Germany and the subsequent reconstruction and mass production of the V-2 in the late 1940s established a vast network of dynamic institutions and trained an entire generation of designers who would later develop the world's first intercontinental ballistic missile. Of course, conceptually and practically, the precise demar-

---

[83] For probably the best narrative summation of the German work based wholly on their reminiscences, see Ordway and Sharpe, *Rocket Team* (cit. n. 24), chap. 17.

cations between matériel and expertise are usually nebulous. In the case of the V-2, for example, the Soviets learned much of what they knew about the rocket from their German colleagues. Yet, the language that Soviet administrators used in the postwar era suggests that they drew a very sharp distinction between the two. Thus, it required no discursive gymnastics for Soviet bureaucrats, engineers, and party apparatchiks to reject the use of Germans on the grounds that they had no place in postwar Soviet science while adopting without reservation the V-2 rocket as the foundation for their new missile program. Soviet chief designer Sergei Korolev was quick to call the R-1, the Soviet version of the V-2, an "original" rocket, despite its being essentially a copy.[84] As the V-2 moved from Germany to Russia, it became less German and more Soviet. It is now the source from which the grand narrative of Russian space exploration emerges: Russian historians still call the R-1 the first "domestic" Soviet rocket. The German scientists who came to Russia have no place in this story.

[84] S. P. Korolev, "Doklad na plenarom zasedanii NTS NII-88 po eskiznomu proektu rakety R-3," in Vetrov and Raushenbakh, *S. P. Korolev i ego delo* (cit. n. 51), 130–40.

Communicating the North:

Scientific Practice and Canadian Postwar Identity

By Edward Jones-Imhotep[*]

ABSTRACT

In the late 1940s, a group of Canadian ionospheric researchers used high-latitude atmospheric research to carve out influence and autonomy for themselves and the nation. Echoing a broader political project that cast the North as a source of Canadian autonomy and power in the postwar world, the group focused its research on the turbulent and anomalous ionosphere above Canada, which disrupted high-frequency communications throughout its northern regions. Those same conditions made ionograms, the key records of ionospheric research, illegible. This paper traces the attempts to demarginalize scientists, the nation, and its nature by remaking the practice of reading in ionospheric research.

> Canadian sensibility has been profoundly disturbed, not so much
> by our famous problem of identity, important as that is, as by a
> series of paradoxes in what confronts that identity. It is less per-
> plexed by the question of "Who am I?" than by some such riddle
> as "Where is here?"
>
> Northrop Frye, *A Literary History of Canada*

INTRODUCTION

In early 1943, as the tide of the Second World War began turning in favor of the Allies, a group of former academics concentrated in Canada's Department of External Affairs began formulating plans to refashion the postwar nation. Early in the war, the members of the group had become increasingly anxious and frustrated by what they saw as the neoimperialism of their allies. After the fall of France in June 1940, it was London's Cabinet War Rooms that seemed to "automatically" determine Canadian policy; following the attack on Pearl Harbor in December 1941, America seemed increasingly to treat Canada like "an internal domestic relationship rather than an international one."[1] Locating the problem in weak and reluctant diplomacy—"the strong glove over the velvet hand"—members of External Affairs began exercising a new muscularity in Canadian foreign policy. As their most imaginative member, Escott

---

[*] York University, STS Programme, Bethune College 218, 4700 Keele Street, Toronto, Ontario, Canada, M3J 1P3; imhotep@yorku.ca.
[1] Norman Robertson, cited in J. L. Granatstein, *A Man of Influence: Norman A. Robertson and Canadian Statecraft, 1929–1968* (Ottawa, Canada, 1981), 118.

© 2009 by The History of Science Society. All rights reserved. 0369-7827/09/2009-0008$10.00

Reid, put it: "If we don't want to be a colony of the United States we had better stop being a colony of Britain."[2]

Leaning over a polar projection map showing the air routes between North America and the Soviet Union in the late summer of 1943, Reid saw the Canadian North—where American violations of Canadian sovereignty had become emblematic—as one critical place to begin breaking that dependency. Political, social, and economic development in the vast region stretching from the 60th parallel to the North Pole would secure the region, improve the lives of northern residents, and provide "an inspiring and somewhat romantic national objective for the people of Canada."[3] Even more important for the diplomatically minded Reid, it would provide a site where the group's emerging core principles of functionalism and internationalism could be put to work. Together those principles would locate postwar power in international bodies such as the United Nations and NATO, diffusing American dominance while claiming significant representation for Canada as a middle-power nation.[4] In the late summer of 1943, though, Reid already glimpsed what his colleague, the historian Hugh Keenleyside, would express after the war: "What the Aegean Sea was to classical antiquity, what the Mediterranean was to the Roman world, what the Atlantic Ocean was to the expanding Europe of Renaissance days, the Arctic Ocean is becoming to the world of aircraft and atomic power."[5] In that potentially apocalyptic polar world, Reid and his colleagues saw Canada as a force for peace among the five Arctic nations, using its northern territories to carve out a neutral space for political collaboration, economic initiatives, communications, and scientific research.

Reid's vision of a polar commons would soon collapse as the cold war took hold, but his idea of the North as a source of power and influence would be taken up by the most iconic Canadian research program of the postwar period. Throughout the late 1940s, a group of Canadian radio scientists wove into that larger political vision for the North their own explorations of the upper atmosphere. Immersed in the wider political discussions about the North, they drew on their wartime research in North Atlantic communications to turn the "Canadian" ionosphere—the site of atmospheric storms and the cause of radio blackouts—into a North writ small, a region that would demarginalize their research and, in doing so, push forward key national aims: territorial and epistemic sovereignty, northern development, international cooperation, distinction, identity, and influence vis-à-vis Britain and the United States. At the core of the group's program were the ionospheric storms and geophysical anomalies that both characterized Canadian ionospheric records and made them unreadable under the dominant interpretive practices of the discipline. This paper explores how the efforts to turn those anomalies into objects of research, to make them "legible," engaged and mirrored the broader political project to remake Canada after the Second World War. It argues, in particular, that reforming the act of "reading" in ionospheric

[2] Escott Reid, "The United States and Canada: Domination, Cooperation, Absorption," cited in Dennis Smith, *Diplomacy of Fear: Canada and the Cold War, 1941–1948* (Toronto, 1988), 18.

[3] Escott Reid to Norman Robertson, 30 July 1943, file A-25-3 (pt. 1), vol. 21, RG 2/18, National Archives of Canada, Ottawa (hereafter cited as NAC).

[4] A. J. Miller, "The Functional Principle in Canada's External Relations," *International Journal* 35 (1979–80): 309–28.

[5] Quoted in Shelagh Grant, *Sovereignty or Security? Government Policy in the Canadian North, 1936–1950* (Vancouver, 1988), 211.

research was part of the broader efforts to demarginalize the nation by recasting it as paradigmatically northern.

The role of science and technology in asserting both national identity and political power has been a lively topic in science studies. Gabrielle Hecht, for instance, has shown how the image of a radiant France buttressed by nuclear technology was central to the redemption and reimagining of the postwar nation.[6] Lisbet Koerner has seen in Linnaeus's taxonomy the dream of Swedish commercial dominance, economic independence, and geopolitical power in an age of colonial expansion.[7] Robert Brain has explored the Parisian linguistics laboratory of the Third Empire to vividly illustrate how the graphic rendering of spoken language formed a vanguard in the battle between French linguistics and German philology, instantiating the linguistic nation and securing the leverage of the metropolis against the provinces and their patois.[8] Again and again scholars have pointed us to the role of science and technology in articulating ideas of the nation on which identities are built.[9]

What interests me here is how the practices of science enable those expressions of national identity and the larger political and cultural projects built upon them. Linking scientific practice and questions of national identity in this way necessarily conjures thoughts of "national style."[10] In its most general form, the national styles argument holds that a specific constellation of characteristically national elements—political, social, economic, institutional, cultural—ultimately shapes a distinctly national way of practicing science or designing and building technology.[11] For all its power and appeal as an explanatory framework, the national-styles argument falls short here. It is not the reflection of an amorphous national culture that I am after, but rather the way that practices themselves embody wider struggles to bring a certain understanding of the nation into being. In her study of interwar Austria, for example, Deborah Coen has argued that after the Treaty of Saint-Germain shrunk Austria's borders in 1919, Austrian meteorologists set out to develop a small-scale climatology ideally suited to their newly diminished nation. By shaping the practice of their sci-

---

[6] Gabrielle Hecht, *The Radiance of France: Nuclear Power and National Identity after World War II* (Cambridge, Mass., 2000).

[7] Lisbet Koerner, *Linnaeus: Nature and Nation* (Cambridge, Mass., 2001).

[8] Robert Brain, "Standards and Semiotics," in *Inscribing Science: Scientific Texts and the Materiality of Communication*, ed. Timothy Lenoir (Stanford, Calif., 1998), 249–84.

[9] For examples, all on the subject of aviation, see Joseph Corn, *The Winged Gospel: Americans' Romance with Aviation 1900–1905* (New York, 1988); Peter Fritzsche, *A Nation of Fliers: German Aviation and the Popular Imagination* (Cambridge, Mass., 1992); Scott W. Palmer, "On Wings of Courage: Public 'Air-Mindedness' and National Identity in Late Imperial Russia," *Russian Review* 54 (1995): 209–18. See also Rudolph Mrazek, "Let Us Become Radio Mechanics: Technology and National Identity in Late-Colonial Netherlands East Indies," *Comparative Studies in Society and History* 39 (1997): 3–34

[10] For two representative studies here, see Mary Jo Nye, "National Styles? French and English Chemistry in the Nineteenth and Early Twentieth Centuries," *Osiris* 8 (1993): 30–49; and Jonathan Harwood, "National Styles in Science: Genetics in Germany and the United States between the World Wars," *Isis* 78 (1987): 390–414. In his classic study of European electrification, Thomas Hughes has managed to avoid the essentializing connotations of national style by examining how specific political and economic structures shaped the structure of electrical systems in different cities. See Thomas P. Hughes, *Networks of Power: Electrification of Western Society, 1880–1930* (Baltimore, 1983).

[11] In explaining the efflorescence of quantum chemistry in the United States, for example, Sylvain Schweber invokes the popular theme of American optimism and pragmatism against the deeply philosophical trepidations of European physics. Schweber has referred to quantum chemistry as the "quintessentially American discipline." See S. S. Schweber, "The Empiricist Temper Regnant: Theoretical Physics in the United States, 1920–1950," *Historical Studies in the Physical and Biological Sciences* 17 (1986): 55–98.

ence around the local—the mountain microclimate but also the city neighborhood—they helped build an alpine national identity for postimperial Austria, as well as a source of expertise that set them apart within the increasingly large-scale meteorology that they felt unable to practice. For them, *Kleinklimatologie* was a way of simultaneously refashioning the nation and the scientific practices that helped articulate its identity and meaning.[12]

Similarly in the case of the Canadian group, making the high-latitude ionosphere a source of authority and influence was itself founded on a rebellion not only against Canadian marginality during World War II but also against the marginalizing practices that helped enforce it. Articulating the nation was as much a question of changing the practice of science—how records were generated, who was authorized to produce them, and how they ought to be interpreted and understood—as it was about changing the products of those practices. The very possibility of expressing a nature and a nation that served the interests of cultural, economic, and political autonomy rested on autonomy and authority in how researchers conducted their work.

To explore those issues, the paper is organized into three parts. The first introduces the political project to remake the postwar nation. It specifically situates that project in relation to the two dominant themes of Canadian postwar identity, communications and the North, and shows how Frank Davies positioned ionospheric research as a solution to its key concerns. The paper then moves into the attempt to make the anomalous conditions of the northern ionosphere into the core of Canadian research, particularly as a way of uniquely contributing to, and gaining autonomy from, British and American investigations. That move relied on a long tradition of exploration and research that normalized the temperate latitudes and organized understandings of both the polar regions and the tropics around notions of the extreme or the anomalous. The final section of the paper then explores how that focus generated problems in reading ionograms, the most important records of postwar ionospheric research. The phenomena at the center of the Canadian group's bid for national autonomy and influence obscured precisely the features of ionograms that defined them as readable. It was only by gradually reforming the act of reading that they were able to make the high-latitude ionogram legible, underwriting other characterizations of Canadian northernness and feeding back into the cultural values that had sustained it in the first place.

NORTHERN COMMUNICATIONS

In the late 1960s, fresh from his search for the deep regulating myths that structured Western literature and furnished the framework for systematic criticism, the renowned critic and founder of archetypal criticism, Northrop Frye, turned his eye to Canadian literary history. "There would be nothing distinctive in Canadian culture at all," he wrote, "if there were not some feeling for the immense searching distance, with the lines of communication extended to the absolute limit, which is a primary geographical fact about Canada and has no real counterpart elsewhere."[13]

---

[12] Deborah Coen, "Scaling Down: The 'Austrian' Climate between Empire and Republic," in *Intimate Universality: Local and Global Themes in the History of Weather and Climate*, ed. J. Fleming, V. Jankovic, and Deborah Coen (Sagamore Beach, Mass., 2006), 115–40.

[13] Northrop Frye, *The Bush Garden: Essays on the Canadian Imagination* (Concord, Mass., 1971), 10.

Frye's statement of Canadian exceptionalism combined three elements that already pervaded thinking about Canadian identity during the Second World War. First, he pointed to the pervasive role of nature in the Canadian imagination: "everything that is central in Canadian writing seems to be marked by the imminence of the natural world."[14] For almost four centuries, it was the supposedly northern character of that natural world that stood out most—long cold winters, barren rock, boreal forests, and flora and fauna that made up what Frye termed the "obliterated environment" in the nation's history.[15] "One wonders," he mused, "if any other national consciousness has had so large an amount of the unknown, the unrealized, the humanly undigested, so built into it."[16] But even as it had furnished the single most potent force in the Canadian imagination, that view of nature had also generated a tone of "deep terror," the second theme. Shifting to metaphysics, Frye elaborated: "It is not a terror of the dangers or discomforts or even the mysteries of nature, but a terror of the soul at something that these things manifest."[17] Underlying Frye's criticism, then, was the paradox that even as the North defined Canada, it also threatened human survival and, through it, the nation. Here, a third theme emerged. What defined Canada was not only the imminence of the natural world and its deeply threatening character but also its conquest "by an intelligence that does not love it."[18] That conquest, Frye explained, had been carried out most effectively through communications technologies. "The enormous difficulties and the central importance of communication and transport, the tremendous energy that developed the fur trade routes, the empire of the St. Lawrence, the transcontinental railways, and the north-west police patrols have given it the dominating role in the Canadian imagination."[19] "It is in the inarticulate part of communication, railways and bridges and canals and highways, that Canada . . . has shown its real strength."[20] For Frye, Canada was a nation at once defined and threatened by northern nature and made possible through communication technologies.

The resurgence of the North within the national imaginary in the 1940s and 1950s—a resurgence that helped shape Frye's own analysis—owed much to a small but influential group of former academics known in government circles simply as "the intelligentsia." Centralist and interventionist in its outlook, the members of the group were heavily concentrated in the Department of External Affairs, where they felt keenly Canada's second-class status in the wartime alliance with the United States and Britain. Drawing on a century-old vision that turned Canada's supposedly barren and hostile northern territories into a source of dynamism and national renewal, the members of External Affairs seized on the North as the basis for Canadian postwar identity, influence, and power.[21] Symbolically, Canada's northernness would distinguish the nation from its allies; substantively, the geopolitical importance of

---

[14] Ibid., 247.

[15] For a discussion of the role of botany and plant biology in characterizing Canada's northern nature, see Suzanne Zeller, *Inventing Canada: Early Victorian Science and the Idea of a Transcontinental Nation* (Toronto, 1987).

[16] Northrop Frye, "Conclusion to the Literary History of Canada," quoted in Carl Mitcham, *The Northern Imagination: A Study of Northern Canadian Literature* (Moonbeam, Ontario, 1983), 222–23.

[17] Ibid., 12, 227.

[18] Frye, *Bush Garden* (cit. n. 13), 224.

[19] Ibid., 171.

[20] Ibid., 222.

[21] Shelagh Grant, "Myths of the North in the Canadian Ethos," *Northern Review* 3/4 (1989): 15.

the region, its mineral wealth, and its scientific value could be used as a fulcrum to elevate Canada's status in the postwar order.

For all its focus on national revival, however, the enthusiasm over the North was driven by deep anxieties about control. Shortly after its entry into World War II, the United States began carrying out meteorological projects in the Canadian North, including one designed by the Massachusetts Institute of Technology and code named "Arctops." The project involved constructing a vast network of radio stations and weather bases in northern Canada and Greenland. Staffed by military radio operators rather than civilian meteorologists, and supplied entirely by either the U.S. Navy or Air Force, the stations produced data that they forwarded directly to the U.S. Weather Bureau. As a complement to the project's meteorological activities, the U.S. Army undertook extensive mapping and aerial photography, experimentation with communications and radar technology (including ionospheric investigations), and scientific studies into Arctic conditions and resources. Those activities would soon be followed by vast U.S. infrastructure projects to help build the northeast and northwest staging routes. By June 1943, at the height of the activities, an estimated 43,000 U.S. military personnel and civilians were in northern Canada, broadly defined.[22]

Those activities raised traditional concerns in Exernal Affairs over Canadian territorial sovereignty. Canada's claims to its northern territories had traditionally been based on the so-called Sector Principle that granted Canada the entire northern area bounded by lines of longitude running from its eastern- and western-most points to the North Pole. Rejecting the validity of the Sector theory, U.S. officials had preferred the more traditional principle of "effective occupation," in which sovereignty was established through the presence of traders, police forces, government officials, and residents. As such, Canadian observers suspected U.S. weather stations might be used as bases to challenge Canadian sovereignty in remote areas. The stations also served as a constant reminder of the difficulties of extending to the North the classic instruments of the state—census collection, welfare measures, and police patrols.[23]

Historically, Canada had relied on communications as a solution to those problems of control.[24] But well into the 1950s, the problem of communications in the North only reinforced those anxieties. Severe radio disturbances, often complete blackouts, across the North often crippled high-frequency radio, even as they left foreign transmissions unaffected. In that context, radio waves would become profoundly symbolic, standing in for the internal and external threats to the nation over the next decade. In the early 1950s, Harold Innis, the renowned political economist and communications scholar, asserted that unreliable northern radio threatened "Canadian national life."[25] His assertions were given visual force in a 1953 report from the Yukon territorial government to its federal counterpart (see Figure 1). Exploiting military cartographic conventions in which the breadth of arrows represented the strength of

[22] The Canadian North is technically defined as the region lying north of 60 degrees latitude, but it more generally refers to the northern portions of the central and western provinces.

[23] James Scott, *Seeing Like a State: How Certain Schemes to Improve the Human Condition Have Failed* (New Haven, Conn., 1999).

[24] The extension of the Canadian Pacific Railway to the West Coast—the formative act of nationhood in Canada—was seen in the 1940s as an attempt to prevent U.S. expansionism in the Canadian west.

[25] Harold Innis, *Changing Concepts of Time* (Toronto, 1952), 18–19.

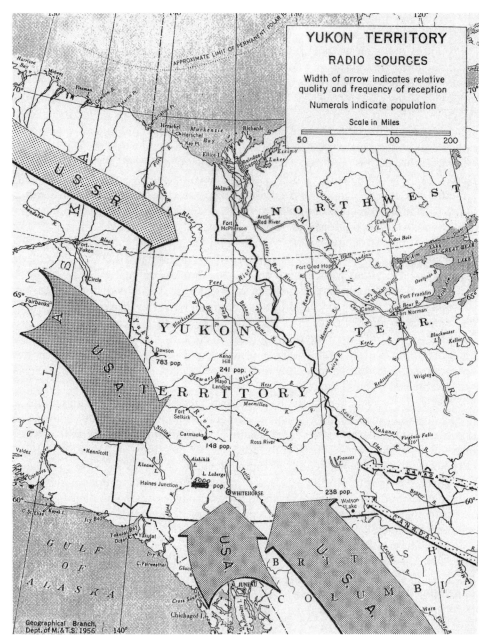

Figure 1. A map showing the reception quality of radio broadcasts entering the Yukon Territory, along with the source of the transmissions (1953). Note the arrows marked "Canada" near the lower-right corner. Representations such as this, common throughout the 1940s and 1950s, pointed to the symbolic nature of electromagnetic waves in postwar Canada. (Source: F. H. Collins, "Radio in the Yukon Territory: A Brief Presented to the Royal Commission on Broadcasting," April 1956, 4–5, part 2, file 5, vol. 127, RG 41, NAC. Courtesy of NAC.)

invading forces, the report featured maps illustrating the source and reception quality of radio broadcasts entering the Yukon, portraying the region as a "battleground of Soviet and American ideologies . . . while Canadian viewpoints are completely absent."[26] The same disruptions responsible for poor public broadcasts were at work in military radio blackouts and fractured government communications. One Canadian Broadcasting Corporation (CBC) source observed that, within this context, the attempt to extend reliable radio to the region was "illustrative of Canada's struggle to survive as a national and cultural entity."[27] Even from their perspective in the later years of World War II, Canadian officials recognized that if the North were to occupy the core of the postwar project of national renewal, it would have to be reclaimed, and reliable communications would play the pivotal role.

THE POLITICS OF ANOMALIES

It was in that context that, in the summer of 1943, a Welsh émigré named Frank Davies began weaving into those larger anxieties and ambitions about the North his own interests in ionospheric research. Educated at Aberystwyth University in Wales, Davies had developed an early and deep interest in meteorology, particularly auroral investigations. After a series of postings on scientific expeditions, and a period as director of the Carnegie Observatory in Huancayo, Peru, he was seconded to the Canadian Navy, where he headed a small group of researchers under Section 6 of its Operational Intelligence Centre (OIC/6). The group's initial task was to aid efforts in code breaking and antisubmarine warfare by investigating the peculiar radio disruptions that plagued North Atlantic shortwave circuits. From two field stations—one outside Ottawa and the other at Churchill, on the shores of Hudson's Bay—OIC/6 began detailing the structure and dynamics of the North Atlantic ionosphere, the ionized regions of the upper atmosphere that make high-frequency (shortwave) radio communications possible. As the situation in the North Atlantic grew more desperate in early 1942, the navy, anxious to secure any advantage it could, expanded the group's activities to include all radio propagation problems in the vicinity of the auroral zone.[28] As the submarine threat in turn subsided in 1943, Davies recognized that his group's research would be much harder to carry out "in the days ahead when money flows less freely from Government pockets."[29] Anxious to secure his group's research, he moved to place it at the core of the emerging postwar vision for the nation.

Over the next three years, Davies's vision would address major anxieties over the place of the North in Canada's postwar plans. In a remarkable foray into political theory titled "The Sector Principle in Polar Claims," for instance, Davies argued that geophysical stations, including ionospheric posts in the vulnerable Arctic archipelago, could stand in for native inhabitants, police, or trading and missionary posts, providing evidence of "effective occupation" while furnishing valuable Arctic data.[30]

---

[26] F. H. Collins, "Radio in the Yukon Territory: A Brief Presented to the Royal Commission on Broadcasting" (April 1956), 4–5, part 2, file 5, vol. 127, RG 41, NAC.

[27] *Current Affairs for the Canadian Forces* 7 (1954), 5.

[28] The northern and southern "auroral zones" are the oval-shaped zones of auroral activity encircling the geomagnetic poles.

[29] Canadian Radio Wave Propagation Committee, "Minutes of First Meeting," 10 April 1944, 1, file HQ 468-3-2, vol. 1, vol. 3413, RG 24, NAC.

[30] Frank T. Davies, "The Sector Principle in Polar Claims," 11 Feb. 1947, file A-25, vol. 46, RG 2/18, NAC.

Throughout late 1943 and early 1944, however, Davies focused more heavily on is-sues of *cognitive* sovereignty. Under wartime arrangements, ionospheric data pro-duced by the Davies group was fed directly to the central ionospheric or "prediction" laboratories in Washington, D.C., and London, where the data was turned into fre-quency prediction charts for shortwave circuits. That flow of data not only subjugated Canadian research to British and American labs but also rearticulated precisely the kinds of historic and potential future colonial relations that Canadian officials feared. Furthermore, as British and American researchers were mostly unconcerned with the peculiar ionospheric conditions that affected Canadian communications, the result-ing prediction charts were largely useless for high-northern latitudes. With this in mind, Davies began echoing a wider sentiment that "information on Canada should come from Canada," rather than from London or Washington, and that the very pecu-liarity of the high-latitude ionosphere, its notorious turbulence and unpredictability, could distinguish his group's research from that of their allies, forming the basis for identity, prestige, and even power after the war.

That emphasis on anomalous northern phenomena as a basis of his group's iden-tity was helped by a curious wartime discovery. Since about 1941, radio researchers, technicians, and telegraphists had observed that long-distance radio circuits from In-dia to England operated efficiently on frequencies much higher than those forecasted. To investigate those effects, the British began operating a new ionospheric station in Delhi in late 1943, the only station covering the enormous stretch between the Middle East and Australia. Within a few months, measurements from Delhi suggested that ionization above Indochina was more intense, and occurred at lower altitudes than at San Juan Puerto Rico, which lay at the same latitude.

The phenomenon—quickly dubbed the Longitude Anomaly—had at least two cru-cial effects. First, it destroyed the spherically symmetrical model of the ionosphere that physicists had worked with until then and which they had used to plan the lo-cation of wartime ionospheric stations around the world. In doing so, the anomaly opened gaps in ionospheric coverage, particularly in the Pacific Northwest, threat-ening operations in the Pacific theater.[31] The best way to seal those gaps, the Allies determined, was to establish a new ionospheric station on the west coast of Canada. Second, researchers tracked more than operational implications across the Pacific. They tracked *explanations*, too. Sir Edward Appleton, Britain's foremost ionospheric physicist, speculated that the elevated ionization and lowered layer heights of the Longitude Anomaly might be explained by their conjugate—the known depressions of critical frequencies and raised layer heights that occurred in high-northern lati-tudes near the auroral zone. One of Appleton's key contributions before the war had been to work out the equations describing the differential refraction of radio waves by the Earth's magnetic field. Noticing that the Delhi station was the most remote from the auroral belt, and knowing that magnetic activity was known to depress the critical frequencies and raise the height of the F2 region, Appleton ventured that the answer to the anomaly might rest with ionospheric investigations in high latitudes.[32] The discovery of the conditions over India therefore linked the two regions of anomalous ionospheric activity, the polar regions and the tropics, suggesting that the anomalous

[31] Central Radio Bureau, "Minutes of Discussion on Ionospheric Problems," 25 March 1944, 6, file N.S. 1078-13-8, vol. 4058, RG 24, NAC.
[32] Ibid.

behavior of radio sets from South Asia to the North Atlantic had their physical origin in the ionosphere above high northern latitudes.[33]

That linking of the northern polar regions and the tropics is interesting here because of what it reveals about how Davies's group positioned their research as the war ended. Historically, notions of both tropicality and nordicity had emerged out of a juxtaposition with the perceived normality of the temperate regions.[34] In his study of European encounters with the New World, Stephen Greenblatt has argued that Europeans engaged the Americas through an essentially improvised but immensely powerful mimetic machinery, centered on the idea of wonder or marvels.[35] The scientific engagement with both the North and the tropics was similarly organized around an idea of nature at the extremes, particularly embodied in notions of the anomalous. In commenting on the tropics, for instance, Alexander von Humboldt explained, "Nature in these climes appears more active, more fruitful, we may even say more prodigal of life."[36] Other commentaries would cut in the opposite direction, characterizing the tropics through narratives of disease that gave tropical nature a "pathological potency that marked them out from milder, more temperate lands."[37] What tied the two together was a notion of aberration: tropical nature was more active, more fruitful, but also more disease ridden and deadlier than in temperate climes. And it was precisely that intensity that gave the tropics a privileged epistemic position. Humboldt wrote, "Nowhere does she [nature] more deeply impress us with a sense of her greatness, nowhere does she speak to us more forcibly."[38] That vision of nature at the extremes similarly structured northern and polar research from the seventeenth century, informing the idea of remote northern regions as "laboratories," used by scientists to turn extreme geographic locations into more central cultural positions.[39] Within that framework, the proper opposite of the polar regions was not the (also anomalous) tropics but rather the comparatively well-behaved temperate latitudes that defined them both.

Geophysical anomalies were precisely what had fascinated Davies from the beginning of his career. With the attention that the Longitude Anomaly drew to the turbulent northern ionosphere, Davies now began to see those interests as a solution to problems of marginality and identity. In his mind, the polar ionosphere became a tool for carving out a distinct geophysical and technological identity for Canada, one that would elevate the work of his team by defining it against British and American research on the "temperate" ionosphere. Starting in late 1943, he began pitching the ionosphere above Canada as quintessentially northern, recasting the problematic

---

[33] Ibid.

[34] By analogy to tropicality, "nordicity" is defined as the state or condition of being northern. Louis-Edmond Hamelin, whom I deal with later in the paper, elaborated the term as a quantitative measure of the degree of northernness of high-latitude regions, based on both human and natural factors.

[35] Stephen Greenblatt, *Marvelous Possessions: The Wonder of the New World* (Chicago, 1991), 22–23.

[36] Alexander von Humboldt, *Cosmos: A Sketch of a Physical Description of the Universe* (London, 1868), 1:3.

[37] David Arnold, *The Problem of Nature: Environment, Nature, and European Expansion* (Cambridge, UK, 1996), 10.

[38] Humboldt, *Cosmos* (cit. n. 36), 13–14.

[39] Patricia Fara, "Northern Possession: Laying Claim to the Aurora Borealis," *History Workshop Journal* 42 (1996): 38. For Iceland as laboratory, see Karen Oslund, "Imagining Iceland: Narratives of Nature and History in the North Atlantic," *British Journal of the History of Science* 35 (2002): 313–34, 324.

relationship between nature and technology in the North Atlantic as a problem that achieved its most distilled and potent form in Canada. Drawing on the language of both arctic and tropical epistemologies, Davies would present the Canadian North as a "natural geophysical laboratory" in which to investigate the high-latitude ionosphere and to trace out the precise links between geophysical phenomena and unreliable radio.[40]

For Davies, all these projects hinged on the creation of a new radio propagation laboratory to house his postwar team. In arguing for it, Davies stressed that Canada needed its own specialized predictions suited to high latitudes and its own laboratory to produce them. This "must be done and be done by a Canadian group," Davies argued, as other groups had only a secondary interest in Canadian domestic communications.[41] The new laboratory would break the subjugation of Canadian ionospheric research by stopping the flow of data at Ottawa, where a proper investigation of high-latitude phenomena would emerge. Those phenomena were particularly crucial for Davies, as they represented the future of the field. "Advances in our knowledge of the ionosphere," he explained to superiors, "will depend largely on analysis of *obscure and exceptional* ionospheric conditions and on improvements in our measurement techniques."[42] As it was transformed into the Radio Physics Laboratory (RPL), OIC/6 became a central part of the Canadian attempt to produce a total science of the northern environment. Aerial mapping expeditions; plant and insect taxonomies; measurements of ice temperature and permafrost thickness; experiments studying the effect of severe cold on materials and metabolism, diuresis, and manual dexterity, tactility, and military tactics were all designed to rigorously document the northern environment and to help indoctrinate and acclimatize humans for military operations in northern conditions.[43] By focusing on the northern ionosphere, Davies's team would implement the vision of External Affairs writ small: it would use the idea of the North as a fulcrum to elevate its status, to underwrite the projects of northern development, to assert autonomy and influence, and to contribute to international cooperation and efforts in the field.

SPORADIC READINGS

To understand the work of the new laboratory, we first have to put aside a common understanding of the kinds of places laboratories represent. Over the past two decades, philosophers, historians, and sociologists of science have given us a number of ways to characterize the rich relations between the laboratory and the myriad graphic products of scientific research.[44] Drawing on recent literature, for example, we might

[40] Frank T. Davies, "DRTE," nd, 5, Frank T. Davies Personal Papers, held at the Communication Research Centre, Ottawa, Canada.

[41] Jack Meek, "A Proposed Frequency Prediction Service for Canada: Report," 18 Jan. 1946, 1, file 468-3-3, vol. 3414, RG 24, NAC.

[42] Canadian Radio Wave Propagation Committee, "Minutes," 28 Feb. 1947, 7, file HQ 468-3-1, vol. 3413, RG 24, NAC (emphasis added).

[43] A. M. Pennie, Foreword to *Defence Research Northern Laboratory, 1947–1965* (Ottawa, Canada, 1966), 1.

[44] The literature on scientific representations is itself extensive. Some representative works include: Bruno Latour, *Science in Action: How to Follow Scientists and Engineers through Society* (Cambridge, Mass., 1987); Caroline Jones and Peter Galison, eds., *Picturing Science, Producing Art* (New York, 1998); Michael Lynch and Steve Woolgar, eds., *Representations in Scientific Practice* (Cambridge, Mass., 1990); Peter J. Taylor and Ann S. Blum, "Pictorial Representation in Biology," *Biology*

place visual representations on par with concepts, theories, social relations, and material processes as the products and facilitators of a more fundamental activity: isolation, interrogation, and intervention into the processes of nature.[45] Or else, images might ground the material, practical and epistemic cultures of the laboratory, serving as critical resources in arguments over the reliability of experimental practice, the validity of instrumental arrangements, and the soundness of theoretical conclusions.[46] Or further still, the laboratory itself might be understood as a site rooted in the transformation into paper of phenomena made manifest within its walls.[47] But whether they see the laboratory as interrogation chamber or inscription factory, or something in between, scholars have argued again and again that its relation to images and to the power they seem to hold derives principally from one thing: the claim of the laboratory to make some physical aspect of the natural world—whether through modeling, manipulation, or mimesis—enter therein.

Historically, radio propagation laboratories had grown out of radio research stations built to investigate issues in radio engineering—antenna and transmitter design, frequency measurement, metrology, electronic standards, and calibration. But throughout the 1930s, they came to focus heavily on the ionosphere as a way of predicting the radio frequencies that would be useful for long-distance communications. In the late 1940s, then, the laboratories could no longer make claims for physical control over their objects of study. What control they did have, they exerted through the production, interpretation, and manipulation of graphic representations of the ionosphere known as ionograms. Laboratories produced the graphs using high-frequency radio echoes displayed on a cathode ray tube and then measured and plotted by equipment operators. Researchers then pored over these records at length, analyzing, measuring, recording, and incorporating their features into larger experimental results and theoretical conclusions.[48] The data were then correlated with other geophysical information—such as sunspot observations, magnetometer readings—to give predictions of useful radio frequencies. It was part of a science carried out in the field but with the laboratory at its core. The graphic manipulation of a nature outside

---

and Philosophy 6 (1991): 125–34; Michael Lynch, "Discipline and the Material Form of Images: An Analysis of Scientific Visibility," *Social Studies of Science* 15 (1985): 37–66; Martin Rudwick, "The Emergence of a Visual Language for Geological Science, 1760–1840," *History of Science* 14 (1976): 149–95. The field of laboratory studies is similarly prolific. For a sampling, see: Bruno Latour and Steve Woolgar, *Laboratory Life: The Social Construction of Scientific Facts* (Princeton, N.J., 1979); Michael Lynch, *Art and Artifact in Laboratory Science: A Study of Shop Work and Shop Talk in a Research Laboratory* (London, 1985); Steven Shapin and Simon Schaffer, *Leviathan and the Air-Pump: Hobbes, Boyle, and the Experimental Life* (Princeton, N.J., 1985); Shapin, "The House of Experiment in Seventeenth-Century England," *Isis* 79 (1988): 373–404; Sharon Traweek, *Beamtimes and Lifetimes: The World of High Energy Physics* (Cambridge, Mass., 1988).

[45] Ian Hacking, "The Self-Vindication of the Laboratory Sciences," in *Science as Practice and Culture*, ed. Andrew Pickering (Chicago, 1992), 29–64.

[46] For the "image tradition" in modern microphysics, see Peter Galison, *Image and Logic: A Material Culture of Microphysics* (Chicago, 1997).

[47] Bruno Latour, "Drawing Things Together," in Lynch and Woolgar, *Representations in Scientific Practice* (cit. n. 44), 19–68.

[48] Bruno Latour has pointed to the broader importance of measurement in two-dimensional images. In his essay "Drawing Things Together," he explains how the result of the merging of image and geometry is "that we can work on paper with rulers and numbers, and combinations of numbers and tables can be used which are still easier to handle than words or silhouettes. You cannot measure the sun, but you can measure a photograph of the sun with a ruler. . . . This is what I call, for want of a better term, the second-degree advantage of inscriptions, or the surplus-value that is gained through their capitalization." Ibid., 46–47.

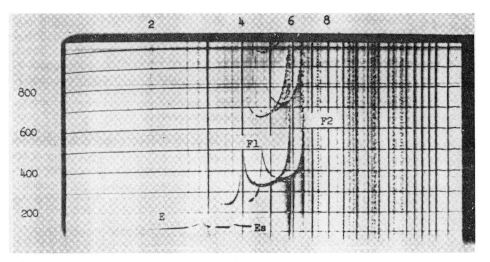

Figure 2. *A "typical" mid-latitude panoramic ionogram. The images were produced by altering the ionosonde's cathode ray tube to display h'f curves directly and then photographing the resulting dynamic image using 16mm cameras. The labels on the record were added after analysis. Note the distinct leftmost traces, known as "ordinary curves." The identification and measurement of these curves formed the basis for the reading regimes of both ionospheric physics and communications engineering. (Courtesy of the Communications Research Centre, Ottawa, Canada.)*

the laboratory and its transformation into other more useful graphic forms were what defined the ionospheric laboratory and its work.

For its part, the ionogram's power lay in its versatility, defined by the reading regimes that had grown up around the image before and during the war. Through a set of techniques developed primarily in the 1920s, an ionospheric physicist learned to read the graphs "backward," so to speak, through magnetoionic theory, thermodynamics, and cosmic ray physics to explain the structures and conditions of the ionosphere at the time of recording. A radio engineer, by contrast, struggling to secure long-distance circuits, projected the images "forward" through the laws of optics and through propagation theorems that converted the lines of the graph into usable radio frequencies and skip distances for communications circuits.[49] But both methods of reading ionograms relied on a third, more basic set of interpretive practices developed by British and American researchers, who had dominated the field since the 1930s and who had based their practices on the relatively simple ionograms from temperate latitudes. Those practices concentrated on identifying a single distinct curve on the diagram known as the "ordinary" trace. (See again Figure 2.) Once the trace had been identified, workers measured two characteristic features: its minimum virtual height (read off the ordinate axis), and its critical frequency (read as the frequency to which the trace became asymptotic). The ionogram's pliability hinged on these measurements, which then entered atmospheric models and radio propagation formulas. The thrust of ionographic analysis, the very act of reading ionograms, and the bulk of the

[49] See, e.g., Newbern Smith, "The Relation of Radio Sky-Wave Transmission to Ionosphere Measurements," *Proceedings of the IRE* 27 (1939): 332–47.

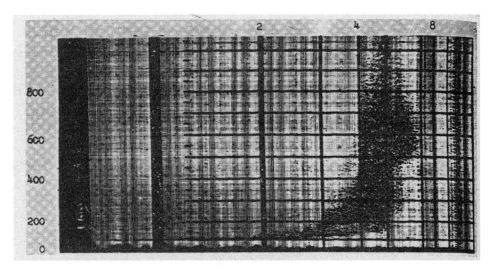

Figure 3. A case of "spread echo." Such phenomena, quite common on high-latitude iono-
grams, rendered the records unreadable according to the standard rules of interpretation.
Baker Lake, 14:00, 4 Aug. 1955. (Courtesy of the Communications Research Centre, Ottawa,
Canada.)

theoretical and practical work built upon it turned on this identification of the ordinary trace and the precise measurement of its characteristic features.

Working with improvised and deteriorating instruments and a constantly changing
workforce, Davies's group had originally struggled to produce ionograms reliably.[50]
To overcome those difficulties, they had adopted a new instrument in 1947—the panoramic ionosonde—which produced ionograms automatically without the difficulties and uncertainties of manual graphing. The new graphs captured the turbulent and
spectacular ionospheric phenomena much more reliably, but they consequently made
the practice of reading ionograms much more difficult. The emphasis on the ordinary
trace meant that, even as Canadian ionograms captured the anomalies and "exceptional conditions" at the center of RPL's research, those same phenomena made the
records unreadable. What, for instance, was an interpretive system based on ordinary
traces and discrete frequency and height measurements to make of something such
as Figure 3, an ionogram produced at Baker Lake, in the Canadian Northwest Territories?

The phenomenon shown there was part of a group of effects, defined by their appearance on ionograms and known as "sporadic" phenomena. Sporadic phenomena
were characterized by the way they obscured the neat graphic elements at the core
of scaling practices, creating what one U.S. researcher admitted was "probably the
most difficult problem in the world, from an ionosphericist's point of view."[51] As
it was used in the late 1940s, the term "sporadic" was ambiguous, as Davies's colleague Jack Meek pointed out, since it referred to two types of effects: ionization that

[50] See Edward Jones-Imhotep, "Nature, Technology, and Nation," *Journal of Canadian Studies* 38
(2004): 5–36.

[51] "Interpretation of Records." Discussion led by Mr. H. W. Wells, Canadian Radio Wave Propagation Conference, Ottawa, Canada, 22–27 Oct. 1945, 59, vol. 3412, RG 24, NAC. The document is
undated, but the context establishes that the discussion took place in the late summer of 1945.

changed rapidly in shape and density, and phenomena that occurred infrequently at middle latitudes. Since those effects were very common in high latitudes, Meek used an article in the *Journal of Geophysics,* to strip the term of everything but its dynamical aspects. Using records taken at Churchill and Baker Lake, Meek then constructed a taxonomy of representative sporadic effects—spread echoes, polar spurs, forked traces, oblique reflections—that he believed characterized ionograms in *"northern* regions."[52] He quickly recognized that the taxonomy left the problem of interpretation and scaling untouched. Gaping omissions in the daily tabulation charts of stations throughout the polar regions testified to their awkward status.[53] If deploying the standard techniques of ionographic analysis in the early 1950s meant producing, to the extent that ionospheric storms and technical difficulties permitted, an authoritative reading of the ionogram, it also meant reinforcing the hold of a taxonomy of ionospheric effects and ionographic appearances that pushed entire classes of records and entire groups of researchers to the margins.

Meek would spend the next five years developing techniques to make high-latitude ionograms readable. His efforts initially focused on creating practices to reconstruct the ordinary trace in the presence of sporadic phenomena. As the techniques became more sophisticated and demanding, however, Meek found them more and more difficult to extend reliably to the field station operators. Mass demobilization in the late 1940s had made experienced military telegraphists, the stock operators of the field stations, a scarce commodity.[54] The itinerant nature of the northern workforce added to these difficulties. Davies remarked in 1947 that seventy operators had passed through three stations in less than two years.[55] Training had to be carried out on site, "sometimes by men who have not had sufficient training themselves. Inaccuracy and serious errors in data have occurred because of this."[56] Meek had been Davies's second-in-command at OIC/6 and the head of the calibration party that toured the ionospheric stations in the late 1940s. In 1947, as the Radio Physics Laboratory was taking shape, he had led a small research team to Manitoba, where they used a converted railway car to make ionospheric measurements at positions across the auroral zone. As the author of wartime and postwar instruction manuals for the station operators, he returned again and again to the same interpretive quandary that now faced him in the early 1950s: adopt a standard reading regime that made interpretation simple and straightforward and risk marginalizing many Canadian records (as the standard techniques had done), or devise a system that accommodated high-latitude ionograms but bred unreliability by requiring technicians to perform interpretive acts beyond their abilities.

[52] Jack Meek, "Sporadic Ionization," *Journal of Geophysical Research* 54 (1949): 343 (emphasis added).

[53] On September 30, 1955, for example, an entire day's soundings (ninety-six ionograms) at Barrow, Alaska, had produced only one reliable critical frequency for the F2 layer, and none at all for the E and F1 layers. See R. W. Knecht, "Statistical Results and Their Shortcomings Concerning the Ionosphere within the Auroral Zone," *Journal of Atmospheric and Terrestrial Physics*, special suppl. (1957): 110.

[54] G. A. Worth, Memorandum to Administrative Deputy, Defence Research Board, 23 July 1947, file 7401-1710, vol. 83–84/167, RG 24, NAC.

[55] Frank T. Davies, Memorandum to Vice-Director General, Defence Research Board, "Assignment of Personnel of Armed Services, and Department of Transport to the Radio Propagation Laboratory," 16 July 1947, 1, file 7401-1710, vol. 83–84/167, RG 24, NAC.

[56] Ibid.

In 1954, Meek hit on what he believed was an optimal solution to the delicate dilemma. He drew up a graph displaying frequency as a function of time. Parsing the abscissa into fifteen-minute intervals (corresponding to the schedule of ionospheric soundings), he began recording information about critical frequencies over a twenty-four-hour period. (See Figure 4.) Distinct frequencies entered the graph as specific points, but (and here lay the force of the device) rather than requiring that all data be entered as discrete values or not at all, as previous techniques had done, Meek allowed the graph to record *ranges* of frequencies corresponding to the actual smeared echo traces on high-latitude ionograms. The *f*-plot, as Meek designated his device, helped change the way RPL approached its study of the high-latitude ionosphere. Whereas the act of interpretation had previously focused on identifying or reconstructing a well-defined trace, it now depended on recording the actual frequency information on the graphs; whereas it had previously been restricted to the single ionogram, it now concentrated on a time series of records illustrating precisely the dynamic properties that Meek believed set the high-latitude ionosphere apart. Researchers soon learned to "see" the formation and disappearance of the sporadic layers and the temporary blanketing of otherwise regular patterns in critical frequency data.[57] As Kenneth Davies, Meek's colleague, explained:

> The emphasis in interpretation [with the *f*-plot] is in estimating the value of a parameter in relation to the sequence of ionograms before and after it, rather than in trying to see something equivalent to an undisturbed echo through the general confusion of a disturbance.[58]

The scheme had at least two immediately crucial implications. The first addressed the problem of skilled technicians that had haunted Meek for years. When confronted with a particularly difficult ionogram, the staff of the ionospheric stations could now ignore the search for the ordinary trace and instead record the information from the record in all its frustrating complexity, as one source put it "with a minimum of prejudgement."[59] The second effect involved the way the graph allowed RPL's researchers to extract and compile quantitative data about high-latitude phenomena such as spread echoes and polar spurs—features that characterized Canadian ionograms but that had previously disqualified the records that contained them.

In the late 1950s, Davies's group began using those quantitative results to correlate ionospheric data with the results of broader investigations of Canadian northernness—unique weather patterns above the Canadian arctic; geologic anomalies that explained the location of the magnetic pole in Canadian territory; models of the solar wind that explained the occurrence of aurora; investigations into tundra cover, permafrost, glacial formations, and soil conditions that affected radio propagation over long distances in the North. As researchers adopted the chart more and more widely, they were able to generate a statistical high-latitude ionosphere (RPL researchers called it the "arctic ionosphere") based on ionization information captured in the charts. Although the precise limits would be disputed, the southern extent of

---

[57] W. J. G. Beynon and G. M. Brown, eds., *IGY Instruction Manual: The Ionosphere,* part 1, *Ionospheric Vertical Soundings*, comp. J. W. Wright, R. W. Knecht, and K. Davies (London, 1956), 105.
[58] Ibid.
[59] Ibid.

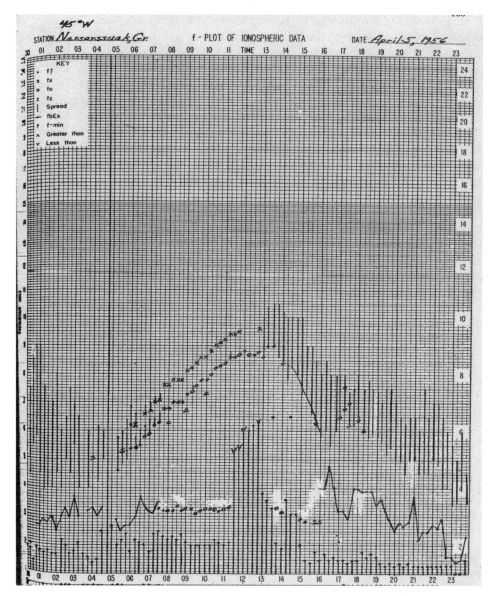

Figure 4. *The f-plot. The legend in the top left demonstrated that scaling no longer required discrete frequencies or even identification of traces. As this example using data from Greenland suggests, the graph would be adopted globally by the late 1950s, particularly for use on high-latitude records. Note the o, x, and z notations and the use of frequency ranges rather than specific values. (Courtesy of the Communications Research Centre, Ottawa, Canada.)*

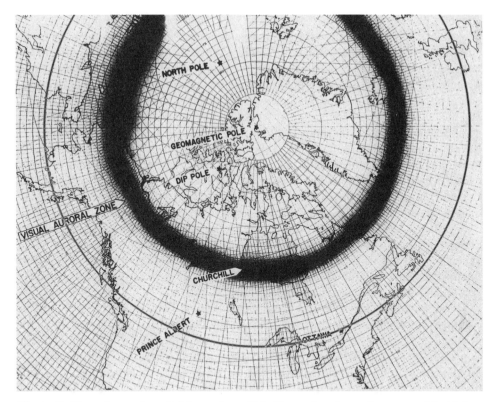

Figure 5. *The polar, or "arctic," ionosphere. This illustration shows the extent of the high-latitude ionosphere (shown by the outer ring running through Ottawa). According to this new construction, all data in Canada was arctic data, and the nation boasted the most "polar" ionosphere in the Northern Hemisphere. (Courtesy of the Communications Research Centre, Ottawa, Canada.)*

the arctic ionosphere zone was generally taken to lie above roughly 57° geomagnetic latitude and was only defined confidently for the northern hemisphere.[60] (See Figure 5.) The boundary was significant since it ran through Ottawa, the site of the most southerly ionospheric station in Canada, making all data collected in Canada "arctic" data and underwriting two identities simultaneously—Canada's characterization as the archetypal northern nation, and RPL's self-image as experts in high-latitude ionospheric investigations.

That quantitative argument for Canadian northernness had parallels outside the laboratory in the late 1950s. One of the most remarkable involved the work of Louis-Edmond Hamelin, professor of geography at Quebec's Laval University. Initially trained in classics and economics, Hamelin spent time in the late 1940s at the Scott Polar Research Institute in Cambridge and in France, where he obtained a doctorate in geomorphology from the Sorbonne in 1951. Throughout the late 1950s, he focused his interests in the northern polar regions toward founding the field of nordology—the study of the cold regions of northern latitudes. At the core of his discipline

[60] This was because the distribution of land mass and the lessened interest in the southern polar regions provided insufficient data to make the determination at the time.

Hamelin placed the concept of "nordicity": the "state or quality of northernness or being north."[61] Unsatisfied with a qualitative definition, he moved to quantify northernness through a "nordicity index," a quantitative measure based on a number of criteria that would help distinguish between a nation's "true" nordicity and false claims to that title. Meteorology, geography, topography, population studies—all helped to define his ambitious index, as did the undifferentiated criterion of "communications." In a move that echoed the premises of the Radio Physics Laboratory's work, Hamelin asserted that (subject to minimum latitude and isothermal requirements) the poorer a region's communications, the more properly northern it was.[62] Combining these features, Hamelin determined that icons of the North such as Iceland, Scandinavia, and even the north of the Soviet Union enjoyed much less "severe nordicity than that of Canada." His calculations allowed Canada's northern area to be estimated at approximately 70 percent of the country. "Given this extent," he concluded, "the territory is no less than a national feature."[63] For Hamelin, the oppositional relationship between communications and northernness in the 1950s was self-evident and presumptive. When he revisited his scheme in the late 1970s, when satellite communications were finally extending reliable communications throughout Canada, he had dropped the communications criteria from his index.

Throughout the late 1950s and early 1960s, the link between northern nature and communications difficulties became elliptical in defining the nation. An article on ionospheric research in the news magazine *Maclean's* included a schematic illustration of the visual auroral zone over the Northern Hemisphere and proceeded to unproblematically and self-evidently connect that icon of Canadian geophysics to radio disruptions in Canadian territory. The links in certain cases became so strong that some researchers would invert the causal chain. When the CBC formed its Northern Service in 1958, its officials conceded that their "north" had no precise boundaries. In a remarkable move that spoke to the identification of shortwave radio with the northern territory, officials turned to "the *inherent nature* of radio transmissions" to define the northern region. The North, they explained, was defined as the region served by shortwave radio, extending "north to the Pole and south *to an imaginary line that would include those listeners who do not receive a consistent and adequate broadcast signal from CBC network stations or private stations located 'outside.'*" By that definition, the territory covered almost two million square miles.[64] A history of the Defence Research Board, which had overseen RPL's work since the late 1940s, continued eliding the boundary between northern nature, problematic communications, and national identity, explaining, "Radio communications in Canada are bedeviled by difficulties which are unique." The report then went on to list not difficulties in radio communications at all but rather geophysical factors whose links to radio disturbances Meek and his colleagues had begun tracing in their research on the North Atlantic:

[61] Louis-Edmond Hamelin, *Nordicité Canadienne* (Montreal, 1975), 35. Hamelin's final system was not presented until the 1970s; there is evidence that the underlying principles of nordicity along with its various criteria were worked out in the late 1950s and early 1960s.

[62] Louis-Edmond Hamelin, "Essai de Régionalisation du Nord Canadien," *North* 11 (1964): 16–19.

[63] Louis-Edmond Hamelin, *About Canada: The Canadian North and Its Conceptual Referents* (Ottawa, Canada, 1988), 22.

[64] CBC, "Northern Service," Minutes of 4th Annual Meeting 16, 1, 1964, NAC (emphasis added).

The North Magnetic Pole lies wholly within Canadian territory; the North Geomagnetic Pole on the north-western edge of Greenland is closer to Canada than to any nation, and the phenomenon of the Aurora Borealis, or Northern Lights, has an adverse affect on radio communications throughout most of the Dominion.[65]

Nor were the resonances confined to official memos and government reports. RPL's investigations took their place alongside the new polar-projection maps, northern news services, film documentaries, and Group of Seven paintings exploring the psychological and spiritual properties of the North.[66] The self-consciously Canadian literature that emerged after the Second World War was suffused with the theme of the North, its malevolence and its capacity to purify through harsh struggle.[67] Documentaries, television specials, radio programs, and magazine articles throughout the 1950s pointed to Canada's supposedly unique northernness while linking auroral displays and magnetic storms to radio disruptions and shortwave blackouts. Like the metrological research of late Victorian Britain, with its links to telegraphy and empire, the ionospheric enterprise carried enormous moral weight.[68] And like that earlier research, its work was both material and symbolic, drawing from and feeding into the image of the nation that had launched its research in the first place.

CONCLUSION

OIC/6 had originally turned to the ionogram as a means of distinguishing the nation, elevating it at a crucial point in its history, and demarginalizing the group's own research as a consequence. Seizing on the record's ability to mediate between peculiar technological effects and unique geophysical conditions, the researchers turned ionograms into even more deeply political objects, seeing in them a way to illustrate, define, and defend claims of Canadian northernness when those claims seemed most urgent. Their intense concerns over the records—their production, circulation, and interpretation; their status as scientific facts; and their transformation into trustworthy evidence—stemmed in part from their role in defining and securing the nation against threats to territorial integrity, epistemic sovereignty, and national identity. The deep irony of their project lay in how the very qualities that characterized the polar ionosphere worked to impede that expression, vitiating efforts to lend the rigor and authority of science to investigations of Canadian distinctiveness, with all this implied for the postwar nation.

Developed by British and U.S. researchers on middle-latitude records, the standard techniques for reading ionograms made no provision for the effects that defined the high-latitude ionosphere and therefore characterized Canadian distinctiveness. We have seen again and again how standards help make the world fit for science, allowing results to circulate beyond the boundaries of the laboratories and institutions that

[65] Donald J. Goodspeed, *A History of the Defence Research Board of Canada* (Ottawa, Canada, 1958), 195.

[66] Douglas Cole, "Artists, Patrons and Public," *J. Can. Stud.* 13 (1978): 69–78.

[67] Margaret Atwood, *Strange Things: The Malevolent North in Canadian Literature* (Oxford, 1995).

[68] Simon Schaffer, "Late Victorian Metrology and Its Instrumentation: A Manufactory of Ohms," in *The Science Studies Reader*, ed. Mario Biagioli (New York, 1998), 457–78.

first produce them.[69] But for the Canadian group, as for others no doubt, those standards needed to be recast to express difference rather than stifling it. By capturing in photographic form the supposed distinctiveness of the Canadian ionosphere, the ionogram only raised questions about how these peculiarities should be treated and understood within an interpretive framework that treated them as noise. The work of the Radio Physics Laboratory in the late 1940s and early 1950s was largely an attempt to rework the reading of the most iconic image of atmospheric research in the interests of both the laboratory and the postwar nation. By rereading the ionogram—and the multiple-frequency penetrations, polar spurs, night E-layers, and Z-modes they identified and measured there—researchers came to possess a reliable point of reference around which arguments about the validity of ionospheric models, the dependability of communications, and the extent of Canadian northernness could be articulated and debated.

In that move against homogeneity, we can glimpse broader histories. Ionograms were profoundly political entities. Internationally, their flow in the 1950s traced out the contours of the cold war political order; their global exchange was the subject of diplomatic negotiations and overtures. Their manipulation drove simultaneously the military projects of nuclear missile detection, the civilian-directed attempts at radio indoctrination, and the political processes of rapprochement. The cold war was waged through and around these images, the knowledge they produced and the technologies they supported. Because of this, we can follow other studies of science and technology to read in the ionogram, in the practices meant to secure its production and interpretation, the homogenizing and dichotomizing effects of that conflict. But we can also look beyond and through those totalizing ambitions to the innumerable acts of resistance and distinction that underlie all identities.

[69] See, e.g., Bruno Latour, "Give Me a Laboratory and I Will Raise the World," in Biagioli, *Science Studies Reader* (cit. n. 68), 258–75; and Schaffer, "Late Victorian Metrology" (cit. n. 68).

The Coproduction of Neutral Science and Neutral State in Cold War Europe:

Switzerland and International Scientific Cooperation, 1951–69

*By Bruno J. Strasser**

ABSTRACT

Neither science nor state has ever been transcendentally "neutral," but they have sometimes been made neutral, together, as this paper shows in the context of cold war Europe. The paper explores how the Swiss government tried to "depoliticize" and "demilitarize" new international research institutions in the fields of high-energy physics (CERN), space research (ESRO and ELDO), and molecular biology (EMBL) in order to make science neutral. Conversely, this paper investigates how participation in "neutralized" scientific institutions supported Switzerland's neutrality policy and strengthened this essential element of its national identity. It thus addresses symmetrically the coproduction of neutral science and neutral state.

INTRODUCTION

During the cold war, setting foot on the Moon or producing an atomic mushroom cloud were as much about nation building as about scientific and technological development. America's national identity was reinforced by its successful landing on the Moon; the French and the Indian national identities were transformed when they detonated their first atomic bombs. These scientific accomplishments expressed American global power, French *grandeur*, and Indian independence.[1] National identity has linked science and state in subtle ways since the scientific revolution, but during the

* Program in the History of Science and Medicine, Yale University, Section of the History of Medicine, PO Box 208015, New Haven, CT 06520-8015; bruno.strasser@yale.edu.

I would like to thank Frederic Joye-Cagnard for his collaboration on an earlier version of this project, John Krige, Dan Kevles, Grace Shen, Ann Johnson, Carol Harrison, three anonymous referees, as well as the participants of the workshop on science and national identity at the University of South Carolina, September 20–22, 2007, for useful comments. I also thank the archivists at the Swiss Federal Archives in Bern, the CERN Archive in Geneva, and the EMBO Archives in Heidelberg for their collaboration. Partial funding for this research is from the Swiss National Science Foundation, project no. 105311–109973.

[1] On the United States' spaceflight program, see Roger D. Launius and Howard E. McCurdy, *Space-flight and the Myth of Presidential Leadership* (Urbana, Ill., 1997). On France's and India's atomic bombs, see Dominique Mongin, *La bombe atomique française, 1945–1958* (Paris, 1997); and Itty Abraham, *The Making of the Indian Atomic Bomb: Science, Secrecy, and the Postcolonial State: Post-colonial Encounters* (New York, 1998). In the case of France, this perspective is taken up most directly and successfully in Gabrielle Hecht, *The Radiance of France* (Cambridge, Mass., 1997). On technology and the United States, see David E. Nye, *America as Second Creation: Technology and Narratives*

© 2009 by The History of Science Society. All rights reserved. 0369-7827/09/2009-0009$10.00

cold war, this relationship grew particularly strong as nation-states became the main patrons of scientific research. Yet more important, to a larger extent than at any time before, the military, economic, and cultural destinies of nations were perceived to rest on advances in science and technology. It should thus come as no surprise that science, along with other social productions, played an essential role in the construction of national identities in this period.

Until recently, scholars have rarely focused on this issue, turning their attention instead on the reverse relationship, namely how national contexts and identities have shaped scientific endeavors. Much of this literature, even in the science and technology studies tradition, has been far from "symmetrical," in the sense that society and culture have been taken as stable entities that can explain the construction of science and technology, but rarely has the reverse been considered.[2] The distinctively national dimensions of "scientific styles" and scientific theories, for example, have been treated as if they emerged from an independent political culture, as in Paul Forman's classic study on quantum mechanics and Weimar culture, or from established national institutions, as in Jonathan Harwood's work on genetics in the United States and Germany.[3]

Here, I would like to adopt both perspectives simultaneously, by focusing on the coproduction of science *and* national identity.[4] Taking the example of Switzerland, I will examine the diplomatic efforts deployed at making international science neutral and, at the same time, look at how science gave substance to Switzerland's neutrality policy.[5] This study focuses on the role of the Swiss Department of Foreign Affairs and Swiss scientific statesmen in the creation of three major institutions devoted to international scientific cooperation in the fields of nuclear physics, space research, and molecular biology, respectively.[6] The paper explores how these different oppor-

---

of New Beginnings (Cambridge, Mass., 2003). More recent examples include cloning research and nation building in South Korea.

[2] Bruno Latour made this point long ago. Bruno Latour, *Science in Action: How to Follow Scientists and Engineers through Society* (Cambridge, Mass., 1987). A rare exception is Hecht, *Radiance of France* (cit. n. 1).

[3] Paul Forman, "Weimar Culture, Causality, and Quantum Theory: Adaptation by German Physicists and Mathematicians to a Hostile Environment," *Historical Studies in the Physical Sciences* 3 (1971): 1–115; Jonathan Harwood, "National Styles in Science: Genetics in Germany and the United States between the World Wars," *Isis* 78 (1987): 390–414.

[4] On coproduction, see Sheila Jasanoff, ed., *States of Knowledge: The Co-Production of Science and Social Order* (New York, 2004), chap. 1.

[5] For a preliminary account of the former argument, see Bruno J. Strasser and Frédéric Joye, "Une science 'neutre' dans la guerre froide? La Suisse et la coopération scientifique européenne (1951–1969)," *Revue Suisse d'Histoire* 55 (2005): 95–112; for the latter, see Strasser and Joye, "L'atome, l'espace et les molécules: La coopération scientifique internationale comme nouvel outil de la diplomatie helvétique (1951–1969)," *Relations Internationales* 121 (2005): 59–72. For a study on the role of Swedish neutrality in Swedish-American relations in the field of guided missiles, see Mikael Nilsson, *Tools of Hegemony: Military Technology and Swedish-American Security Relations, 1945–1962* (Stockholm, 2007). More generally on scientific cooperation as a tool and foreign policy, especially in the United States, see John Krige and Kai-Henrik Barth, "Introduction: Science, Technology, and International Affairs," *Osiris* 21 (2006): 1–21; Clark A Miller, "'An Effective Instrument of Peace': Scientific Cooperation as an Instrument of U.S. Foreign Policy, 1938–1950," *Osiris* 21 (2006): 133–60; Ronald E. Doel and Kristine C. Harper, "Prometheus Unleashed: Science as a Diplomatic Weapon in the Lyndon B. Johnson Administration," *Osiris* 21 (2006): 66–85.

[6] For an overview of European scientific cooperation, John Krige and Luca Guzzetti, eds., *History of European Scientific and Technological Cooperation* (Brussels, 1997); and John Krige, "The Politics of European Scientific Collaboration," in *Science in the Twentieth Century*, ed. John Krige and Dominique Pestre (New York, 1997): 897–918.

tunities for international scientific cooperation in Europe were perceived by the state not just as scientific, economic, or military opportunities but also as tools for reinforcing the central pillar of Switzerland's national identity, namely its proclaimed neutrality. The alleged neutrality, universality, and objectivity of science[7] were a perfect illustration of what political neutrality stood for—if only science could be made to conform to these ideals in the midst of the cold war. This was obviously not a simple task because, as recent historiography has shown, science was, during this period, more often highly politicized and militarized than neutral. As John Krige has argued, for example, the United States used science as a powerful political and cultural weapon in postwar Europe.[8] At the same time, the military establishments in Europe and in the United States embraced science for its relevance to the key technologies of the cold war, even in fields seemingly far removed from immediate practical applications.[9] Thus, if science were to play a role in Switzerland's national identity, it had to be *made* neutral.

Scholars no longer consider national identities as natural entities—as was the case when they used the Enlightenment's concepts of the "spirit of nations" and the "national genius"—nor as abstract political ideologies or essential human attributes.[10] National identities are now understood as artifacts, the products of cultural and political processes, and as tools to attain certain political and economic goals, even long after nations have become stabilized entities. To be sustained, national identities need to be constantly reproduced and reinterpreted to adapt to changing environments, while at the same time giving a sense of permanence, reflecting the nation's past.[11] National identities can thus been seen to lie at the intersection of collective memories of a shared past and of wishful projections of a community's future. Understandably, in modern states, governments have played an essential role in crafting and sustaining national identities to support their current political goals. Indeed, the resulting national identities, as "imagined communities," to take Benedict Anderson's classic formulation,[12] have been essential to the very existence of the nation and to the state's power.

---

[7] On neutrality, universality, and objectivity, see Robert N. Proctor, *Value-Free Science? Purity and Power in Modern Knowledge* (Cambridge, Mass., 1991); Bruno Latour, *Pandora's Hope—Essays on the Reality of Science Studies* (Cambridge, Mass., 1999); Lorraine Daston and Peter Galison, *Objectivity* (Brooklyn, N.Y., 2007).

[8] John Krige, *American Hegemony and the Postwar Reconstruction of Science in Europe* (Cambridge, Mass., 2006).

[9] Paul Forman, "Behind Quantum Electronics: National Security as Basis for Physical Research in the United States, 1940–1960," *Historical Studies in the Physical and Biological Sciences* 18 (1987): 149–229; Stuart W. Leslie, *The Cold War and American Science: The Military-Industrial-Academic Complex at MIT and Stanford* (New York, 1993); Amy Dahan and Dominique Pestre, eds., *Les sciences pour la guerre, 1940–1960* (Paris, 2004); and the life sciences, Angela N. H. Creager and Maria Jesus Santesmases, "Radiobiology in the Atomic Age: Changing Research Practices and Policies in Comparative Perspective," *Journal of the History of Biology* 39 (2006): 637–47.

[10] For a historical sociology of the concept, see Anthony D. Smith, *National Identity: Ethnonationalism in Comparative Perspective* (Reno, Nev., 1991); and Smith, *Nationalism: Theory, Ideology, History* (Malden, Mass., 2001). For a recent overview of the topic, see Gerard Delanty and Krishan Kumar, *The Sage Handbook of Nations and Nationalism* (London, 2006). For the point about human attributes, see Ernest Gellner, *Nations and Nationalism: New Perspectives on the Past* (Ithaca, N.Y., 1983), 6.

[11] On the creation of traditions, see Eric J. Hobsbawm and Terence Ranger, eds., *The Invention of Tradition* (Cambridge, UK, 1983).

[12] Benedict Anderson, *Imagined Communities: Reflections on the Origin and Spread of Nationalism* (London, 1983).

This conception applies particularly well to the case of Switzerland. The people who were subsumed into the modern federalist Swiss state as it was created in 1848 did not all share a single language, religion, or culture. They represented various legal, social, and political traditions, so much so that they often had more in common with the people of neighboring nations than they did with each other. Thus, building and sustaining an "imagined community" was crucial to cementing such a diverse population into a national whole. Neutrality, federalism, and direct democracy were promoted as the political foundations of Switzerland's national identity. Neutrality became so essential to the Swiss national identity that by 1957 Switzerland's foreign minister could claim that "neutrality is for the Swiss a phenomenon as natural as that of water flowing along a riverbed."[13] In particular, neutrality in foreign relations was a political necessity for the state, given the diverse cultural and political allegiances of its citizens.[14] Foreign policy was a means for shaping the perception of Switzerland's national identity not only among foreign political elites but also among a broader public domestically. The origins of Switzerland's neutrality were often presented as coinciding with the joining together in 1291 of the first states that eventually formed Switzerland. In fact, Swiss neutrality was only recognized in international law at the Congress of Vienna in 1815. And since then, it had been permanently reinvented to adapt to changing foreign and domestic environments. The political value of neutrality, originally about the avoidance of war, came to incorporate crucial diplomatic and commercial functions as well, allowing Switzerland to play a disproportionately important role in world affairs and to sustain continued commercial relationships with all parties, even in wartime.

After World War II, the idea that Switzerland had been neutral during the conflict was widely believed in the country and provided a morally acceptable explanation as to why it had escaped the destruction that still afflicted its neighbors.[15] In international affairs, this idea was used to justify Switzerland's intense commercial relationships with Nazi Germany during the war, all the while positioning the country unambiguously on the side of the winning allied nations. As cold war tensions mounted, neutrality was again perceived as a political and commercial opportunity in the context of the new world order. It would guarantee the independence of a small state surrounded by powerful neighbors, allow it to play a privileged role in international affairs, and widen the range of its potential commercial partners, even as the country aligned itself, particularly after the 1950s, squarely with the "free world."[16]

The notion of neutrality, however popular in Switzerland among the general population and government officials, was heavily criticized abroad in the immediate postwar period. The United States, for example, interpreted Switzerland's neutrality as a cover-up for its sustained commercial relationships with Nazi Germany and as an excuse for engaging in trade with communist countries while benefiting from its geo-

[13] Max Petitpierre, "Conférence donnée le 9 novembre 1957 à Milan," 9 Nov. 1957, *Documents Diplomatiques Suisses* (available from http://www.dodis.ch) (hereafter cited as DoDiS), 14037.

[14] Max Petitpierre, Switzerland's foreign minister from 1945 to 1961, made this point in Max Petitpierre, "La neutralité Suisse," 28 May 1953, DoDiS, 14036.

[15] Hans Ulrich Jost, *Le salaire des neutres: Suisse, 1938–1948* (Paris, 1999); Georg Kreis, *Switzerland and the Second World War* (New York, 2000).

[16] Dominique Dirlewanger, Sébastien Guex, and Gian Franco Pordenone, *La politique commerciale de la Suisse de la seconde guerre mondiale à l'entrée au GATT (1945–1966)* (Zürich, 2004).

graphic position and cultural ties with the Western alliance.[17] The United States and other countries continued to criticize Swiss neutrality in the following two decades. In 1957, Switzerland's foreign minister put it mildly when he observed, "The word 'neutral' is not by itself very attractive. . . . Neutrality does not have a very good reputation."[18] Indeed, as George W. Ball, the U.S. undersecretary of state for economic and agricultural affairs in the Kennedy and Johnson administrations, recalled about the early 1960s: "In my view, Sweden and Switzerland defined 'neutrality' to suit their own purposes, and I had no sympathy for such casuistry."[19] Building a neutrality policy that would be credible abroad and at home was one of the great challenges of the Swiss Department of Foreign Affairs after 1945.[20]

To achieve this goal, those responsible for foreign policy resorted to boundary work: they constructed a pragmatic distinction in international affairs between "political" domains, in which Switzerland would abstain from participating, and "apolitical" domains, in which the country could fully take part. Max Petitpierre, Switzerland's foreign minister from 1945 to 1961, created the category of "technical" organizations and purposes to cover all aspects of foreign relations that were neither military nor political. In 1947, for example, speaking before the parliamentary commission debating Switzerland's possible candidacy for the United Nations, he made this point clear: "Neutrality remains our guiding principle and for this reason we cannot, at least for now, envision joining the planned world security organization. However, we should, as of now, make clear our interest in collaborating with technical (economic, social and humanitarian) organizations."[21] Indeed, Switzerland did not join the United Nations but did become a member of most of its affiliated organizations, such as the World Health Organization (WHO) and the United Nations Educational, Scientific and Cultural Organization (UNESCO). The notion of technical domains was convenient in that it was flexible enough to adapt itself to different contexts, while still carrying the idea that it was necessarily nonpolitical. Switzerland thus made every effort to play a leading role in these "technical" international organizations, so as to compensate for its absence from the others, and to give a positive meaning to "neutrality." Bern put a special emphasis on humanitarian aid through the International Committee of the Red Cross, for example, and cultural cooperation was promoted in the name of "solidarity." This notion had also been developed by Max Petitpierre as the other side of the neutrality coin, in order to respond to the criticism that neutrality amounted to isolationism and was nothing more than a pretext to escape international, moral, and political responsibilities.[22]

---

[17] Daniel Trachsler, *Neutral Zwischen Ost Und West? Infragestellung und Konsolidierung der Schweizerischen Neutralitätspolitik durch den Beginn des Kalten Krieges, 1947–1952* (Zürich, 2002); Mauro Cerutti, "La Suisse dans la guerre froide: La neutralité Suisse face aux pressions Américaines à l'epoque de la guerre de Corée," in *Guerres et paix*, ed. Michel Porret, Jean-François Fayet, and Carine Fluckiger (Geneva, 2000), 321–42.

[18] Petitpierre, "Conférence donnée" (cit. n. 13).

[19] George W. Ball, *The Past Has Another Pattern: Memoirs* (New York, 1982), 219.

[20] Daniel Möckli, *Neutralität, Solidarität, Sonderfall: Die Konzeptionierung der Schweizerischen Aussenpolitik der Nachkriegszeit, 1943–1947* (Zürich, 2000).

[21] Quoted in Antoine Fleury, "La Suisse et le défi du multilatéralisme," in *La Suisse dans le système international de l'après-guerre, 1943–1950*, ed. Georg Kreis (Basel, 1996), 68–83. On Petitpierre, see Louis-Edouard Roulet, ed., *Max Petitpierre: Seize ans de neutralité active: Aspects de la politique etrangère de la Suisse, 1945–1961* (Neuchâtel, 1980).

[22] Petitpierre, "Conférence donnée" (cit. n. 13).

Recent historiography of Swiss neutrality has emphasized the role of Switzerland's participation in international cultural, social, and humanitarian collaboration, while ignoring the importance of international scientific cooperation.[23] By focusing on Switzerland's involvement in different projects of international scientific cooperation in the name of its neutrality policy, one can understand how science can play a role in defining national identities and at the same time how national identities can shape the institutional dynamics of science—what I like to call the coproduction of neutral science and neutral state.

ATOMIC PHYSICS IN A GLASS HOUSE

The creation of CERN, the European organization for nuclear research, in 1954, constituted the first major accomplishment of the cold war in terms of European scientific collaboration. Much has been written about the complex processes that led to the birth of CERN; most of that historiography focuses on the role of the European scientists and science administrators who promoted the project, the interests of the major European states involved, and the context of the European integration movement. Before examining closely the role of Switzerland in this process, I will give a brief overview of the course of events.[24]

In December 1949, at the European cultural conference in Lausanne, Switzerland, the French science administrator Raoul Dautry had a resolution to study the creation of a "European center for atomic research" passed.[25] Six months later, at the annual UNESCO conference in Florence, the American physicist and statesmen Isidor I. Rabi proposed a resolution encouraging the creation of regional laboratories in Europe; the participating states adopted the proposal. These proposals gained momentum as they were taken up by several nuclear and cosmic ray physicists and by science administrators from different European countries. Dautry and the French cosmic ray physicist Pierre Auger took leading roles in defining a project focused on building on the European continent the largest particle accelerator in the world. A proposal was eventually submitted by a group of scientists led by Auger to an intergovernmental conference sponsored by UNESCO in December 1951. Support from the member states led three months later, at a meeting in Geneva, to the creation of a temporary organization and the signing, in June 1953, of a convention establishing CERN. Almost three-quarters of the funds were provided by France, Germany, and Great Britain, proportional to their gross national products, to build large particle accelerators. This convention was eventually ratified by eleven Western European countries plus the nonaligned Yugoslavia.[26] In May 1954, the construction of CERN began just outside the city of Geneva, in an agricultural field.

The Swiss government played a distinctive role in defining the political contours of CERN. Even though the country's financial contribution promised to be modest and its expertise in nuclear physics was not on par with that of France or Great

[23] Möckli, *Neutralität* (cit. n. 20); Trachsler, *Neutral Zwischen Ost Und West?* (cit. n. 17).

[24] Armin Hermann, John Krige, Ulrike Mersits, and Dominique Pestre, eds., *History of CERN*, 4 vols. (Amsterdam, 1987–96).

[25] Dominique Pestre and John Krige, "Some Thoughts on the Early History of CERN," in *Big Science: The Growth of Large-Scale Research*, ed. Peter Galison and Bruce Hevly (Stanford, Calif., 1992), 78–99.

[26] Belgium, Denmark, France, Germany, Greece, Italy, the Netherlands, Norway, Sweden, Switzerland, the United Kingdom, and Yugoslavia.

Britain, the fact that Switzerland had been a likely site for the future laboratory had given the Swiss government considerable leverage from the beginning. Together with most scientists involved in the project, the Swiss government pressed hard to "depoliticize" and "demilitarize" the CERN project to bring it into conformity with the ideal of scientific neutrality. Of course in the tense period of the cold war, building an immensely expensive laboratory devoted to research in one of the militarily most strategic fields of science could hardly be expected to be brought about from scientific interests alone. Yet in the end, CERN became an international nuclear physics laboratory devoid of military influence, relatively independent of the participating nations' particular political agendas, and dominated by the physicists' scientific goals as the result of a sinuous political and social process in which neutral countries, such as Switzerland, played a key role.

The Swiss government became officially involved in the CERN project in August 1951, when the foreign minister, Max Petitpierre, received an invitation to have Switzerland represented at the conference convened by UNESCO.[27] Most of the community of Swiss physicists consulted by Petitpierre favored the project of an international laboratory, largely because it would offer research opportunities they could not have dreamed of in the national context. One of them remarked that "the future of physics in Switzerland depends immensely upon the realization of this project."[28] Other physicists outlined the possible economic advantages of Switzerland's participation, which could constitute an "excellent deal," because the Swiss industry could sell equipment for the future laboratory, perhaps at a total price even higher than the government's contribution to the project.[29]

A few physicists were more hesitant, however. In addition to any moral reservations they might have had, they feared that if the project came to embody the military and political interests of some larger participating states, it would be difficult to build the intergovernmental consensus necessary to carry out CERN's mission.[30] In the unlikely event this hurdle could be overcome and a laboratory to pursue these military or political agendas was finally built, they warned, it would be of only limited scientific interest. They would have lost their autonomy in setting the laboratory's scientific goals, and because of the military relevance of their research, they would be restricted in freely sharing their results, a practice they deemed to be essential for the production of scientific knowledge. They did not want CERN to resemble the American national laboratories, which they perceived as being embedded in a culture of secrecy and national security concerns. These scenarios were not too farfetched in the European case, because a number of European nations, including Switzerland, had ongoing programs to develop atomic bombs and often skillfully maintained an ambiguous dividing line between civilian and military research.[31] The president of

---

[27] James Torres-Bodet to Max Petitpierre, 31 Aug. 1951, E 2001-04 (-) -/6, vol. 39, Swiss Federal Archives, Bern, Switzerland (hereafter cited as BAR).

[28] "Laboratoire européen de recherche nucléaires, Annexe 3: Procès verbal de la réunion de Lucerne, 2 October 1951," 2 Oct. 1951, 8, E 2001-04 (-) -/6, vol. 39, BAR.

[29] Ibid., 11.

[30] Ibid.

[31] On the Swiss atomic bomb project, see Fréderic Joye-Cagnard, *La construction de la politique de la science en Suisse: Enjeux scientifiques, stratégiques et politiques (1944–1974)* (PhD diss., Univ. of Geneva, 2007), chap. 3. The Swiss government was opposed only to military research carried out in collaboration with other countries, since that would violate its neutrality policy, not to research carried out independently.

the Swiss society for physics thus insisted vehemently that the CERN project should "carefully avoid any military aspect" and be "open to any nation."[32] The openness to all countries, whatever their political allegiances, was a means to ensure that the laboratory did not pursue military goals. The Swiss physicists also believed that locating the laboratory in Switzerland, a country with a "neutrality tradition," could help prevent the possible militarization and politicization of the laboratory.[33]

Other Swiss physicists had different concerns. They feared that the international project would compete with financial resources available for research at the national level.[34] Paul Scherrer, for example, had opposed an earlier and more ambitious project precisely on these grounds.[35] Given Scherrer's stature as the leading Swiss physicist and a high-profile public figure, his position was crucial to the Swiss government. Other European physicists were well aware of Scherrer's influence. As Pierre Auger wrote to a colleague: "Among the personalities that we need to have with us in the enterprise, there is Scherrer. You know perhaps that he is not favorable to the project. But we must convince him, because he determines the attitude of the federal government, and because the Dutch and Swedish opinions depend on it to some extent."[36] Switzerland's participation in the project seemed to be of prime importance to Auger, most likely because it would constitute the best public statement and long-term guarantee that the project was not being manipulated by some of the large European states' military or political establishments. Opposition to the project from other leading European physicists led Auger's group to propose a more modest accelerator as a first step, and only later a larger one, a scheme that then received Scherrer's full support. This constituted the proposal submitted to the UNESCO conference of December 1951.

Given the Swiss physicists' support, the government decided to participate in the UNESCO conference and gave precise instructions to its delegation, such as proposing Geneva as the site of the future laboratory.[37] But the Department of Foreign Affairs made clear that there was one nonnegotiable condition to Switzerland's participation: "The organization must be open to all European nations, including Eastern Europe, [and] its activities cannot be secret in any way, and should aim only at scientific and civilian goals."[38] With these instructions, the Swiss government followed precisely the key concerns voiced by the physicists one month earlier.

At the UNESCO conference, Alfred Picot, a local politician from Geneva and member of the Swiss delegation, was particularly attentive to the issue of CERN's neutrality. To avoid any military interpretation of the project, he suggested, for example, the addition of some broad lines indicating that CERN would "benefit the

---

[32] "Laboratoire européen de recherche nucléaires" (cit. n. 28).

[33] Ibid., 11.

[34] On the opposition to CERN from the older generation of European physicists, such as Paul Scherrer, see John Krige, "Scientists as Policymakers: British Physicists' 'Advice' to Their Government on Membership of CERN (1951–1952)," in *Solomon's House Revisited: The Organization and Institutionalization of Science*, ed. Tore Frängsmyr (Sagamore Beach, Mass., 1990), 270–91.

[35] "Procès-verbal du Conseil fédéral," 6 Nov. 1951, 1004.1 (-) -/1/, vol. 535, DoDiS 8137, BAR.

[36] Pierre Auger to Victor Weisskopf, 5 July 1951, CERN Archives, Geneva, reproduced in Hermann et al., *History of CERN* (cit. n. 24).

[37] The Swiss delegation comprised three physicists (Paul Scherrer, Peter Preiswerk, and André Mercier), a psychologist (Jean Piaget), a member of the state of Geneva's government (Albert Picot), and a member of the Department of Foreign Affairs (Bernard Barbey). "Décision du Conseil federal," 6 Nov. 1951, E 1004 1951, vol. 535, DoDiS 8137, BAR.

[38] "Instructions pour la délégation Suisse," 27 Nov. 1951, E 1004 1951, vol. 535, BAR.

progress of medicine and hygiene."[39] More important, he opposed the French delegation, which wanted to continue the negotiation within an ad hoc intergovernmental group and insisted that the future discussions about CERN be held under the auspices of UNESCO. The Swiss delegation unanimously supported this proposition to save the "principle of neutrality" of CERN. Indeed, as Picot explained: "On the one hand, with UNESCO the door remains open for new candidates from the Eastern countries. On the other side, this institution is a glass house, and one cannot suspect it of hiding military secrets."[40] The participation of Eastern European countries would make clear, according to Picot, that CERN was "not an American project" as some had implied because it had been initiated in part by Isidor I. Rabi, as a representative of the United States to UNESCO in 1950.[41] Picot tried to convince the Swiss public and the parliament alike that this was not the case and that CERN was first and foremost an initiative of the European scientific community. The openness of the laboratory was also a crucial point for Picot, because it hinged on the possible militarization of the laboratory, as the foreign minister had already pointed out. To the parliament, he acknowledged that "the words 'nuclear energy' evoke for many 'atomic bombs', and consequently certain fears. [But] atomic bombs are constructed in extremely secret places," which would not be the case with CERN, Picot argued.[42] Making sure the laboratory became a "glass house" was thus a means to prevent a possible militarization of CERN, because secrecy was a prerequisite for any research related to national security.

The participation of communist countries in CERN was a contentious issue among the delegations represented at the UNESCO conference.[43] For the Swiss government, the inclusion of Eastern European countries was key to the neutrality of the laboratory; for others, it represented an unacceptable opening toward their rivals in a bipolar world. Two years earlier, in an internal memo, the Swiss government had already been concerned with its position in this regard:

> The cooperation in the field of atomic research with nations of the Western alliance could be interpreted to some extent by the Eastern countries as a cooperation in the field of military armament. The explanation that for us it involves only the exploitation of atomic energy for economic purposes will not be taken seriously and considered as bare camouflage.[44]

Thus, the Swiss Department of foreign affairs began to worry when its delegation reported that "the French and Italian delegations have insisted, in corridor conversations, that the future laboratory be exclusively open to Western European nations. They don't want the laboratory to offer the Eastern countries precious sources of information without equivalents for Western physicists, and they are especially worried about adverse public opinion" in these countries.[45] This was bad news for Max Petitpierre, who confessed that he was "very concerned by the question of the laboratory's

---

[39] Albert Picot to Max Petitpierre, 17 Jan. 1952, E 2001-04 1970/346, vol. 209, BAR.

[40] Albert Picot to Max Petitpierre, 24 Dec. 1951, E 2001-04 (-) -/6, vol. 39, BAR.

[41] On the place of CERN in U.S. foreign policy, see Krige, *American Hegemony* (cit. n. 8), chap. 3.

[42] "Procès verbal de la séance de la commission du Conseil des Etats chargée d'examiner le message du Conseil fédéral du 4 avril 1952," 4 April 1952, E 2001-04 1970/346, vol. 209, BAR.

[43] See John Krige's contribution to Hermann et al., *History of CERN* (cit. n. 24), chap. 8.

[44] DMF to DPF, DEP, and DFI, 13 Feb. 1950, E 5155 (-) 1971/202, vol. 65, BAR.

[45] Jean Piaget, "Notice," 25 Jan. 1952, E 2001-04 1970/346, vol. 209, BAR.

universality."[46] In a moment of discouragement, he reflected that "there is a rift so wide between East and West that it makes all collaboration absolutely impossible in the atomic domain."[47] In this context, he feared that the laboratory might become "a new organization of the Western bloc."[48] If this were to be the case, he was ready to "renounce that the laboratory be hosted in Switzerland."[49] Petitpierre thus instructed the Swiss delegation to remain extremely careful concerning the location of the future laboratory.

To make matters worse, the United Kingdom proposed not only excluding Eastern European nations but also opening up CERN to the Commonwealth (including Australia and Canada) and the United States.[50] The British delegation even succeeded in having the word "European" suppressed from the categorization of potential participating nations to CERN; among the participants, Switzerland alone voted against the measure.[51] This event shows the very different meanings that the political configuration of CERN would take for various countries. For the United Kingdom, the Atlantic positioning of CERN would make clear that the project was not related to the ongoing European integration movement, a process toward which the British remained extremely skeptical, and thus it would look less politically contentious and well in tune with its "special relationship." For Switzerland, the opposite was true: if CERN embodied an alliance between Western Europe and the United States, it could be the sign of an Atlantic alliance, incompatible with Swiss neutrality policy. Thus, the Swiss argued, only by opening CERN up to Eastern European counties could it be considered truly neutral—neither Atlantic nor linked to the European integration movement.

A careful positioning of CERN was not just a foreign policy requirement for Switzerland; it was a domestic policy necessity as well, in view of the critical comments expressed "in numerous journals of all tendencies" about the compatibility of CERN with Switzerland's neutrality, as Petitpierre pointed out. These reactions worried him all the more in that he expected them to have an impact on the Swiss parliament, which would have to approve Switzerland's participation in CERN.[52] Furthermore, the Communist Party of Geneva had launched a referendum against the establishment of CERN,[53] arguing that the project "embroiled Switzerland deeper in the bloc of the imperialist warmongers, in the bloc fighting communism and Soviet Russia."[54] Those behind the CERN project, a Communist Party leader claimed, "took advantage of Switzerland's neutrality to protect an institute serving the United States by carrying out military research."[55] In the case of new conflict, Geneva would thus become a target for the belligerents, he argued. With a referendum pending and criticism

[46] Ibid., 2.
[47] Ibid.
[48] Ibid.
[49] Ibid.
[50] On the position of the United Kingdom, see Hermann et al., *History of CERN* (cit. n. 24), chaps. 12–13.
[51] Ibid., 251.
[52] Max Petitpierre to Paul Scherrer, 8 Feb. 1952, E 2001-04 1970/346, vol. 209, BAR.
[53] Bruno J. Strasser, *La fabrique d'une nouvelle science: La biologie moléculaire à l'âge atomique (1945–1964)* (Florence, 2006), chap. 1.
[54] Intervention by Léon Nicole, *Mémoriaux du Grand Conseil du Canton de Genève*, 30 May 1953, on 628. On the referendum against CERN, see Strasser, *La fabrique* (cit. n. 53), chap. 1.
[55] Jean Vincent, "Rapporteur de la minorité," in *Mémoriaux du Grand Conseil du Canton* (cit. n. 54), 612.

mounting, in not only the communist but also the conservative press, that CERN posed a threat to Switzerland's neutrality, the Swiss Department of Foreign Affairs was under pressure to find a way to demonstrate that this was not the case.

A small space for compromise opened up in the negotiations over CERN when someone realized that Switzerland's neutrality imperative did not require the actual participation of Eastern European nations to CERN, only the possibility of their participation. France, which was not favorable to the British plans, found the compromise formulation—in which candidacy to CERN was open to any country, but conditional on approval by every current member state, i.e. a negative vote by any member was tantamount to a veto.[56] As such, CERN would, in theory, be open equally to Poland and to the United States, but each member state could veto either candidacy. As John Krige put it in a nutshell, this solution "preserved the appearance of openness while masking the reality of exclusivity."[57]

Switzerland's diplomatic efforts were successful in bringing the laboratory to Geneva and, in conjunction with the interests of the leading scientists, isolating it from the most powerful military and political interests.[58] The scientists' decision to concentrate on a particle accelerator and not a reactor (which would have been much more closely tied to military applications, due to the necessary production of plutonium), brought Switzerland's mission within reach. As a result, a neutral laboratory in high-energy physics was created at the pinnacle of the cold war. Its convention stated clearly that CERN "shall have no concern with work for military requirements and the results of its experimental and theoretical work shall be published or otherwise made generally available."[59]

The neutrality of CERN's science not only made Switzerland's participation possible but also reinforced the credibility of its neutrality policy at a time when it was under strain. Indeed, in 1951 Switzerland reluctantly signed the Hotz-Linder Agreement with the United States, severely limiting Switzerland's exports to Eastern Europe, and drawing criticism at home that it had submitted itself to American interests.[60] The CERN negotiations constituted an opportunity to reposition Switzerland's diplomacy as more independent from the United States and open to relationships with Eastern European countries. Switzerland's position toward CERN was a manifestation of its concerns not only about finding its place in an increasingly bipolar world but also about its position toward the European integration movement, another challenge to its neutrality policy. The Swiss federal government had watched warily the development of the European Council in 1949 and of the European Coal and Steel Community in 1951, not joining either one.[61] But CERN gave the Swiss authorities a

[56] On France's position about the United States' joining CERN and toward the United Kingdom, see Dominique Pestre's contribution to Hermann et al., *History of CERN* (cit. n. 24), chap. 9; and "Note pour l'Ambassadeur de France à Londres," 29 April 1953, CHIP 10022, CERN Archives; "Minutes of the Session," 15 Dec. 1953, A151, CERN Archives.

[57] Hermann et al., *History of CERN* (cit. n. 24), 1:252.

[58] At the same time locating CERN in Switzerland was part of the concerns with the "German problem." See Krige, *American Hegemony* (cit. n. 8), chap. 3.

[59] "Convention for the Establishment of a European Organization for Nuclear Research," July 1953, article 2, CERN Archives.

[60] Cerutti, "La Suisse dans la guerre froide" (cit. n. 17). On the case of Sweden, see Nilsson, *Tools of Hegemony* (cit. n. 5).

[61] Antoine Fleury, "La Suisse et le Conseil de l'Europe," in *Jalons pour une histoire du Conseil de l'Europe*, ed. Marie-Thérèse Bitsch (Bern, 1997), 151–65; Fleury, "La Neutralité Suisse à l'Epreuve de l'Union Européenne," in *Neutrality in History*, ed. Jukka Nevakivi (Helsinki, 1993), 188–99.

chance to demonstrate their openness to a European project, reiterating their willingness to cooperate with their European neighbors, in areas considered nonpolitical and nonmilitary, or "technical" in the Swiss Department of Foreign Affairs' terminology. High-energy physics, thus adjusted, was made to fit this agenda. CERN became a symbol of Switzerland's neutrality and its role in mediating East-West relations during the cold war. On numerous occasions, the Swiss Department of Foreign Affairs used the case of CERN's neutrality, exemplified by the fact that Russian and American scientists met there, to illustrate the inherent neutrality of Switzerland.[62] Neutral science and neutral state had been efficiently coproduced.

By the end of the decade, when a new opportunity for international scientific cooperation presented itself, the Department of Foreign Affairs reacted very favorably. The same issues presented themselves once again, but under a slightly different light, due to the changed international context and the evolution of Switzerland's neutrality policy.

SPACE RESEARCH AND THE ROCKET PROBLEM

In February 1959, when CERN's massive proton synchrotron was about to go into operation for the first time, two of the physicists who had made it possible, Pierre Auger and Edoardo Amaldi, were in Paris discussing the creation of a similar organization devoted to space research.[63] The timing of this discussion should not come as a surprise. In October 1957, the first artificial satellite Sputnik began to orbit the earth, and as a response, the United States created NASA in October 1958.[64] By the end of the decade, space research had become a priority in science policy for a number of countries, including the United Kingdom, France, and Italy, all of which had their own national civilian or military space programs or both. In broad terms, space research was like atomic research, in that it was considered to involve "dual-use" technologies, military and civilian.[65] Atomic research could lead to building bombs and civilian power plants, just like space research was important for building missiles that could carry warheads and satellites. The launcher that carried Sputnik into orbit is a case in point, since it was also Russia's first nuclear ballistic missile.

The Paris discussion between Auger and Amaldi resulted, through a complex process similar to the one that led to CERN, in the creation in 1962 of two organizations, the European Space Research Organization (ESRO) and the European Space Vehicle Launcher Development Organization (ELDO). The fact that two distinct organizations were created instead of one illustrates the tension between the civilian and mili-

---

[62] See, e.g., "Essai de définition d'une politique dans le domaine de la collaboration scientifique future," 29 Aug. 1960, DoDiS 16953; and Petitpierre, "Conférence donnée" (cit. n. 13).

[63] On the history of space research in Europe, see John Krige and Arturo Russo, *A History of the European Space Agency, 1958–1987* (Noordwijk, Netherlands, 2000); for the case of Switzerland, see Büro für Weltraumgelegenheiten, ed., *Die Schweiz, Europa und die Raumfahrt, Abenteuer und Notwendigkeit* (Lausanne, 2000); Stephan Zellmeyer, *A Place in Space: The History of Swiss Participation in European Space Programmes, 1960–1987* (PhD diss., Univ. of Basel, 2007); Peter Creola, *Switzerland in Space—a Brief History* (Noordwijk, Netherlands, 2003).

[64] Roger D. Launius, John M. Logsdon, and Robert W. Smith, eds., *Reconsidering Sputnik: Forty Years since the Soviet Satellite* (London, 2002). The International Geophysical Year from July 1957 to 1958 also included an important component of atmospheric and space research.

[65] John Krige, "What Is 'Military' Technology? Two Cases of U.S.-European Scientific and Technological Collaboration in the 1950s," in *The United States and the Integration of Europe*, ed. Francis H. Heller and John R. Gillingham (New York, 1996), 307–38.

tary dimensions of space research.[66] This tension would prove particularly difficult to resolve and make compatible with Switzerland's neutrality policy.

In January 1960, the Swiss federal government was asked by a parliamentary representative from Geneva if "it should not take the lead in creating a European organization for space research."[67] Max Petitpierre, still foreign minister at the time, deemed the question "quite urgent" and created a delegation of scientists, including Paul Scherrer and Marcel Golay, director of the Geneva Observatory, to represent Swiss interests.[68] The reason for moving ahead so promptly, and for enrolling scientists to represent national interests as early as possible in the process, was the hope of the Department of Foreign Affairs that the Swiss scientists could shape the future organization significantly: "It is only by acting quickly that we will be able to influence the shape and the activities of the new organization and that we can avoid the risk of facing projects developed within organizations that do not correspond to our institutions (we think in particular of NATO)."[69] Indeed, NATO had, as early as 1957, outlined plans to develop international cooperation in space research on a European basis.[70] The director-general of CERN, Cornelis Jan Bakker, warned the Swiss Department of Foreign Affairs that "the essential difficulty will consist in creating an organization that can conduct its activities without the interference of the military. One needs to prevent them from having any means of influencing the research of an organization that should solely respond to scientific and peaceful concerns."[71] In this respect, the European scientists behind the project could not agree more with Switzerland's position, since they too wanted to prevent political and military control over the organization.

Switzerland became a particularly valuable ally for the European scientists when, in April 1960, the United Kingdom proposed linking the project of a European space organization to their obsolete ballistic missile Blue Streak, which they hoped to convert to civilian uses.[72] Some scientists feared that this proposition could give a military flavor to the European project. After this threat of militarization, another politicization menace began to loom over the project. The Organization for European Economic Co-operation (OEEC) proposed to host the secretariat of commission in charge of developing the Auger and Amaldi project at a time when the OEEC was just about to become a more Atlantic organization by including the United States and Canada (becoming the OECD in 1961). Auger and Amaldi opposed both propositions, as they had resisted the earlier propositions of NATO to organize European

---

[66] The enormous difference in cost in developing a rocket or building a satellite was also a crucial issue. I thank John Krige for drawing my attention to this point.

[67] "Question Borel posée au Conseil fédéral le 25 janvier 1960," 25 Jan. 1960, E 2003 (A) 1971/44, bd. 94, BAR.

[68] "Note dictée par Max Petitpierre," 27 Jan. 1960, E 2003 (A) 1971/44, vol. 94, BAR.

[69] "Proposition du Département politique fédéral au Conseil fédéral," 30 May 1960, E 2003 (A) 1971/44, vol. 95, BAR.

[70] John Krige and Lorenza Sebesta, "US-European Co-Operation in Space in the Decade after Sputnik," in *Big Culture: Intellectual Cooperation in Large-Scale Cultural and Technical Systems*, ed. Giuliana Gemelli (Bologna, Italy, 1994), 263–85.

[71] Samuel Campiche, "Note en vue de l'entretien avec les savants suisses sur une organisation européenne pour la recherche spatiale," 14 March 1960, 1, E 2003 (A) 1971/44, vol. 94, BAR.

[72] The Royal Society, "Western European Space Research Meeting," 29 April 1960; "Draft Minutes," 30 April 1960, E 2003 (A) 1971/44, vol. 94, BAR. On Blue Streak, see Krige, "What Is 'Military' Technology?" (cit. n. 65); and Krige and Sebesta, "US-European Co-Operation" (cit. n. 70).

space research.[73] A member of the Swiss delegation made clear what the new orientation of the OEEC might mean for European space research:

> The possible presence of the United States and Canada could compromise the political neutrality of the OEEC. Given the strategic dimension of space research and given the fact that it is particularly well developed in those countries, there might be unpleasant interference from their side on our European research.[74]

A member of the Swiss delegation also expressed his concerns that Austria, another neutral country, had not been invited to participate, raising further suspicions about the politicizing of the project.[75] And Switzerland was not the only neutral country concerned by these developments. Sweden also feared for its neutrality policy and insisted that any plans for a European organization be officially communicated to the Soviet Union.[76]

These irreconcilable positions clashed during the discussions between the delegations, bringing the negotiations to a dead end until Marcel Golay informed the assembly that he had just been authorized by the Swiss government to offer its help in convening an intergovernmental conference.[77] Understandably, the proposition was very well received by European scientists such as Auger, who pointed out that Switzerland could play a role for space research similar to the one that UNESCO had played for nuclear physics.[78] In the following meeting, Switzerland, with the support of the Norwegian, Swedish, and Danish delegations, was chosen as the host country for the intergovernmental conference.[79] According to Marcel Golay, "[A]ll participants seemed to be aware that it would be in their interest to put forward the name of Switzerland as a symbol of the non-political activities of the future organization."[80] What Golay did not mention was that Switzerland's interest in keeping the future organization neutral would leave the scientists in charge and protect them from interference by the nation-states in the shaping and running of the organization, but at the cost, as the future would show, of more limited national participation in ELDO.

In July 1960, the Swiss government took over the negotiations that the scientists had handled up to that time.[81] It examined more carefully than ever the consequences of a European space research organization for the country's neutrality policy. In a detailed report, it outlined once again the convergence of interest between the scientific community and the Swiss authorities in preventing political and military interference:

---

[73] Krige and Russo, *A History of the European Space Agency* (cit. n. 63), 44; John Krige, "NATO and the Strengthening of Western Science in the Post-Sputnik Era," *Minerva* 38 (2000): 81–108.
[74] Marcel Golay, "Recherche spatiale: Rapport relatif à la conférence de Londres," n.d. [May 1960], annexed to his letter to Max Petitpierre, 5 May 1960, E 2003 (A) 1971/44, vol. 94, BAR.
[75] Ibid.
[76] Royal Society, "Western European Space Research Meeting" (cit. n. 72).
[77] Ibid.
[78] Bernard Barbey to Jean de Rham, 18 May 1960, E 2003 (A) 1971/44, bd. 94, BAR.
[79] "Rapport sur la réunion de Paris des 23 et 24 juin 1960," 29 June 1960, E 2003 (A) 1971/44, bd. 94, BAR.
[80] Ibid., 5.
[81] Département politique fédéral to Président de la Société helvétique des sciences naturelles, 13 July 1960, E 2003 (A) 1971/44, bd. 94, BAR.

> The preference to date of the scientists for a model similar to CERN corresponds to our position. This organization succeeded, in the field of fundamental research on the atom, to demilitarize and depoliticize a number of scientific activities that were previously the prerogative of the military. Today, thanks to CERN, atomic science is in large part in the public domain. An impressive number of American and Soviet scientists meet there. However—and this is essential—it is the fact that CERN was active on our territory that made it possible for this institution to play this role of intermediary between East and West.[82]

Thus according to the Swiss government, not only could its participation in an international scientific cooperation project be made compatible with its neutrality policy, but also the latter would serve scientific cooperation well by guaranteeing that it was depoliticized and demilitarized—neutral, in other words.

Switzerland was not alone in holding these views. The Swedish and Austrian delegations, for example, two other neutral countries, reasoned along the same lines. In the fall of 1960, the United Kingdom was still trying to promote the use of its recycled launcher Blue Streak, stripped of its military characteristics, but experienced only incomplete success.[83] Sweden refused the proposition outright,[84] aware of the political risks inherent in international cooperation in the field of launchers as a dual-use technology, and Switzerland opposed the proposition, too. Switzerland's decision was also motivated by the fact that Swiss industry believed the investments in space launchers to be insufficiently profitable, and the military had no interest in developing missiles, preferring to keep in touch with space research merely "to appreciate the potential threat coming from space."[85] As a result of Switzerland's, Sweden's, and other countries' opposition to the British plans, the delegations decided to create two distinct organizations. The European Space Research Organization (ESRO) would be devoted exclusively to space research and would include ten European countries.[86] The European Space Vehicle Launcher Development Organization (ELDO), by contrast, would be devoted to the development of a launcher and would include only six European countries, plus Australia. Austria, Sweden, and Switzerland, all small neutral countries, were absent from ELDO.

Switzerland's efforts at depoliticizing and demilitarizing space research were not as successful as they were in the case of high-energy physics. On the one side, the size of the investments and the military and commercial importance of launchers stood in the way of Switzerland's and other neutral countries' diplomacies. Moreover, these particular aspects of space research remained too tied to political and military interests to be made compatible with Switzerland's neutrality policy. The Swiss government therefore tried to preserve the country's participation in European space research by advocating a split between launchers (ELDO) and satellites (ESRO) in

---

[82] "Rapport au Conseil fédéral sur la nécessité d'une organisation européenne pour la recherche spatiale et sur l'intérêt pour la Suisse d'y participer," 22 July 1960, 11, E 2003 (A) 1974/52, vol. 153, BAR.

[83] The Swiss government received a confidential memo on this topic 2 Sept. 1960; "Note pour le dossier," 15 Sept. 1960, E 2003 (A) 1971/44, vol. 95, BAR.

[84] "Note pour Monsieur le Ministre de Rham," 18 Oct. 1960, E 2003 (A) 1971/44, bd. 94, BAR.

[85] "Rapport du DMF du 8 Mai 1962 à l'appui de la proposition du DPF du 30 Avril 1962," 8 May 1962, E 2003 (A) 1971/44, bd. 94, BAR.

[86] Belgium, Denmark, France, Germany, Italy, the Netherlands, Spain, Sweden, Switzerland, and the United Kingdom.

two separate organizations. In this way, Switzerland could at least join the neutralized ESRO. On the other side, the fact that satellites, unlike nuclear research, were not perceived as dual-use technology meant that it was unnecessary to publicly affirm their neutrality by locating any of the ESRO institutions on Swiss ground. Indeed, Switzerland could not land any of the ESRO research laboratories. However, at least it succeeded in ensuring that European space research did not take place under the framework of NATO or the European Communities, something it had feared early on, and which would have made Switzerland's participation in any way impossible.[87]

Switzerland was satisfied with the neutrality of ESRO, as it embodied three mechanisms to prevent militarization of any kind. In April 1962, Switzerland's new foreign minister, Friedrich T. Wahlen, shared his thoughts about this delicate issue with his Swiss ambassador in Paris:

> It is obvious that with the new weapons, the scientists and the military work in the same fields and as a result, a definition and exact delimitation of their programs is a necessity. In the case of CERN, it is inconceivable that their programs could be inspired by military considerations, since all the results of studies on the atom must be published. CERN is, as it is customary to say, a glass house. . . . In the case of ESRO, any state who thinks that the program diverges from the objectives of the conventions can veto the budget, since it has to be adopted unanimously. It is true that it could be difficult in practice to determine if one experiment or the other is really devoid of military preoccupations. We cannot exclude that certain powers—even against the opinion of their scientists—could use this European organization to increase their strategic potential, through indirect ways difficult to detect. Our representatives must thus be vigilant, and a way to achieve this is through the possibility and the right of every nation to have its own citizens participate in any experiment carried out by ESRO. But besides this assurance, there is another guarantee—the most efficient perhaps—against such tendencies, namely the publication of all the experimental results. The secrecy required by military experiments would be irreconcilable with this publication, which is a principle of ESRO. The necessary vigilance is the common concern of the member states most directly concerned, Switzerland, Sweden and Austria.[88]

Thus space research involving satellites, and not linked to launchers, could be made neutral using the criterion of open publication as a benchmark for basic "nonmilitary" research, as with CERN.[89] In addition, the absence of the United States and of Eastern European countries and the USSR, and their potentially polarizing effect, as well as the presence of neutral countries, offered a good chance that ESRO would be exclusively following scientific goals, as the European scientists, and the Swiss authorities, had wished. This solution avoided "the unpleasant intervention" of the United States in European space that the European scientists had feared because of the "strategic importance of this field." At the same time, it fulfilled their desire to keep "a complete independence from the Soviet Union in order not to compromise the essentially scientific goals" of ESRO.[90] European scientists and Swiss federal authorities could not agree more completely.

The country's participation in space research thrilled the Swiss media. The media saw it as particularly in tune with several aspects of Switzerland's national identity.

---

[87] "Rapport au Conseil fédéral" (cit. n. 82), 10.

[88] Friedrich T. Wahlen to Agostino Soldati, 14 April 1962, E 2804 1971/2 bd. 44, BAR.

[89] On the flexible definition of what counts as a military technology, see Krige, "What is 'Military' Technology?" (cit. n. 65).

[90] Golay, "Recherche spatiale" (cit. n. 74).

For example, space research would involve the development of high-precision minia-
turized devices to be embarked on satellites, a special field of expertise of the Swiss
watchmaking industry and a subject of national pride.[91] Switzerland's participation
in European space research was also understood as an unusual opportunity for its
neutrality—that is, "independent of both the American and the Soviet blocs," thus
making uniquely European collaboration possible.[92] Space research caught the imagination of the media, which came to envision "Swiss citizens on the moon!"[93] The
Swiss authorities' diplomatic efforts in the field of space research thus helped position the country's national identity not only in foreign relations but also domestically,
as the extensive and enthusiastic media coverage demonstrates.[94]

MOLECULAR BIOLOGY: AN EASY CASE FOR INTERNATIONAL COOPERATION

The explosion of the first atomic bomb over Hiroshima in 1945 and the launch of
Sputnik in 1957 played decisive roles in setting the science policy agenda in the in-
dustrialized nations. The impact of these two events proved all the greater in that both
were linked to warfare technologies and commercial opportunities. Physicists were
most successful in mobilizing these different interests and translating them into massive support for physical science research.[95] CERN and ESRO/ELDO are just two
examples reflecting these changing science and technology policy priorities in the
cold war.

Until the 1960s, there were no events comparable to Hiroshima or Sputnik that
reoriented the national scientific priorities in favor of the life sciences. Only decades
after James Watson and Francis Crick's 1953 discovery of the DNA double helix was
this event remembered as the starting point of a new scientific discipline.[96] However,
in the early 1960s, it was becoming increasingly perceptible that a deep transformation was taking place in the understanding of life and disease, as a new discipline
called "molecular biology" was progressively redefining the avant-garde of biological research.[97] Significantly, a number of researchers working under this banner were
physicists converted to the study of life. Beginning in 1958, a growing number of
local research institutions created new departments of molecular biology.[98] In 1962,
the Nobel Prize in Chemistry and the Nobel Prize in Physiology or Medicine were
awarded to five researchers in the field of molecular biology, three of whom were
physicists, adding to the growing prestige of this new discipline.

In December 1962, on their way back from Stockholm where they had received

---

[91] Ibid.

[92] "Le Centre européen de recherche spatiales est un projet pacifique auquel la Suisse participe," *La
Tribune de Genève*, 23 June 1960, 21.

[93] "Des Suisses . . . sur la lune!," *Le Genevois*, 10 Dec. 1960, 1.

[94] For a detailed analysis of the press in the Swiss-German context, see Zellmeyer, *A Place in Space*
(cit. n. 63).

[95] Daniel J. Kevles, *The Physicists: The History of a Scientific Community in Modern America*, 2nd
ed. (Cambridge, Mass., 2005).

[96] Bruno J. Strasser, "Who Cares about the Double Helix?" *Nature* 422 (2003), 803–4.

[97] On molecular biology in postwar Europe, see Bruno J. Strasser and Soraya de Chadareian, eds.,
Studies in History and Philosophy of Biological and Biomedical Sciences 33 (2002); Soraya de Cha-
darevian, *Designs for Life. Molecular Biology after World War II* (Cambridge, UK, 2002); Jean-Paul
Gaudillière, *Inventer la biomédecine: La France, l'Amérique et la production des savoirs du vivant,
1945–1965* (Paris, 2002); Strasser, *La fabrique* (cit. n. 53).

[98] Bruno J. Strasser, "Institutionalizing Molecular Biology in Post-War Europe: A Comparative
Study," *Stud. Hist. Phil. Biol. Biomed. Sci.* 33 (2002): 533–64.

their Nobel Prizes, James Watson (physiology or medicine) and British crystal-lographer John Kendrew (chemistry) stopped by CERN.[99] In a discussion among Watson and Kendrew and American physicist Leo Szilard and CERN director Victor Weisskopf, Szilard suggested that European molecular biologists also try to convince European governments to fund an international laboratory modeled on CERN. In September 1963, a small group of European researchers founded the European Molecular Biology Organization (EMBO) to foster molecular biology in Europe by, for example, creating an international laboratory modeled on and located close to CERN.[100] However, for reasons to be explored later, this venture was not received with great enthusiasm by European governments, lacking as it did the political urgency and military resonances of high-energy physics and space research. The Swiss physicist turned molecular biologist Eduard Kellenberger, a student of Paul Scherrer's and the Swiss representative to the newly born EMBO Council, persisted in bringing molecular biology to the Swiss political agenda.[101] He took the initiative in making EMBO a private organization under Swiss law, which would later facilitate negotiations with the Swiss federal government. Indeed, the Swiss Department of Foreign Affairs decided to support EMBO's plans, and in 1966, Switzerland convened an intergovernmental meeting that led to the signing, in February 1969, of a convention creating the European Molecular Biology Conference (EMBC), including representatives of the twelve Western European nations.[102] In 1974, the EMBC came to an agreement to fund the European Molecular Biology Laboratory (EMBL), which eventually opened in Heidelberg in 1978.

In April 1964, the new Swiss foreign minister, Friedrich T. Wahlen, received a document with the puzzling title "What Is Life? A New Organization for Biological Research in Europe."[103] It had been sent by Eduard Kellenberger, as the Swiss representative to the EMBO Council, who informed the minister about the plans for his new organization. Wahlen took an immediate interest in the EMBO plans[104] and proposed that Switzerland take the diplomatic lead to support it.[105] Wahlen, with his background in agricultural science, had a keen interest in biology. Furthermore, between 1949 and 1958, he had been director of the Agriculture Department of the Food and Agriculture Organization (FAO), a United Nations affiliated entity, giving him a significant experience in international organizations. The FAO represented the kind of organization that Max Petitpierre, Wahlen's predecessor, had defined as "technical" and which was therefore compatible with Switzerland's neutrality. The Department of Foreign Affairs also justified its interest in EMBO by pointing to the rapidly changing scientific priorities of the cold war:

[99] On the history of EMBO and EMBL, see John Krige, "The Birth of EMBO and the Difficult Road to EMBL," *Stud. Hist. Phil. Biol. Biomed. Sci.* 33 (2002): 547–64; Bruno J. Strasser, "The Transformation of the Biological Sciences in Post-War Europe," *EMBO Reports* 4 (2003): 540–43.

[100] Krige, "Birth of EMBO" (cit. n. 99).

[101] On Kellenberger, see Bruno J. Strasser and Jacques Dubochet, "Eduard Kellenberger (1920–2004)," *Nature* 433 (2005), 817.

[102] Austria, Denmark, France, Germany, Greece, Italy, the Netherlands, Norway, Spain, Sweden, Switzerland, and the United Kingdom.

[103] Edouard Kellenberger to Friedrich T. Wahlen, 27 April 1964, E 2003 (A) 1987/29, vol. 185, BAR.

[104] Edouard Kellenberger to Max Perutz, 4 July 1964, EMBO Archives, Heidelberg.

[105] "Minutes of the Meeting of Council held at CERN, Geneva, on 12 July 1964," 12 July 1964, EMBO Archives, Heidelberg.

> Since atomic energy has entered the stage of industrial accomplishments and the NASA "moon-crash-program" will most likely be carried out, all responsible parties should, as of now, make efforts to reduce, or at least find ways to continue employing the scientific workforce of the country. From this perspective, the EMBO initiative could be very useful for keeping pace for once with the USA.[106]

Other European nations were much less enthusiastic about a centralized laboratory for the life sciences. They cast some doubt on the necessity of centralizing resources in the field, because unlike high-energy physics or space research, molecular biology did not require any unique pieces of heavy equipment, such as cyclotrons or rockets, that would be too expensive for a single smaller country to develop. An international laboratory might also deplete the country of its elite scientists.[107] Finally, molecular biology did not seem very promising commercially or militarily. This last point, perceived as a disadvantage of the field by some nations, was thought of as a great opportunity by Switzerland. The absence of military or political interests in molecular biology would ease greatly Switzerland's efforts at inscribing EMBO's plans into its current neutrality policy.

Switzerland supported EMBO without any of the hesitations that had characterized its earlier involvement in international scientific cooperation. However, the Swiss government was not the only political institution to identify the possible advantages of collaborating in the field of molecular biology. As a representative of the OECD would put it in May 1965, cooperation in the field of the life sciences "is relatively easy, because, on one side it doesn't raise delicate political or economic problems, and on the other, its development—even through international cooperation—does not require very substantial investments."[108] UNESCO, which had played a leading role for CERN, proposed in 1965 to reiterate its efforts for EMBO and convene an intergovernmental meeting the following year. At the same time, the European Council decided to do the same. The OECD and the WHO considered taking similar action, as did Switzerland.[109]

The EMBO Council was wary about these different organizations and declared its willingness "to maintain a complete independence of UNESCO and other similar organizations on political as well as administrative questions."[110] They emphasized the risks resulting from the patronage of international organizations, such as "political pressure on the choice of scientific personnel and lack of interest in a research program in fundamental biology."[111] Privately, the EMBO Council members were very favorable to Switzerland's initiative, even if they decided not to show their preferences publicly or to turn down the other offers.[112]

Swiss diplomacy was confronted, on the surface at least, with strong competition from international organizations that had the advantage of already having governmental-level representations from the EMBO member states. Switzerland made serious efforts to keep the lead in this process, especially since, as in previous

---

[106] G. Pnett, "Mitteilung," [mid-Sept. 1965], E 2003 (A) 1978/29, vol. 185, BAR.
[107] Krige, "Birth of EMBO" (cit. n. 99).
[108] OECD, *Problems of International Scientific Co-Operation—International Co-Operation in Biology—The Problem of Choice*, 27 May 1965, 28, EMBO Archives.
[109] Strasser, "Transformation of the Biological Sciences" (cit. n. 99).
[110] "Minutes of the Meeting of Council held at CERN" (cit. n. 105).
[111] "Meeting of EMBO Council at Geneva on 2 Feb. 1964," 2 Feb. 1964, EMBO Archives.
[112] "Minutes of the Meeting of Council held at CERN" (cit. n. 105).

collaborative efforts, it had a precise idea about what political contours the organization should take to conform to Switzerland's neutrality policy of the 1960s. Indeed, by that time, Switzerland had clearly found its place in Western Europe, even though it remained highly suspicious of any convergence with Atlantic positions.[113] It was also making significant efforts to become closer to its European neighbors in the European Economic Community, without considering joining the Community, however, especially after the failure of the larger European Free Trade Association in 1959.[114] In 1963, for example, Switzerland joined the European Council, after having refused to do so in 1949.[115]

As a result of Switzerland's new understanding of its neutrality policy, it privileged international cooperation with Western European nations, and actively tried to avoid any other political configuration. It opposed UNESCO, for example, which favored an organization open to all European nations, East and West, reversing the position it had adopted earlier in the case of CERN.[116] It also asked the EMBO Council not to seek financial support from the Ford Foundation or other American funding agencies to prevent any sign that it was an Atlantic organization.[117] The Swiss Department of Foreign Affairs vigorously defended its position to strictly limit membership to Western Europe, thus excluding Israel, even though Israeli scientists had been founding members of EMBO.

In 1966, an independent event reminded the Swiss authorities that they had much to lose politically if they were unable to guide European scientific cooperation in the field of molecular biology along their own agenda: the Italian government proposed to reactivate an older NATO project for a European institute of science and technology modeled after MIT[118] and designed to include a department of molecular biology. This proposition provoked a stir in the Swiss Department of Foreign Affairs. An internal note spelled out some of the political consequences of the Italian proposition for Switzerland's neutrality policy:

> The motivation for this action—besides the strengthening of NATO—lies in the correct conclusion that the widening technological gap between the United States and Europe could also widen proportionally the political differences between the two continents. A preliminary inspection reveals several adverse prospects. The neutrals will have to be very careful in order not be caught by surprise and pushed aside by politically inspired scientific plans. The danger lies in a repetition of the experience of the European Economic Community, in a different universe, but with the same divisive result (there: at-

[113] Jakob Tanner, *Grundlinien der schweizerischen Aussenpolitik seit 1945* (Bern, 1993).

[114] Antoine Fleury, "La Suisse: Le projet de grande zone de libre échange et la création de la CEE," in *Il rilancio dell'Europa e i trattati di Roma*, ed. Enrico Serra (Milan, 1989), 355–76. On NATO in science, see Krige, *American Hegemony* (cit. n. 8), chap. 8; and Krige, "NATO and the Strengthening of Western Science" (cit. n. 73).

[115] Fleury, "La Suisse et le conseil de l'Europe" (cit. n. 61), 151–65.

[116] At the time of the CERN negotiations, Switzerland was still considering including countries of Eastern Europe, in line with UNESCO's universalist perspective; but by the early 1960s, the country had clearly aligned itself with Western Europe, making UNESCO's proposal politically unacceptable.

[117] Jeffrey Wyman, "Notes on Discussion at Swiss Department of Foreign Affairs," 17 Jan. 1966, Eduard Kellenberger private archives. On the Ford Foundation in Europe, see Krige, *American Hegemony* (cit. n. 8), chap. 6.

[118] Giuliana Gemelli, "Western Alliance and Scientific Diplomacy in the Early 1960s: The Rise and Failure of the Project to Create a European MIT," in *The American Century in Europe*, ed. R. Laurence Moore and Maurizio Vaudagna (Ithaca, N.Y., 2003), 171–92.

tempts to unify Europe through the economy, here: attempt to strengthen NATO through science).[119]

As a consequence of this reasoning, the Department of Foreign Affairs decided to move ahead more energetically than ever, taking the risk of clashing with the international organizations, such as UNESCO, which had similar plans. Switzerland rushed to consult all European governments and issued an invitation for an intergovernmental conference, to take place in September 1966, on international cooperation in the field of molecular biology. Only countries already members or observers at CERN were invited to participate. In a letter of protest, the director of UNESCO "deplored" the action of the Swiss government and expressed regret that the invitation had only been addressed to CERN members.[120] The Swiss foreign minister was somewhat relieved by this reaction, as he had feared a worse, even a "hostile," reaction of UNESCO.[121] By 1969, the conference organized by Switzerland led to the signing of a convention by twelve Western European countries creating the EMBC, an intergovernmental funding body for EMBO, and leading to the creation, in 1974, of the European molecular biology laboratory in Heidelberg.

Even though Switzerland did not succeed in obtaining the central laboratory, it was able once again to shape the political contours of the organization and to bring it in line with its neutrality policy of the 1960s. As with its leadership in the development of CERN and ESRO, Switzerland's active role in the creation of the EMBC and its participation in all three organizations demonstrated better than any discourse could that neutrality was neither an isolationist policy nor a pretext to escape international responsibilities. As an internal memo of the Foreign Department made clear, this was precisely Switzerland's agenda before embarking on the EMBO projects: "We should seize the great opportunity of a European science policy initiative [the EMBO project] to refute our alleged selfish isolationism."[122] The participation in international scientific cooperation in the field of molecular biology was an ideal way to dispel the impression that neutrality amounted to political isolationism. Neutrality would be pursued as a coherent policy to create a space free from military or political interference, a space that would benefit not only international negotiations aimed at conflict resolution, for example, but also the pursuit of scientific research in an international setting. Thus with the EMBO project, the Swiss authorities strove to sustain the neutrality of science and, at the same time, to reinforce the neutrality of its national identity. Admittedly, the case of molecular biology was far easier than that of atomic physics or space research, as molecular biology was not linked to sensitive military technologies. Nevertheless, the EMBC could have been structured in a quite different political framework had it been shaped by organizations such as UNESCO and the OECD. Although a different structure would not have forbidden Switzerland's participation, it would have at least made this organization much less useful as an expression of the current meaning of Switzerland's neutrality policy.

[119] "Projekt Fanfani (wissenschaftlich-technologischer Marshall-Plan)," 29 Oct. 1966, E 3375 (A) 1992/25, vol. 9, BAR.

[120] Alfred Rappard to Division des Organisations Internationales, 7 April 1966, E 2003 (A) 1978/29, vol. 185, BAR.

[121] Ernst Thalmann to Max Perutz, 9 Dec. 1966, EMBO Archives.

[122] "Einladung der Schweiz zur Gründung einer europäischen Organisation für Molekular-Biologie," 30 Aug. 1966, E 2003 (A) 1978/29, vol. 185, BAR.

CONCLUSION

The neutrality of science and the state have never been givens; rather they represent a process of negotiation taking place in historically specific contexts, aimed at deflecting particular political forces. In cold war Europe this context included the political and ideological commitments of the nation-states in the divide between East and West, and their interest in building strong military defense programs underpinned by science and technology. States such as Switzerland defined being neutral as being permanently engaged in the process of finding a path along the delicate line balancing the necessity of active involvement in international affairs and the refusal to commit to political alliances aimed at shifting balances of power. Given the paramount importance of political alliances of all kinds in international affairs, the Swiss government was hard pressed to find domains that could be made to fit this agenda. Scientific cooperation was one of these domains, along with cultural, humanitarian, and social cooperation. By actively attempting to depoliticize and demilitarize these international scientific organizations, Switzerland could affirm publicly how much it cared about its neutrality policy. These actions were not cynical manipulations of the neutrality idea. The Swiss believed them to be expressions of what neutrality historically stood for and attempts to ground it in tangible institutions. Thus it was precisely when science was the least neutral, as in the case of high-energy physics and space research, that it would best serve the assertion of the country's national identity built around the idea of neutrality.

The three examples of scientific cooperation discussed here illustrate how in the decades following the Second World War, the Swiss authorities considered science an opportunity to participate in international affairs and reinforce the credibility of the nation's neutrality policy. Science could serve this purpose only if it were made neutral, something that proved relatively simple in the case of molecular biology, more difficult for nuclear physics, and almost impossible for space research, except when a sharp institutionalized distinction could be made between satellites and launchers. It is a measure of Switzerland's success that the complex negotiations between scientists and statesmen from different countries finally led to the creation of institutions that have, in fact, been considered neutral by all parties and dominated by the scientists' agendas. This was a remarkable achievement in the cold war, during which state support for large-scale research, for example in the United States national laboratories, was generally subservient to the attainment of practical goals of military or economic interest. Nobel prizes awarded to scientists working for CERN and EMBL demonstrate that these organizations hosted fundamental research at the highest level of scientific excellence. In the same time period, the scientific failure of EURATOM, created in 1957 to stimulate research on nuclear energy among members of the European Economic Community, showed where an excessive politicization of scientific research could lead.[123]

Switzerland was not alone in defending the neutrality of European scientific cooperation, as other neutral and non-neutral countries shared similar views, but it did play a leading role in this respect. Its strongest ally throughout the negotiations re-

---

[123] Michel Dumoulin, Pierre Guillen, and Maurice Vaïsse, eds., *L'Energie nucléaire en Europe: Des origines à euratom* (Bern, 1994). EURATOM's scientific failure does not imply it was not a successful political weapon; see John Krige, "The Peaceful Atom as Political Weapon: Euratom and American Foreign Policy in the Late 1950s," *Historical Studies in the Natural Sciences* 38 (2008): 5–44.

mained the European scientists themselves, for whom the fight for neutrality was also a fight for scientific leadership in these organizations. During the CERN negotiations, a Swiss politician could acknowledge that "these great physicists are diplomats far more skillful than we are."[124] The convergence of interest between European scientists and neutral states was perhaps the strongest driving force that led to the depoliticizing and demilitarizing of these scientific organizations.

Public debates about Switzerland's participation in international scientific organizations, in the press as well as in the parliament, often revolved around the question of Switzerland's political neutrality. Indeed, Switzerland's neutrality was not only a useful fiction employed by the government for defending its foreign and economic policies but also a central tenet of the nation's identity. "Neutrality is the people's business, not the government's or the parliament's," emphasized Max Petitpierre in 1957.[125] By participating in international scientific cooperation and defending the neutrality of science, the Swiss government could demonstrate to other nations, as well as to its own citizens, what political neutrality stood for. Swiss scientists took pride in the fact that their country hosted these international scientific organizations and, moreover, that these institutions were neutral and devoted to peaceful research only, following the wishes of the scientific community. CERN, in particular, along with the International Committee of the Red Cross, became for Swiss citizens part of their country's national identity, embodying its neutrality, and at the same time, its participation in world affairs. When the Swiss government presented the ESRO agreement in front of the parliament in 1962, it made the link clear—"our participation in ESRO follows our solidarity policy"—an essential part of the country's neutrality posture for Petitpierre.[126] Neutrality, its supporters argued, was intrinsic to Switzerland's openness toward the rest of the (free) world, and not just a self-serving policy aimed at defending a small country's economic and political interests. The putatively neutral, universal, and objective value of science was ideally suited to make this point.

[124] Albert Picot, "Le laboratoire scientifique européen de recherches nucléaires," 21 April 1952, E 2001-04(-) 1970, vol. 346, BAR.

[125] Petitpierre, "Conférence donnée" (cit. n. 13).

[126] *Feuille Fédérale* 2 (1962): 325–51, on 341.

"Signs of the Times":
Medicine and Nationhood in British India

By Pratik Chakrabarti[*]

ABSTRACT

Medical practice and research in colonial India historically had been an imperial preserve, dominated by the elite members of the Indian Medical Service. This was contested from the 1900s on by the emerging Indian nationalism. This essay studies debates about the establishment of a medical research institution and how actors imposed the political identities of nationalism on British colonial practices of medical science. At the same time, Indian nationalism was also drawing from other emerging ideas around health and social welfare. The Indian nationalists and doctors sought to build the identities of the new nation and its medicine around their own ideas of its geography, people, and welfare.

INTRODUCTION

On April 18, 1923, Lt. Col. S. H. Burnett, an officer of the Indian Medical Service (IMS), traveled from 29 Pembridge Square, London, along Hyde Park to the India House, Whitehall, to meet Edward J. Turner, undersecretary of state for India, to explain his reasons for premature retirement from his post of surgeon superintendent of St. George's Hospital in Bombay.[1] Burnett had two main grievances against the medical service in India. The first concerned the financial disincentives: lack of decent salary and the loss of private practice. The second grievance was an immediate one: "I find myself now confronted with a totally new and unexpected position in that a committee of three Indians appointed by the Bombay University are to visit and, virtually, inspect the European General Hospital reporting on it with a view to its affiliation as a field for clinical study and instruction of postgraduates among whom will, doubtless, be Indians."[2]

In December 1921, Fardunji M. Dastur (registrar, University of Bombay) had written to the government of Bombay's surgeon general that a university committee had

---

[*] School of History, Rutherford College, University of Kent at Canterbury, Canterbury, CT27NX, UK; p.chakrabarti@kent.ac.uk.

I am grateful to the Wellcome Trust for funding the research for this project. My thanks to Carol E. Harrison and Ann Johnson and the other participants in the *Osiris* conference for their comments on the previous drafts.

[1] "Lt Col SH Burnett, I.M.S., Reasons for taking leave preparatory to Retirement including Question of Examination of European Patients by Indian Medical Students," C&R 1784, 1923, L/E/7/1156, Asia, Pacific and Africa Collections, British Library, London (hereafter cited as APAC).

[2] Burnett to the Personal Assistant to the Surgeon General, Govt. of Bombay, 28 Dec. 1921, 1–2, L/E/7/1156, APAC.

© 2009 by The History of Science Society. All rights reserved. 0369-7827/09/2009-0010$10.00

been appointed to inspect St. George's Hospital.[3] Burnett, in charge of the hospital, saw a more sinister plan behind this inspection: the university was encroaching upon a British medical institution; the proposed recognition of this European General Hospital as a study field for certain postgraduates was likely to be followed by the inclusion of Indian postgraduates also.[4] The proposal of inspection by Indians, he suspected, was designed to convert the hospital into a site of regular study for Indian medical students. This was unacceptable to Burnett, as the hospital would then not remain an exclusively European institution. He concluded, "[T]he visit of the Committee I regard as signs of the times which, in conjunction with the drawbacks I have already represented, prompt me to ask for leave pending retirement."[5]

In London, Burnett urged Turner to ensure that the European hospital in Bombay should not be turned into a training ground for Indian students. Turner forwarded Burnett's letters to the government of India (GOI) and the registrar of Bombay University and urged that the case be taken up by the director general of Indian Medical Service (DGIMS) as well.[6] But he was informed that there was little the imperial government could do in this matter, as medical administration, including hospitals and provisions for medical education, following the Indian reforms of 1919, was a "transferred subject" in which only the governor of Bombay and the provincial Council of Ministers were responsible.[7] Lord William Peel, the secretary of state, wrote to Governor George Lloyd, who replied that he considered Burnett's objection "a grave one."[8] Soon, Lloyd informed Peel that the surgeon general of Bombay had assured him there was no move to make the St. George's Hospital an institution of medical instruction for university graduates. He had in fact withdrawn this facility from the university itself, and now the hospital was available only to the military assistant surgeons.[9] Lloyd reassured Peel, "I need scarcely add that I should never dream of allowing European patients in our hospitals out here to be used as clinical material for the study of Indian medical students."[10]

Although the matter seemed resolved, this was indeed a sign of the times in India. With the rise of nationalism, medical research and education were becoming contested territory. The contestation was occurring in three areas. One was the culture of medical instruction for Indian doctors. The second was the enrolment of more Indians in such institutions, and the third was the more general arena of the state and welfare.

There was growing professional pressure from Indian university-based medical faculties and students who formed the nongovernment independent medical profession (non-IMS) that had led to the antagonism and distrust of the IMS toward this group. Related to this was the question of facilities of clinical study and research for Indian doctors and students. Burnett's case also represented the increasing discontentment among the IMS officers about their service conditions, salaries, living

---

[3] Fardunji M. Dastur to the Surgeon General, Govt. of Bombay, 13 Dec. 1921, L/E/7/1156, APAC.
[4] Burnett to the Personal Assistant to the Surgeon General, Govt. of Bombay, 8 Jan. 1922, 1, L/E/7/1156, APAC.
[5] Ibid., 2–3.
[6] Turner's note to the Government of India (GOI), 1 May 1923, L/E/7/1156, APAC.
[7] Note from Dawson to Turner, 8 Aug. 1923, L/E/7/1156, APAC.
[8] Draft paragraph of the private letter from Lord Peel to Sir George Lloyd, 14 May 1923, 1, L/E/7/1156, APAC.
[9] Lloyd to Peel, Turner, and Hirtzel, 8 June 1923, L/E/7/1156, APAC.
[10] Ibid.

standards, and loss of private practice. There was also a rising anxiety among the IMS about losing control in the country's medical matters, as after 1919, public health, hospitals, and medical administration had passed to provincial legislative control.

Burnett's objection to the Indian inspection of a British institution reflected a general British fear and anxiety about Indian encroachment. What was that encroachment? Nationalism in its attempts to create its own identities and spaces had challenged some of the established norms of medical tradition that the British had so carefully established in India, a process that engendered political, physical, moral, and institutional encroachment. This had also created a disjuncture in the established ideas of medicine and the state. British ideas of public health, from the mid-nineteenth century on, were based on the Chadwickian notion that people's health was a matter of public and state concern.[11] The British had sought to introduce this notion in India as well, albeit on a limited scale. But it was precisely this convergence of the "public" and the "state" that was now being contested in India. The nationalist movement, which was growing in popularity, was challenging the authority of the colonial state to represent public concern and welfare as part of its struggle for state power. Thus in India, the question of science and national identity had become a political one. David Potter has suggested that one of the main reasons for decolonization in the postwar period was the shortage of manpower in the British Empire.[12] This crisis, he argues, developed independent of the nationalist movement. I would like to suggest that this crisis of manpower was indeed a political issue, sparked by nationalist politics.

The essay will also demonstrate that in this period, despite the obvious differences, there was also a convergence in the practices of science and medicine in the developmental frameworks of emerging nation-states. This was within a broader concomitance of industrialism and development, the Soviet model of planned economy, and the political ideology of socialism. The health planning of India in the 1930s and the 1940s, as pursued by the nationalists, was driven by the same faith and idealism that had shaped science-based development of the emerging nations, particularly following developmental plans of the Soviet Union and the intellectual realignment between science and society following the Second International Congress of the History of Science in London in 1931. This was an important alliance of science and medicine in planning national development, a plan which shared a problematic relationship with Gandhi.

RESEARCH AND MEDICINE IN COLONIAL INDIA

The university-government divide in medical research that Burnett's episode alludes to holds a key to the history of medicine in India, as well as to the history of Indian nationalism. University education and teaching had been the mainstays for the emergent Indian middle class, and universities had historically been an important site for Indian intellectuals, for their political struggles. The first generation of Indian scientists who became prominent at the turn of the twentieth century were all products of the Indian universities.

There are some notable features in this enmeshed history of science and medi-

[11] Christopher Hamlin, "State Medicine in Great Britain," in *The History of Public Health and the Modern State*, ed. Dorothy Porter (Amsterdam, 1994), 134–44.

[12] David C. Potter, "Manpower Shortage and the End of Colonialism: The Case of the Indian Civil Service," *Modern Asian Studies* 7 (1973): 47–73.

cine. David Arnold has pointed out that although there were several Indian scientists working in physics and chemistry, there were very few working in medicine.[13] A related phenomenon was that even the engagement in physics and chemistry tended to be more in fundamental science than in applied science. As noted by Y. Guay, between 1907 and 1926, while British chemists in India devoted their time exclusively to applied research, Indian scientists were more interested in pure and theoretical research.[14]

Part of the explanation for this, as argued elsewhere, was that for many Indian scientists, research was a moral and fundamental quest, part of their search for nationhood and identity in the modern world.[15] Besides, university education in India, as initiated by the British, was based on the principles of the Victorian educator Thomas Arnold, which stressed "character building." The British saw these principles as particularly relevant for Indians, whom they perceived to be in need of the virtues of science and rationality.[16] The Indian university-based science curriculum tended to focus on fundamental sciences, physics and chemistry, and mathematics.

The other reason was the British monopoly in medical sciences in India beginning in the mid-eighteenth century. This dominance was institutionalized through the formation of the IMS, a body of medical men emerging from the eighteenth-century military traditions of the English East India Company.[17] As the company's territorial control over the Indian subcontinent expanded by the middle of the nineteenth century, British doctors of the essentially military establishment had larger civilian practices as well, in the hospitals, dispensaries, and research institutions springing up in the various parts of the country. This dual role of the IMS, in military and civilian health care, was unique and crucial to its survival and influence. With the assumption in India of crown rule in 1858, public health became an important concern of the government, thereby further increasing the sphere of activity of the British medical officers. The other important characteristic of the IMS was its strong metropolitan links. The entrance exams for recruitment were held only in England, and candidates received their training almost entirely in British universities. By contrast, in the Indian university curriculum, courses in medical science and research remained rudimentary, and very few Indians joined the profession until the turn of the twentieth century.

The history of the British initiative in medical research has also to be seen against the backdrop of the lack of government investment in either science laboratories or technical education. The only investment in research by the state was in medical science, in the bacteriological laboratories, pioneered by the IMS cadres in the 1890s.

[13] David Arnold, "Colonial Medicine in Transition: Medical Research in India, 1910–47," *South Asia Research* 14 (1994): 10–35, 27.

[14] Y. Guay, "Emergence of Basic Research on the Periphery: Organic Chemistry in India, 1907–1926," *Scientometrics* 10 (1986): 77–94, 87–88. On the other side of the spectrum was the nationalist movement for technical education and institutions: Aparna Basu, "The Indian Response to Scientific and Technical Education in the Colonial Era, 1820–1920," in *Science and Empire: Essays in Indian Context, 1700–1947*, ed. Deepak Kumar (Delhi, 1991); and Basu, "Technical Education in India, 1900–1920," *Indian Economic and Social History Review* 4 (1967): 361–74.

[15] Pratik Chakrabarti, *Western Science in Modern India: Metropolitan Methods, Colonial Practices* (New Delhi, 2004), 146–214.

[16] Ellen E. McDonald, "English Education and Social Reform in Late Nineteenth-Century Bombay: A Case Study in the Transmission of a Cultural Ideal," *Journal of Asian Studies* 25 (1966): 455–60.

[17] For a study of the emergence of the IMS, see Mark Harrison, *Public Health in British India, Anglo-Indian Preventive Medicine, 1859–1914* (Cambridge, UK, 1994), 6–35.

Throughout the colonial period, hardly any facilities for research in physics or chemistry existed outside the small university laboratories. Appeals for research facilities by nationalists and scientists remained unheeded. In 1890, following the visit by the prominent physicist J. C. Bose, some distinguished scientists in the United Kingdom wrote to the secretary of State for the establishment of a central laboratory in Calcutta.[18] The GOI, however, refused to consent, citing "the present state of finances."[19] This was indeed part of a long tradition of refusing such requests.[20]

Apprehension about the uncertain consequences of the practice of science by the natives, particularly in the context of increasing nationalism, perhaps motivated this reluctance as well. In 1890, an editorial in the Anglo-Indian newspaper *Civil and Military Gazette*, while commenting on the introduction of modern science in India, mentioned that "one has to be careful because the crooked native mind could transfer any knowledge, as it had done with modern education, rather than accepting as an idea of reason, into an ideology of dissent."[21] Medical research, by contrast, was considered to be the stronghold of the imperial regime and thus a safer investment. The *British Medical Journal* supported the suggestion put forward in the Indian Medical Congress of 1894 that some of the government funds for education be used for building medical research institutes, adding that the "educated Bengali babus" were only interested in becoming "disloyal and seditious agitators."[22]

This had created a two-tiered medical profession in India. Prior to the First World War, IMS officers had manned most of the medical research institutes and university professorships, apart from the military posts, while the Indian graduates occupied the subordinate posts. Until 1913, Indians composed only 5 percent of the IMS; by 1921 their numbers had risen to compose 6.25 percent.[23] The nonofficial medical workers had formed their own organizations, such as the Bombay Medical Union (BMU) and the Calcutta Medical Club. They mainly directed their activities toward "enhancing the status and dignity of the Indian medical profession."[24] They also formed what was commonly known as the "university group."

Such a system generated areas of conflict. The growth of the independent medical profession was not totally undesired by the government as it promised more cheaply trained doctors at a time when doctors were increasingly in demand in the empire. For example, in 1899, George Hamilton, the secretary of state, had urged the GOI to

[18] "Memorial from Certain Distinguished Scientists Advocating the Establishment of a Central Laboratory at Calcutta," 12 May 1897, File 723, L/PJ/6/445, APAC.

[19] Reply from A. Godley, India Office, 21 March 1898, File 460, L/PJ/6/473, APAC.

[20] In 1898, a similar request was refused by the secretary of state on the ground that the required sum of six lakhs of rupees could not be spared as it was the time of the Afghan wars. *Hitavadi*, a vernacular newspaper, commented, "It is a wonder that a man in the position of the Secretary of State has not hesitated to make such a reply. A Government unable to spare six lakhs for a beneficial object is wasting crores in foreign wars!" *Hitavadi*, 1 April 1898, in *Report on Native Papers of Bengal Presidency*, 1898, File 345, L/R/5/24, APAC. Bose's requests for assistance for his institute in 1917 were refused as well: see "Application of Dr. J. C. Bose, CSI, CIE, for Certain Grants of Money to Enable Him to Carry Out His Scheme for a Research Institution in Furtherance of the Work in Which He Is Engaged," Department of Education, A, Proceedings, July 1917, Nos. 9–11, National Archives of India, New Delhi (hereafter cited as NAI).

[21] *Civil and Military Gazette,* 17 April 1890, 4.

[22] "A Bacteriological Department for India," *British Medical Journal (BMJ)* 1 (1897): 31–32.

[23] Roger Jeffery, "Recognizing India's Doctors: The Institutionalization of Medical Dependency, 1918–1939," *Modern Asian Studies* 13 (1979): 301–26, 311.

[24] As quoted in Mridula Ramanna, *Western Medicine and Public Health in Colonial Bombay, 1845–1895* (New Delhi, 2002), 3.

sponsor the independent medical profession in India for provincial civilian services. But the plans were dropped as British residents opposed any reduction of European medical assistance for their well-being.[25] However, by the turn of the century, two developments had become apparent: the government had realized that the growth of the IMS needed to be reduced for reasons of cost, and members of the IMS were growing anxious about the loss of their lucrative civilian practice.

These developments had followed several decades of protests by the independent medical professionals. Since the late nineteenth century, the BMU, in partnership with the Indian National Congress (INC), had been demanding an end to the monopoly of the IMS.[26] In 1913, BMU had sent its representations to the Royal Commission on the Public Services in India demanding equal status, privileges, and emoluments for the independent medical men, especially those in higher grades.[27]

The same year, the *Indian Medical Gazette*, a publication by the IMS officers, complained about the political interventions and subsequent decline in IMS recruitment in Britain: "The unrest in India, the treatment of that unrest by the authorities and the political developments of the present day, have made men hesitate before embarking on an Indian career." Thus, "civil practice is not what it was, little money can be made in many stations; moffussil life is less attractive than it used to be."[28] The British Medical Association (BMA), representing the IMS in England, also sent a memorandum deploring the present conditions of the service. According to them, the decreasing number of British men in the IMS posed a structural as well as a "grave moral" question for the future welfare of India. They stressed that India still needed the healing hands of imperial medicine: "Those who know the Indian most intimately, and who admire most intelligently his many excellent qualities as a profession man, cannot blind themselves to the fact that his standards are still far from being those of his British brother."[29] The reference here was to the unprecedented legislative control over British medical practices in India that was being imposed following the agitation by Indians, which had brought the elite service within the contemporary political spectrum. According to the BMA, the government was responsible for restricting private practice, but the opposite had occurred due to a movement that emanated "from the educated Indians who have been trained in our colleges."[30] The situation also led the British medical personnel to demand centralization and a reapportioning of responsibilities. The memorandum suggested the creation of a provincial medical service, which through its own (mostly Indian) medical officers could carry the "blessings" of Western medicine to the remote parts of India, leaving the IMS officers to concentrate on more central questions of research and public health policy.[31]

The contemporary political context was indeed becoming important. The early decades of the twentieth century in India had seen the struggle for political power. These had led to several constitutional and legislative concessions. The post–World

[25] Jeffery, "Recognizing India's Doctors" (cit. n. 23), 310.

[26] Ramanna, *Western Medicine and Public Health* (cit. n. 24), 217–21.

[27] *Representation of the Bombay Medical Union to the Royal Commission on the Public Services in India* (Bombay, 1913), 1–4.

[28] *Indian Medical Gazette* (*IMG*), Oct. 1913, 396–99.

[29] "Memorandum on the Present Position and Future Prospects of the Indian Medical Service," 1913/14, British Medical Association, Medical Appeal Board, 1, L/S&G/8/305, APAC.

[30] Ibid.

[31] Ibid., 11.

War I period provided the great push for the Indianization of medical services and further accentuated political tension. In 1919, the British introduced the Montague-Chelmsford Reforms, which prompted provincialization. The reforms introduced the concept of "dyarchy," transferring functions such as education, health, and agriculture (referred to as "transferred" subjects) to provincial legislative bodies, while retaining others, such as finance, revenue, and home affairs, as "reserved" or "imperial." This indirectly increased the number of elected Indian members in district boards and municipal corporations, since the authority to regulate local government bodies was placed in the hands of the popularly elected ministers, whose constituents naturally wanted more devolution of power.[32] The reforms also provided a further incentive for centralization of British medical involvement in India, as the IMS and medical research remained largely under central governmental control.[33]

RESEARCH AND PRIVILEGE

These changes had highlighted for the IMS officers that for them the future site of involvement in India could be, not in provincialized public health institutions, but in the centralized research institutes. This realization came with a certain sense of regret, as the British considered the introduction of modern public health and hygiene to India as one of their greatest contributions. In 1927, GOI public health commissioner J. D. Graham, in his lecture on the "Medical and Sanitary Problems of India," elaborated that these political reforms interfered with the essential and necessary "evolutionary" process of India's public health policy.[34] Graham now wanted more control to be vested in the hands of the central government, as in the old days, and defended the need for European medical men, particularly in research.[35] In another lecture Graham stressed the need for British control over medical research, especially following the creation of the Bacteriological Department within the Indian government in 1906 to staff the medical research organizations.[36] The 1919 devolution of power had made such a choice more obvious, financially and institutionally. The department was reorganized in 1922, to "make it more attractive."[37] When the Retrenchment Committee[38] proposed the reduction of expenses in research and in recruitment for the department, J. B. Smith (IMS, retired), medical advisor to the secretary of state, opposed the reduction, arguing that research opportunities became a new rationale for recruiting IMS cadres in the UK: "appointments in the Bacteriological Department are held out as an inducement to men entering the Indian Medical Service, and it is a breach of the promise made if the majority of these appointments are withdrawn."[39]

This had been the new journey of the IMS, from public health and hygiene, its proud heritage in India, to the laboratories of research. Beginning in the early de-

---

[32] See Harrison, *Public Health in British India* (cit. n. 17), 60–98.

[33] Arnold, "Colonial Medicine in Transition" (cit. n. 13), 24.

[34] *Seventh Congress of the Far Eastern Association of Tropical Medicine, Souvenir, the Indian Empire, Being a Brief Description of the Chief Features of India and Its Medical and Sanitary Problems* (Calcutta, 1927), 55.

[35] Ibid., 58.

[36] "Medical Research and Organisation," *Seventh Congress of the Far Eastern Association of Tropical Medicine* (cit. n. 34), 102.

[37] Ibid., 102–3.

[38] Headed by Lord Inchcape (1922–23).

[39] "Retrenchment in Expenditure on Medical Research in India," Dept. of Education, Health, and Lands, A, May 1925, Nos. 17–25, 9, NAI.

cades of the twentieth century, medical research in India became the site of privilege and preserve. One important aspect of that privilege was in the location of the institutes. In the British habitation of colonial India, location had been a fundamental imperial concern. From the middle of the nineteenth century, this had driven them to the salubrious hills, in an attempt to retain their racial and climatic distance from the dusty tropical plains. In the thickly wooded hills and swirling mists of the hill stations, the British had sought to build around themselves a replica of English life.[40] Anglo-Indian medical ethics had followed the same trend, and in seeking to avoid the heat and dust of tropical research it had established most of its laboratories in the hills, such as the Pasteur Institutes (from 1900) and a Central Research Institute (CRI, 1907).

The second feature of privilege in research was in funding and personnel. Pardey Lukis (DGIMS) wanted a group of trained men and resources that would not require the sanction of the secretary of state or the legislative assemblies to fund medical research and so created the autonomous Indian Research Fund Association (IRFA) in 1911.[41] As pointed out by Mark Harrison, through the IRFA, which came to symbolize "imperial efficiency" and humanitarian reform, "research-oriented medical men" managed to create a niche for themselves in colonial medicine and administration in India in the interwar period.[42] The IRFA also created a skewed financial distribution, as it contributed almost exclusively to government research institutes and hardly anything to the universities.

The IRFA had an imperial character, it had the Scientific Advisory Board to advise on technical matters, but the real control was vested in the Governing Body, initially consisting only of the DGIMS and the sanitary commissioner of India.[43] The issue of preserving medical research for the IMS officers had become important by the 1920s, and research organizations were becoming the new sites of hope for the British recruits. The retired IMS officer and an expert on Indian medical affairs, Leonard Rogers wrote in the *BMJ*, while discussing the present problems of the IMS, "There remains one feature of the position the importance of which for the future of India and the IMS can hardly be exaggerated. The development of the research department, so far from being checked by recent difficulties, has received a definite impetus during the past few months."[44] On another occasion, Rogers mentioned that one of the main attractions of the IMS was the IRFA and the research department.[45]

A CENTRAL MEDICAL RESEARCH INSTITUTE

These pressures and the inadequacies faced during the First World War made the GOI look into the matter of medical research more closely. Plans were being made for a Central Medical Research Institute (CMRI) as the facilities of CRI in the hill station of Kasauli were felt to be inadequate. In 1920, the DGIMS felt it to be "a matter of

[40] Dane Kennedy, *The Magic Mountains: Hill Stations and the British Raj* (Berkeley, Calif., 1996).
[41] Helen J. Power, "Sir Leonard Rogers FRS (1868–1962): Tropical Medicine and the Indian Medical Service" (PhD diss., Univ. of London, 1993), 148–49.
[42] Harrison, *Public Health in British India* (cit. n. 17), 165.
[43] Arnold, "Colonial Medicine in Transition" (cit. n. 13), 14.
[44] *BMJ* 1 (1929): 1168–69. Rogers was the Indian correspondent for *BMJ* between 1898 and 1929.
[45] "Notes on the Indian Medical Service, 1930," ROG/C.19/22, Leonard Rogers Papers, Manuscripts, Wellcome Library, London (hereafter cited as Rogers Papers).

urgency to establish an additional research institute in a central position."[46] The same year, Professor E. H. Starling of University College, London, sailed for India to advise the GOI on the location and establishment of the new institute. *BMJ* reported that the IRFA was to fund it, and Delhi was being considered as the possible site.[47]

Starling drew up a detailed plan and proposed the erection of an all-India medical research institute at Delhi, the imperial capital since 1911.[48] The GOI accepted the scheme, and the secretary of state gave his approval.[49] But following the recommendations by the Retrenchment Committee, the plan was postponed indefinitely.[50] In 1927, the issue was raised again, and this time Walter Fletcher, secretary of the Medical Research Council (MRC), was appointed the head of another committee to look into the matter.

Fletcher's arrival in India was significant. One of the staunchest advocates of the primacy of medical research in contemporary Britain, Fletcher, as the secretary of the MRC from 1914 to 1933, had played a crucial role in the organization of medical sciences in Britain.[51] A distinguished laboratory physiologist, Fletcher became a prominent statesman and administrator in the interwar period. As a strong believer in research, he had stressed that medical practice had to be based on research as much as in hospital wards.[52] The interwar period in Britain saw a struggle between scientists and medical practitioners about the nation's medical policy, and Fletcher was engaged in ensuring the supremacy of MRC.[53] Fletcher's clash with Lord Dawson, president of the Royal College of Physicians, regarding the status of the MRC within British medical practice was an important episode in this.[54] Fletcher was also a keen advocate of autonomy in research and was responsible for putting the MRC on the same footing as the Department of Scientific and Industrial Research (DSIR), relatively free from governmental control.[55]

Fletcher's agendas of prominence and autonomy of medical research, which had shaped contemporary British clinical medicine, corresponded perfectly with the interests of the IMS officers in India, although the motives were very different here. Fletcher suggested some important modifications to Starling's plans. For Fletcher, the question of location was fundamental for autonomy in research in India. His *Report of the Committee on the Organization of Medical Research*, issued in 1929, suggested that seclusion was paramount, however, in the colony; the agenda of seclusion was also linked to the efforts to escape from the heat, dust, and chaos of the tropical and political plains.[56] The report created a new link between climate and research: "it

---

[46] "Proposed All India Research Institute" *BMJ* 1 (1920): 344.

[47] Ibid.

[48] *Report of the Committee on the Organization of Medical Research under the Government of India* (Calcutta, 1929), 15–16.

[49] Ibid., 15.

[50] "Retrenchment in Expenditure on Medical Research" (cit. n. 39), 7.

[51] Joan Austoker, "Walter Morley Fletcher and the Origins of a Basic Biomedical Research Policy," in *Historical Perspectives on the Role of the M.R.C.: Essays in the History of the Medical Research Council of the United Kingdom and Its Predecessor, the Medical Research Committee*, ed. Joan Austoker and Linda Bryder (Oxford, 1989), 23–33.

[52] Ibid., 24.

[53] Ibid., 29–31.

[54] Ibid., 31–32.

[55] Ibid., 28–29.

[56] *Report of the Committee on the Organization of Medical Research* (cit. n. 48), 43.

is equally certain that basic research can best be done in a climate favourable to the energy and mental acuity of the workers and moreover, in a climate where delicate technical processes and procedures in the laboratory as well as animal experiments, are not interfered with by extreme heat."[57] Delhi, which had been recommended by Starling, was rejected as it was considered too hot.[58] Bombay was rejected as it already had a small provincial laboratory. While rejecting the other main cities, the report gave the same reason: "What we have said about Bombay applies also to Calcutta," and "Our remarks on Bombay and Calcutta are applicable to Madras."[59] It is important to mention that large urban centers in India had become the main sites of the nationalist movement, particularly during the Non-Cooperation Movement (1921–22), which under Gandhi's leadership had shaken the British government.[60]

Fletcher's choice was Dehra Dun, in the salubrious Garhwal foothills of northern India.[61] A small cantonment town, Dehra Dun had been built by the British as a site for imperial institutions, including the Imperial Forest Research Institute and the Geodetic Branch of the Survey of India, and was also the projected site of the Royal Indian Military College, the Railway Institute, and Telegraph Headquarters. Colonial officials favored Dehra Dun as a retirement site.[62] According to the report, these institutions provided an ideal atmosphere for scientific discourse. It also had a "Leper Hospital for 80 beds," which was considered adequate for providing clinical materials for the proposed CMRI.[63]

Fletcher's agenda of autonomy in research thus translated itself in the colony into an urge to retain control of research in British hands. Before leaving for India, Fletcher, who had little prior knowledge about the country, had met Leonard Rogers in London, and the latter had updated him about the "Indian affairs." Fletcher wrote to Rogers gratefully: "Thank you indeed for the confidential memorandum about Indian affairs that you have been good enough to send. Now I hope you will pray for me in the very anxious task I see ahead, for which ignorance seems to be my only qualification."[64] Rogers had made Fletcher apprehensive of Indian involvement. In December 1927, Victor Heiser, the Rockefeller Foundation's International Health Board's director for the East, met Fletcher in a "confidential conference" in Calcutta. Heiser found that Fletcher "fears greatly that the laboratory may soon pass

---

[57] Ibid.

[58] Ibid., 44–45.

[59] Ibid.

[60] Sumit Sarkar, *Modern India, 1885–1947* (New Delhi, 1989), 204–25.

[61] *Report of the Committee on the Organization of Medical Research* (cit. n. 48), 45.

[62] John F. Richards, "Environmental Changes in Dehra Dun Valley, India: 1880–1980," *Mountain Research and Development* 7 (1987): 299.

[63] *Report of the Committee on the Organization of Medical Research* (cit. n. 48), 45.

[64] Fletcher to Rogers, 9 Nov. 1927, Medical Research Council, Adelphi, London, "Correspondence with Sir Walter Fletcher, Medical Research Council 1927–33," ROG/D.5/1–11/1, Rogers Papers. Rogers was a very influential figure. An IMS officer with a long medical career in India, a leading figure in tropical medicine and the founder of the Calcutta School of Tropical Medicine, he was also the staunchest critic of the Indianization of IMS. After his retirement in 1921, he was appointed a member of the India Office Medical Board in 1922 and its president in 1928, with collateral duty as medical adviser to the secretary of state. He used this latter position along with his status as the Indian correspondent of *BMJ* to resist any move by the government or Indians to provide concessions to non-IMS medical practitioners. George McRobert, "Rogers, Sir Leonard (1868–1962)," *Oxford Dictionary of National Biography* (Oxford, 2004).

into Indian hands unless special means are devised to change present tendencies."[65] Thus, Fletcher's other emphasis was on recruitment of staff. The report suggested the appointment of a committee, ostensibly to secure a closer liaison between medical research organizations in India and in Britain. In effect, it was to foster closer links between the metropolis and the empire by making the issue of recruitment a metropolitan one, superseding the political voices in India. The committee was to be composed of a representative of the India Office, of the Ministry of Health (UK), of the MRC, and of the Royal Society. It would advise on general recruitment policy for Indian medical research workers as well. The Fletcher report also recommended that twenty-three out of thirty posts be reserved for IMS officers.[66]

Fletcher's report reflected an interesting convergence of contemporary metropolitan and imperial concerns of laboratory research. The colonial scientists saw in the MRC and its arguments for autonomy in research a model for research in the colony. They found strong parallels between the MRC and the IRFA, particularly in their political insulation. When there were demands to include more legislative and Indian members into the IRFA, Rogers countered by showing how the IRFA had been "politicised" and "liberalised" over the years, unlike the MRC.[67] Rogers also wanted the MRC to play a more active role in funding imperial research in India and Africa.[68] The Fletcher report gave a formal shape to such trends.[69]

With the submission of the report, the mood among the British medical men was confident. Rogers supported the idea that "suitable research men will in future have to be recruited mainly in Great Britain."[70] The GOI acted quickly on the proposals and passed the financial approval through a Standing Finance Committee on August 29, 1928, soon after the report was submitted, not waiting for it to be discussed in the Legislative Assembly.[71] Edward Turner wrote to Arthur Hirtzel, undersecretary of state, in June 1929, that the secretary of state had approved of the formation of a selection board in India and a consultative board in England for the appointment of the scientists in the institute.[72]

While the government moved ahead with the plans, the Indian Legislative Assembly was questioning the motives behind them, particularly those concerning lo-

[65] Diary of Dr. Heiser's World Trip, 17 Oct. 1927 to 1 May 1928, 53–54, Rockefeller Archive Center, New York. Fletcher had forged a strong relationship between MRC and the Rockefeller Foundation, and in the interwar period the RF provided Britain with £2,500,000 for medical teaching and research. Austoker, "Walter Morley Fletcher" (cit. n. 51), 28.

[66] Report of the Committee on the Organization of Medical Research (cit. n. 48), 48–62.

[67] "Public Health Institute Calcutta—Founded with Assistance from Rockefeller Foundation Central Medical Research Institute. Scheme in Abeyance of Fletcher Committee Report on the Organization of Medical Research," File 7B, 148, L/E/9/610, APAC.

[68] Rogers to Graham, 17 Nov. 1927, ROG/A/55/76, Rogers Papers.

[69] However, A. V. Hill, in his famous report on scientific research in India, was careful to point out the limits of this comparison. While mentioning that the MRC in the UK "has an extremely free hand" and that the "Indian Research Funds Association (IRFA) plays a rather similar role in India" he added: "Its [IRFA's] funds, however, are very scanty . . . Because of its poverty it cannot take the same broad view of its functions as the MRC does: for example, to maintain even a single clinical research unit in one of the medical colleges would be financially out of the question until ampler funds are available." A Report to the Government of India on Scientific Research in India (London, 1944), 20–21.

[70] Rogers, note, "Public Health Institute Calcutta" (cit. n. 67), E&O 6102, 1928, 6, File B, L/E/9/609.

[71] Extract from Volume VIII, No. 1 of Standing Finance Committee, 29 Aug. 1928, L/E/9/610, APAC.

[72] Turner, undersecretary of state, to Fletcher, India Office, 7 June 1929, 3449/28 E&O 6163, L/E/9/609, APAC.

cation and personnel, two important areas of nationalist contestation. Members of the assembly raised questions about whether the Fletcher report had been reviewed before approval and about the choice of Dehra Dun as the site.[73] In March 1929, M. S. Aney, leader of the INC and member of the Central Legislature, asked, "Are the Government aware that there is a strong feeling in the Indian medical profession against the recommendation of the Fletcher Committee to reserve a very large proportion of the cadre of the appointment in the Central Medical Research Institute for members of the Indian Medical Service?"[74] In June, the BMU wrote a note protesting the plans for the CMRI. They objected to two aspects of the Fletcher report: the choice of Dehra Dun—"a far-away place, removed from all facilities for clinical work"—and the reservation of almost all posts for IMS cadres.[75] The president of BMU, Dr. G. V. Deshmukh, in a letter to the secretary of the Department of Education, Health, and Lands, alleged that the government was trying to rush the matter without a proper debate.[76] He added that the institute should be located in a university town and that the members of the independent medical profession wanted to participate in the debate. The isolated and imperial charters of Indian hill stations were increasingly subject to the pressures of political criticism in this period. This came as much from outside by leaders from the plains, as it did through the emergence of nationalist awareness among the resident population of shopkeepers, skilled workers, and the laboring poor of the hill stations.[77] The critique of Dehra Dun as the site of the institute was part of this movement.

Thus, while for the British scientists the issue of location was about isolation, for the nationalists it was about inclusion. The *Calcutta Medical Journal*, a publication of the Calcutta Medical Club, protested the selection of Dehra Dun, saying that such a central research institute would have to deal with the problems of the whole of India.[78] It stressed that medical research also needed to be closely linked to the "great sites of learning" in India.[79] The *Bombay Chronicle*, a nationalist newspaper, complained that while plans were going ahead for the establishment of a medical research institute for some time, "it is surprising that . . . no attempt has been made to secure the confidence and invite the co-operation of the public, including the independent medical profession in the country."[80] The newspaper also criticized the government's moves to secure funding with such "unseemly haste," without discussions in the Legislative Assembly. About Dehra Dun, the paper was sarcastic. "Dehra Dun is a nice, cool station, affording to the presumably European staff to be engaged for such investigations a perpetual holiday there on the adjacent hills."[81] The *Bombay Chronicle* supported the BMU demand that the new institute should be in a university town. The

[73] *Extract from Official Report of the Legislative Assembly Debates*, 14 March 1929, 1843, L/E/9/610, APAC.

[74] Ibid.

[75] "Copy of Extract from 'The Servant of India,' Poona, Thursday, 13 June 1929," 3449/28, 1–2, L/E/9/609, APAC.

[76] G. V. Deshmukh to Secretary of Dept. of Education, Health, and Lands, E&O 5586 1929, 16 May 1929, L/E/9/609, APAC.

[77] Pamela Kanwar, *Imperial Shimla: The Political Culture of the Raj* (Delhi, 1990).

[78] "Medical Research in India," *Calcutta Medical Journal* 23 (1929): 613–14, 613.

[79] Ibid., 613–14.

[80] "Medical Research in India," reprinted from the Editorial, *Bombay Chronicle*, 4 June 1929, 2, E&O 5586/1929, L/E/9/609, APAC.

[81] Ibid.

issue soon became part of the Indian university movement, where nationalists had a long-term presence. The All-India Inter-Universities Conference of 1929 in Delhi passed resolutions asking the government to appoint a committee, with representatives of the medical faculties of Indian universities, to report on the most suitable university centre for locating the proposed research institute.[82]

A prominent doctor and dean of Gordhandas Sunderdas (GS) Medical College of Bombay, Jivraj N. Mehta, wrote to Dr. Drummond T. Shiels, undersecretary of state, that the existing institutes in Dehra Dun, such as the Geodetic Branch of the Survey of India, the Railway Institute, and the Imperial Forest Research Institute, had little to contribute intellectually to medical research. The real motive, he suspected, was privilege and autonomy, with the institute set to become the "Eton or Harrow" of India.[83] Mehta reiterated the nationalist demand that medical research had to go "hand in hand" with medical education.[84] He also suggested that it had been possible to pass such a proposal through the IRFA because almost all its members were high government officials, which was not the case in the Legislative Assembly.[85]

In London, Leonard Rogers remained dismissive of this nationalist challenge and advised the government not to pay much attention to these "unofficial" medical men.[86] He urged the GOI to go ahead with the plans, which had been developing since the end of the war.[87] Rogers prepared another note on recruitment. He pressed for a committee formed in England for the recruitment of the proposed institute as recommended by Fletcher.[88] Rogers suggested that the committee should have representatives from the Royal Society as they had recently started a Tropical Diseases Committee. "Such men as Professor Nuttall of Cambridge would be very suitable," as well as experts from the London School of Hygiene and Tropical Medicine (LSHTM). Rogers also wanted university scholars from Edinburgh, "and I know that a first class man is willing to serve on it if asked to do so."[89]

Such a response sparked intense protests in India. In the Bombay Council, in October 1929, Dr. M. K. Dixit of the Surat Medical Union moved a resolution against Dehra Dun as the site.[90] Ebrahim H. Jaffar moved a resolution in the Shimla Council in September, warning the government "that they cannot, they dare not, accept such a recommendation if they desire to retain the confidence of the medical profession and the general public."[91] The government, however, still seemed to be in an uncom-

[82] "Statesman," 7 Nov. 1929, "Medical Research in India; I.M.S. Recruitment," New Delhi, 2359/21, L/E/9/609, APAC.

[83] Mehta to Shiels, 27 June 1929, 3–4, L/E/9/609, APAC.

[84] Ibid., 8.

[85] Ibid., 9–10.

[86] Leonard Rogers, note on the petition of an unofficial Bombay Medical Institute to postpone for further consideration the eleven-year-old scheme for a Central Research Laboratory of the Research Fund Association of India, 9 July 1929, 1, L/E/9/609, APAC.

[87] Ibid., 1–2.

[88] "Recruitment of Medical Research Workers, Recommendation of Fletcher Committee Regarding Setting up of Consultative Board in This Country," 25 Oct. 1929, E&O 6159, 1–2, L/E/9/609, APAC.

[89] Leonard Rogers, note, "Proposal of the Fletcher Committee to Form a Consultative Committee at the India Office," 21 Oct. 1929, 1–2, L/E/9/609, APAC.

[90] Extract from Official Report of the Bombay Legislative Council Debates, 8 Oct. 1929, 590, L/E/9/609, APAC.

[91] Extract from the Council of State Debates, vol. 2, no. 2, "Hon'ble Sir Ibrahim Haroon Jaffer's resolution in connection with the location of the Central Medical Research Institute. The Council met in the Council Chamber," Simla, 17 Sept. 1929, E&O 9492, 2–3, L/E/9/609, APAC.

promising mood and responded, "The Government of India . . . see no reason to alter their decision that the Institute should be located at Dehra Dun as recommended by the Fletcher Committee."[92] Plans for the elite committee in England went ahead. The secretary of state, following Rogers's suggestions, had written to the Medical Research Council, the Royal Society, and the LSHTM to send their representatives to the committee, and they had all warmly accepted.[93]

Meanwhile, the nationalists continued to press forward with their agenda. M. R. Jayakar[94] moved another resolution that Dehra Dun as the site for the institute should be reconsidered and a committee be appointed for that purpose.[95] Discussions in the Legislative Assembly were intense, and Frank Noyce, undersecretary to the GOI, suggested a compromise: a conference would be convened consisting of representatives of the medical faculties of Bombay, Calcutta, Lahore, Lucknow, and Madras to discuss the issue. The government promised to give "their fullest and most careful consideration" to the conference's recommendations.[96] Jayakar accepted the suggestion on the condition that three issues—location, recruitment, and funding—were included in the agenda.[97] He ended on an ominous note: "I hope the Government will realise that the days of isolated hilltops are gone for ever."[98]

Rogers reacted sharply to the proposal of the conference, calling it a "surrender."[99] For him the greatest blunder was in opening the recruitment issue for discussion, which had grave moral and physical implications. "[L]iterally millions of lives, now annually sacrificed to such scourges as cholera, malaria, plague etc., are at stake, as nothing but the best qualified research workers the world can produce are required in India at the present time."[100] He stressed that the funding and recruitment issues should not be sanctioned for discussion.[101] Hirtzel agreed with Rogers, describing these political interferences in a matter of science as a "gross example of that prostitution of every other consideration to political considerations."[102]

The conference was held at Shimla on July 21–22, 1930. Noyce was the chairman; among the members were J. W. D. Megaw (DGIMS), J. D. Graham, and S. R. Christophers (director of CRI). The Indian contingent had a strong university representation: T. Krishna Menon Avargal (Madras University), M. N. Saha (Allahabad University), and Dr. J. N. Mehta (medical faculty, Bombay University). Zia Uddin Ahmad was the representative of the Legislative Assembly. Significantly, the two agendas to be discussed were the site and constitution of the IRFA; the recruitment issue was left

[92] "Location of the Central Medical Research Institute at Dehra Dun," 7 Jan. 1930, E&O 9492/92, L/E/9/609, APAC.

[93] 5 Dec. 1929, E&O 9038; 6 Dec. 1929, E&O 9068; and 10 Dec. 1929, E&0 9168, L/E/9/609, APAC.

[94] Mukund Ramrao Jayakar (1873–1959), barrister, member of Bombay Legislative Council and vice chancellor of Poona University.

[95] Extract from the Legislative Assembly Debates, vol. 1, no. 12, 8 Feb. 1930, New Delhi, E&O 2919, 1, L/E/9/609, APAC.

[96] Ibid., 3–4.

[97] Ibid., 7.

[98] Ibid., 8.

[99] Roger to the Secretary of State, 29 April 1930, E&O 2919, 1, L/E/9/609, APAC.

[100] Ibid., 4–5.

[101] Ibid., 5.

[102] Hirtzel, handwritten note, 7 May 1930, E&O 2919, L/E/9/609, APAC.

out.[103] Noyce explained that the question of recruitment was considered settled in a 1928 press communiqué by the secretary of state, who saw no need to raise it again. The Indian delegates decided to discuss the issue nevertheless.

The conference led to a protracted, often fractious, discussion between the Indian medical men and the British experts. The main objective of Indian scientists concerning location was that the institute should be located in a university town in the plains—in Calcutta, Bombay, or Madras—which would ensure a closer link between research and education. Mehta even quoted from the earlier Starling committee, which had rejected the concept of a hilltop laboratory in favor of Delhi.[104] The agenda of location of the institute was intrinsically linked to the other issue of recruitment. An institute in a university town could recruit and train more Indian graduates. The final resolution, on which Christophers and Graham as advisory members did not vote, reflected the nationalist agenda. It was resolved that a Central Medical Research Institute should be located at a university center.[105] Large-scale changes in the governing body of the IRFA were also proposed, with more members from the Department of Education, Health, and Lands and the Legislative Assembly, and from among nonmedical scientists and the medical faculties of universities.[106] There was an addendum about recruitment: "the question of the reservation of posts in the Medical Research Department for the Indian Medical Service should have been referred to this Conference."[107]

These debates in India, to some extent, reflected another debate in contemporary medicine: that between research scientists and medical practitioners, between laboratories and hospitals, regarding the appropriate approaches to public health. In colonial India, to the IMS officers—the proud inheritors of the traditions of public health now faced with the political challenges—research and autonomy seemed the more attractive option. To Indian medical practitioners, by contrast, research needed to be linked to the hospitals and the universities of the country.

"AN IMPASSE HAS NOW BEEN REACHED"

Forwarding the details of the conference to the government, J. D. Graham attached little credibility to the conference's resolutions. He described them as "both anomalous and confusing," because a university group had debated upon a nonuniversity issue.[108] He urged the government to keep medical research separate from the university matters. He alleged that it was, in fact, the Indian universities that were the real sites of insularity regarding the issues of public concern in India, "nor did these gentlemen have any clear idea of the needs of India as a whole in this respect."[109]

What had bothered the IMS officers most was the political intervention in medi-

---

[103] "Proceedings of the Conference Held at Simla on the 21 and 22 July 1930 to Discuss the Location of the Proposed Central Medical Research Institute and Other Matters," Calcutta, 1931, ii, V/27/850/4, APAC.

[104] Ibid., 9.

[105] Ibid., 109.

[106] Ibid.

[107] Ibid.

[108] "Notes on the Present Position of the Indian Medical Research Association in View of Recent Proposals for Reorganization," 6 June 1931, L/E/9/609, APAC.

[109] Ibid.

cal research. Rogers stressed that the Indian contingent in the conference was more a *political* outfit than a scientific one. "*Not a single one of the voting members had ever engaged in medical research work.*"[110] They were, what he called, the "politically minded university group."[111] The conference thus could not be regarded as being representative either of public opinion or of people connected with medicine. The conference and its resolutions to him were a "calamity" that had grave implications for the millions of Indians:

> An impasse has now been reached, through the sanctioning of this ill-advised conference, in which the Government of India and the Secretary of State must face, once and for all, the responsibility of deciding whether the efficiency of medical research in India, on which the future health of 350 million souls largely depends, is to be sacrificed to political expediency or not . . . Real courage will now be required to avert the calamity which must inevitably result, if the hitherto efficient administration of the medical research department of the Govt of India by a scientific Governing Body is subjected to the ruinous political control of the non-research university representatives; as demanded by their majority group at the recent conference.[112]

This dislike of politics, as evident in Rogers's words, went along with a preference for "efficiency" and excellence. This particular inclination had a complex lineage. Part of it was the product of the national efficiency movement in British Edwardian politics, which sought to revitalize the country in the face of its loss of position economically to rising powers such as Germany and the United States.[113] Part of it was that modern science and development itself represented a language of efficiency against the seemingly chaotic muddle of politics. For the British, this language and politics of efficiency made particular sense in an increasingly politicized India. It had become the rationale behind arguments for preserving their control over medical research and other aspects of Indian civil life. Defending the older constitution of the IRFA with its British character, Rogers, in a 1932 note to the secretary of state, reiterated the link between efficiency and morality: "if the Secretary of State should be forced to conclude that efficiency must be sacrificed to political considerations, that sacrifice shall be made in clear knowledge of the facts, and with a full sense of responsibility for the inevitable resulting future loss of life from epidemic and other preventable diseases in India."[114] In London, meanwhile, Fletcher had learned about the Shimla conference and its resolutions. In a letter to Rogers, he expressed his contempt for Indians and their politics: "It was the authentic chattering of the bandar-log."[115] Fletcher suggested that now British involvement in Indian science and medicine was in question. "If, in a perfectly straightforward piece of scientific business like this, the Government of India show such vaccilation [sic], and play such a feeble

[110] Rogers, Minute Paper, Economic and Overseas Department, 1 May 1931, 2, L/E/9/609, APAC (emphasis in original).

[111] Ibid., 1–2.

[112] Rogers, note, 1 May 1931, 5, L/E/9/609, APAC, 9–10.

[113] G. R. Searle, *The Quest for National Efficiency: A Study in British Politics and Political Thought* (Berkeley, Calif., 1971).

[114] Rogers to Secretary of State, 16 Feb. 1932, File 7B, 147, L/E/9/610, APAC.

[115] Fletcher to Rogers, 8 May 1931 [Private, confidential], ROG/D.5/1–11, Rogers Papers. *Bandar-log* is "monkey-people" in Hindi, a racially derogatory term used by the British for Indians. Fletcher was known for his strong language and opinions and was sometimes "too trenchant in his denouncement of others." Quoted in Austoker, "Walter Morley Fletcher" (cit. n. 51), 24.

game against ignorant and divided but sharp-witted antagonists, it makes one doubt whether we have any business to be in India at all."[116]

Rogers's words and Fletcher's sentiments proved to be prophetic. In 1932, a group of government medical personnel wrote to the secretary of state opposing most of the resolutions of the Shimla conference. According to them, the composition of the conference was biased toward nonofficial members, which made the resolutions prejudiced.[117] They also opposed the conference's propositions for the reorganization of IRFA.[118] Most important, they noted that a new institute funded by the Rockefeller Foundation was being planned in Calcutta, so the plans for the central research institute could be postponed. The current financial crisis, too, made the plans untenable.[119] The CMRI was never to be established in British India, and the Rockefeller funded institute was to become a site mainly for training, not research.

The impasse that Rogers had referred to was an important one. It demonstrated that at certain points, accommodating the colonial identities of science and medicine to their emergent national ones had become untenable. This was the changing identity of science in India at this moment of political transformation. The hilltop laboratories, the IMS cadre, the IRFA, and the moral imperatives of an "evolutionary" public health represented the identity of the imperial science. The transformation described above was not just from imperial to national science. The spaces of negotiation had broken down also because British imperial medicine increasingly found itself in a changing world. While nationalist ideology questioned the motives, aesthetics, and ethics of imperial research, international health initiatives and funding, from the interwar era, had fast changed the norms of colonial health policy and research. Already by the 1920s, the schools of public health of American universities such as Johns Hopkins, Yale, and Harvard had become the centers of international public health instruction, attracting large numbers of foreign students. Significantly, A. V. Hill, in 1944, while drawing up plans for the future national scientific institutions of India, mentioned that the postindependence all-India medical center should be an "Indian Johns Hopkins."[120] In addition, organizations such as the Rockefeller Foundation were increasingly determining tropical health policies and research funding in Asia and Africa.[121] Moreover, as we shall see next, the rationales of imperial medicine had become untenable in another respect. The identity of the new nation and its medicine was to be built around new ideas of state, people, and welfare.

"A REAL AND PRACTICAL IDEALISM"

While there was a stalemate reached at Shimla, there was new hope emerging elsewhere. The 1930s and 1940s were also a period of new dreams and visions around medicine and public health, particularly for the emerging nation-states. The main exposition of this was in the ideas of "socialized medicine," which defined much of In-

---

[116] Fletcher to Rogers, 8 May 1931, ROG/D.5/1–11, Rogers Papers.

[117] P. Willingdon, W. Chetwode, G. Raine, J. Crerar, G. E. Schuster, B. L. Mitter, Ahmad Said, and J. W. Bhore to Sir Samuel Hoare, 18 Jan. 1932, E&O 585, File 7B, 153–7, L/E/9/610, APAC.

[118] Ibid., 156.

[119] Ibid., 154–55.

[120] Hill, *Report to the Government of India* (cit. n. 69), 17.

[121] John Farley, *To Cast out Disease: A History of the International Health Division of the Rockefeller Foundation (1913–1951)* (New York, 2004).

dia's public health planning of this period. Henry Sigerist played an important part in promoting this new ideology around health and development. A prominent critic of American health care, Sigerist stressed the need for a national health service and a socially equitable distribution of health care with funding by the state. He stressed that "all the people should have medical care, irrespective of race, creed, sex, or economic status, and irrespective of whether they live in town or country."[122] The model behind his ideas was the Soviet Union, and in his *Socialised Medicine* (1937), he promoted its structure of health care to be adopted by other countries. According to him, the Soviet Union was the first country to socialize medicine and recognize that protecting the health of all the citizens was the responsibility of the state.[123] These were not just issues of health and well-being; they were also overtly political and ideological issues. According to Sigerist, socialized medicine could exist in its true form only as an integral part of a completely socialized state. Here Sigerist differed from the contemporary proponents of "social medicine," who stressed the more general social application of medicine and a convergence of medical and social sciences. For Sigerist, socialized medicine meant "socialist medicine."[124] His ideas and his writings in the history of medicine had their ideological links with those of J. D. Bernal, who following his Soviet inspiration criticized social inequalities of science, developed plans for socializing science, and stressed the interaction between scientific, technical, and economic development. It is also important to mention that in the interwar period, the BMA and the medical profession of Britain had been challenged by a small group of radical socialist physicians under the Socialist Medical Association.[125] This group too had been deeply influenced by the developments in the Soviet Union and the principles of socialist medicine.[126] This new ideology thus required a new politicization of the question of health care and a fresh alignment between the public and the state.

Despite ideological opposition from several sections in the West, Sigerist's *Socialised Medicine* had become immensely popular among Indian medical personnel. Soon after its publication, Sigerist had come into contact with Indian doctors and administrators who were impressed by his work. In 1941, Dr. Kamala Ghosh, who had worked in India for eight years as part of the Women's Medical Service and then pursued further studies in England, traveled to the United States for a few weeks. It was here that she found a new direction after reading *Socialised Medicine*. She wrote to Sigerist from New York: "reading your book on 'Socialised Medicine' a few weeks ago, all the problems in public health & education that I have been facing, were presented in an entirely new aspect, I became suddenly capable of solution."[127] She decided to extend her stay in the United States to pursue a master's program at

[122] Henry E. Sigerist, "Medical Care for All the People," *Canadian Journal of Public Health* 35 (1944): 253–67.

[123] Henry E. Sigerist, *Socialised Medicine in the Soviet Union* (London, 1937).

[124] Dorothy Porter and Roy Porter, "What Was Social Medicine? An Historiographical Essay," *Journal of Historical Sociology* 1 (1998): 92–93.

[125] The SMA was responsible for instituting the postwar National Health Service in Britain, despite opposition from the BMA. See John Stewart, *"The Battle for Health": A Political History of the Socialist Medical Association, 1930–51* (Aldershot, UK, 1999).

[126] Stewart, "Socialist Proposals for Health Reform in Inter-War Britain: The Case of Somerville Hastings," *Medical History* 39 (1995): 338–57.

[127] Ghosh to Sigerist, 23 March 1941, "Correspondences with Dr Kamala Ghosh, 1941–43," folder 176, box 5, Henry Sigerist Papers, Manuscripts and Archives, MS 788, Yale University Library, New Haven, Conn. (hereafter cited as Sigerist Papers).

Johns Hopkins under Sigerist. Sigerist helped her to develop her research scheme, and Ghosh focused on the activities of local district boards and municipalities in India.[128] When she finished her course, she wrote to Sigerist, filled with enthusiasm for her future work in India. "This whole year has meant a very great deal to me— an entire new world of thoughts & ideas has opened out, something I was totally unaware of before; & I know it is going to make a difference in my work when I get home."[129] Although, tragically, Ghosh was killed on her way back to India, in Sigerist and in socialized medicine she had discovered a new vision for the future. As her sister Bimala Wallis wrote: "she found, for the first time in her life, a real and practical idealism."[130]

Sigerist visited India in December 1944, as an invitee to the Bhore Committee. (The committee had been set up in 1943, following nationalist demands for better health planning in India, under the chairmanship of Joseph Bhore, a lawyer and an ICS officer.) Sigerist realized that he was a familiar figure among Indians who knew about his knowledge of Arabic and Sanskrit and their classical texts. But it was his book on Soviet medicine that had been more widely circulated and read. "In Lucknow I was told that there was only one copy in town—it was out of print—but that it had been circulated and that every doctor had read it."[131] Sigerist's paper on medical education had been mimeographed by the Bhore Committee, and copies had been sent to all the members of the committee and to the deans of all the medical colleges. The Bhore Committee accepted a number of his recommendations and incorporated them in its final report.[132]

For India, Sigerist suggested the same doctrine of socialized medicine, in which medicine and socialist developments were to be integral. "It is quite obvious that a health plan for India can hold a promise of success only if it is an integral part of a general economic and social plan."[133] Socialist planning with industrialization was to be the basis of the new regime of health, "the electrification of the country must be the backbone of any health programme." Electrification would lead to better agricultural productivity and higher income for the rural population. It would also provide power for the development of industries. The formula of development was straightforward: "[a] rising material and cultural standard would decrease the death rate and also the birth rate."[134]

Under Sigerist's influence, the Bhore Committee report became a documentation of socialized medicine in its insistence that public health was the fundamental responsibility of the state. The report made a case for a national health service, making references to his work on Soviet medicine. It was to also become the blueprint of India's future medical infrastructure.

In this stress on medicine and planned social development, the Bhore report had a precedent in India. The National Planning Commission (NPC) in 1939, created by the nationalists under Jawaharlal Nehru's leadership, advocated a state-sponsored

---

[128] Ghosh to Sigerist, 23 July 1941, folder 176, box 5, Sigerist Papers.
[129] Ghosh to Sigerist, 19 Oct. 1942, folder 176, box 5, Sigerist Papers.
[130] Bimala Wallis to Dr. Charlotte Silverman, 9 July 1943, folder 176, box 5, Sigerist Papers.
[131] *Report on India*, folder 74, 1, box 35, Sigerist Papers.
[132] Ibid., 2.
[133] Ibid., 9.
[134] Ibid., 10.

planned development following the Soviet model.[135] The INC established the NPC in 1938 to draw up the blueprint of the social and economic reconstruction of postindependence India, marking the beginning of socialist planning in India. While developing the scheme of the NPC, around the same time that Sigerist published *Socialised Medicine*, Nehru was drawing equally from the developmental experiences of the Soviet Union.[136] The NPC had a subcommittee charged with drawing up the provisions of public health and medicine, with Sahib Singh Sokhey, an Indian IMS officer, in charge.[137] The committee had recognized poverty as the main cause of ill-health, and the cornerstone of the scheme was a community health worker for every 1,000 village population.[138] Maintaining that the health of the people was the responsibility of the state, it stressed the need to integrate curative and preventive functions under a single state agency. However, through the Second World War, the Quit India movement (1942), and Indian Independence (1947), the report remained unimplemented and the final version was not published until 1947. Like these proponents of the Planning Commission, Sigerist too had his strong differences with Gandhi, whom he described as a "reactionary," seeking to take India back into the Middle Ages.[139]

Sokhey, the main man behind the 1939 health planning, was also influenced by Sigerist and the Soviet Union.[140] After independence, Sokhey became part of Nehru's influential coterie of scientist-statesmen and was also a nominated member of the parliament. He continued his correspondences with Sigerist throughout the 1950s. One of Sokhey's main interests was in the convergence of industrialism and health care, highlighted by his own involvement in the industrial manufacture of penicillin in India in the 1950s.[141] In pursing these ideas he was confronted with the inequalities in international development, which hindered India's medical infrastructure. While attending the WHO meeting in Geneva in 1953, he wrote to Sigerist about the importance of social and economic parity for better health care and how these international bodies did not appreciate the needs of the poorer countries. "[W]e fail to attach due weight to the fact that the world is very unevenly developed. We tend to graft Western practices on to an economic system which cannot carry them."[142] Reiterating the conclusions of his own report of a decade ago, he added: "the inescapable fact remains that if we would improve the health status of a people we can do so only by improving their economic and cultural status. . . . Improvements in health conditions can materialise only if the economic basis for them is prepared at the same time."[143] Sokhey

[135] Sunil S. Amrith, *Decolonizing International Health: India and Southeast Asia, 1930–65* (Basingstoke, UK, 2006), 61.

[136] Bidyut Chakrabarty, "Jawaharlal Nehru and Planning, 1938–41: India at the Crossroads," *Modern Asian Studies* 26 (1992): 277–78.

[137] *National Planning Committee, Subcommittee on National Health Report* (Bombay, 1948).

[138] Debabar Banerji, "The Politics of Underdevelopment of Health: The People and Health Service Development in India; A Brief Overview," *International Journal of Health Services* 34 (2004): 127.

[139] He ended his report on India saying, "[W]e may expect great events when Gandhi dies and when two million soldiers return from the war." Report, 15, folder 74, box 35, Sigerist Papers. For a study of the differences between Gandhi and Nehru over the NPC see, Chakrabarty, "Jawaharlal Nehru and Planning" (cit. n. 136), 282–85.

[140] "Correspondences with Sahib Singh Sokhey, 1951–54," folder 808, box 22, Sigerist Papers.

[141] Nasir Tyabji, "Gaining Technical Know-How in an Unequal World: Penicillin Manufacture in Nehru's India," *Technology and Culture* 45 (2004): 331–49.

[142] Sokhey to Sigerist, 25 Sept. 1951, folder 808, box 22, Sigerist Papers.

[143] Ibid.

visited the USSR in 1953 to witness the development strategies around health and industry and was deeply impressed.[144] He visited Moscow, Leningrad, and Stalingrad, went through the Volga dam and the industrial sites of Rostov, Kiev, and Tashkent, and he described the postwar reconstructions as well as the medical infrastructure to Sigerist; the large number of qualified doctors, the great stress on "Health Education" of the people, and the "abundance" of its medical infrastructure.[145] Sokhey was awarded the Stalin Peace Prize in 1953, the same year as John Desmond Bernal.

Although this is not the place for a discussion of India's national development projects and its links with those of Soviet Union, it is important to highlight the similarities that these two new nations experienced in such pursuits. The Soviet Union was the first of Europe's multiethnic states to promote national consciousness over a wide variety of ethnic populations and to establish the institutional forms of the modern nation-state. The Bolshevik government, to defuse nationalist sentiment, trained new national leaders, established national languages, and financed the industrial infrastructures of the modern nation-state. To that extent, its nationalist developmental experience was comparable to that of India.[146]

Another Indian doctor deeply inspired by Sigerist was Dr. Mahendra Bhatt, who passed his MBBS from Bombay University in 1941, and in 1945 he was awarded a Watumull Foundation fellowship for two years' study in the United States. He worked at the Department of Tropical Medicine, Tulane University, New Orleans. Following that he traveled in the United States and Canada, studying public health and medical care programs as arranged by the Harvard School of Public Health and sponsored by the Watumull Foundation.[147] Throughout this period, Bhatt was in regular contact with Sigerist, and following the latter's suggestion, he and his wife set off on a tour of Europe as well to study public health institutions in different countries. He also planned to visit the USSR but could not secure a visa during the war. In November 1947, soon after Indian independence, Bhatt set out for India, full of ideas and experiences he had encountered during his sojourn and enthusiasm for the new nation. He wrote to Sigerist on board ship in the Mediterranean, "So we are heading towards new India!"[148]

In India, he applied himself in serving local health in the new state of Saurashtra, in western India. He helped with the creation of a new Department of Public Health, started rural health centers, child welfare centers, a nutrition and school health organization, antimalaria and filarial units, and mobile health education units in the shape of vans.[149] In 1950, he was appointed director general of Health Services of the Indian government as a special officer of health education and was posted to the capital Delhi.[150] However, he returned to Saurashtra a few years later to continue his work there. He wrote to Sigerist about the need he had felt to work in the interior of the country: "[I]f we want to improve and build up our health services the young

[144] Sokhey to Sigerist, 10 Aug. 1953, folder 808, box 22, Sigerist Papers.
[145] Sokhey to Sigerist, 14 Oct. 1953, folder 808, box 22, Sigerist Papers.
[146] Terry Martin, *The Affirmative Action Empire: Nations and Nationalism in the Soviet Union, 1923–1939* (Ithaca, N.Y., 2001), 2.
[147] "Correspondences with Dr. Mahendra Bhatt," 1947–48, folder 317, box 8, Sigerist Papers.
[148] Bhatt to Sigerist, 25 Nov. 1947, folder 317, box 8, Sigerist Papers.
[149] *Annual Report for the Department of Public Health, United State of Saurashtra, Rajkot, for the Year Ending 1st March 1949*, 6–7, folder 317, box 8, Sigerist Papers.
[150] Bhatt to Sigerist, 25 Dec. 1950, folder 317, box 8, Sigerist Papers.

people with ideas and ability have to go to the interior and work in the field. I feel I will be able to contribute more by doing comprehensive public health work in these areas."[151] Despite these local engagements, Bhatt had continued his interest in international health movements from where his inspiration had come. He had contacted George Rosen of Columbia University following Sigerist's suggestions in developing his projects on health education in India.[152] In 1954, when Sigerist was planning to visit the USSR, Bhatt expressed a desire to join him to study the public health setup there.[153] Sigerist probably never made this trip due to failing health; he died in 1955.

This movement of the 1930s, which followed through to postindependence India, was an important aspect of Indian's nationalist engagement with the question of health, which developed in congruence with the ideas of India's scientific industrialism and development planning. While the ideas and identities of imperial health care were being challenged, there was a new alignment taking place between politics and medicine. Sigerist's scheme had appealed to Indian nationalists as it was a health plan with a clear political inspiration, something that the imperial and British models had been opposed to. The stress on local health, as evident in the cases of Ghosh, Bhatt, and Sokhey, was also distinct from the erstwhile British involvement in Indian local health and the constitutional provincialization of medical infrastructure on 1919, which was more an administrative issue deriving from the British traditions of local health care. The inspiration for these people had come from a different political ideology of grassroot activism and equitable distribution of resources. At the same time, there was a new legitimization of the state; the nation-state was to be the inspiration behind and provider of the health care of its people. Thus a new link between public health and state was envisioned. The three individuals studied above show three different aspects of internationalism, socialism, and the question of Indian health practices.

CONCLUSION

This essay has demonstrated how the identity of Indian nationhood and its choices of scientific models and infrastructure were intrinsically linked. Although the government of India had abandoned plans for a central research institute, the debates changed the discourse of medical research in India. In the short term, the British increasingly lost their grip over recruitment issues, and hostile questioning over the salaries and other benefits of European medical researchers continued in the Legislative Assembly.[154] The Bhore Committee recommended the establishment of a national medical center.[155] The foundations were laid in 1952, and the All India Institute of Medical Sciences (AIIMS) started functioning in Delhi in 1956.

In the longer term, there has been the mixed heritage of Indian medical science and national identity. The twin critiques of location and personnel, which was so crucial against imperial privileges and essential to Indian nationalism, have remained

[151] Bhatt to Sigerist, 13 May 1953, folder 317, box 8, Sigerist Papers.
[152] Sigerist to Bhatt, 3 Jan. 1952, and Bhatt to Sigerist, 13 May 1953, folder 317, box 8, Sigerist Papers.
[153] Bhatt to Sigerist, 28 Sept. 1954, folder 317, box 8, Sigerist Papers.
[154] Arnold, "Colonial Medicine in Transition" (cit. n. 13), 29–30.
[155] Report of the Health Survey and Development Committee (Delhi, 1946).

important issues in postcolonial India. Through their criticism of the IMS, Dehra Dun, and IRFA, the nationalists and the medical men had urged for more provincialized and rural health care and more economically and socially accountable institutions. The Indian Council of Medical Research (ICMR), as the IRFA was renamed in 1949, actively promoted research in medical colleges and universities rather than in research institutes by bringing them in direct contact with research workers and with the outstanding research problems facing the country.[156] From 1953, with the help of Rockefeller Foundation fellowships, the ICMR funded research in medical colleges and established several research units in university colleges in Agra, Bombay, and Calcutta.[157] A report by the ICMR declared in 1957 that a new chapter had been opened and medical research was no longer confined, as in the days of the IRFA, to the research institutes in hill stations. "It has been brought to where it belongs—to centres of teaching and learning."[158]

Indian nation building, however, has not been without its own institutions and sites of privilege and the marginalization of the questions of politics. The Bhore Committee report, while being critical of the British medical research initiatives and urging for more widespread medical infrastructure in India, also supported the continuance of centralized research and disregarded the rural medical plans of the Sokhey report of 1939. Medical infrastructure, despite substantial investment in rural sectors through planned economy, has tended to remain urban oriented. Urban centers and metropolises have become the sites of real privilege and power in postcolonial India. In the preoccupation with building the major projects and medical infrastructure for the country, the smaller sanitary projects around drains and clean water had been overlooked, a movement that the British did initiate in India and Gandhi championed over large research institutes and hospitals. "The science of sanitation is infinitely more ennobling though more difficult of execution, than the science of healing."[159]

The legacy of "socialized medicine" lives in India in the People's Health Movement (parallel to the People's Science Movement) with the goal to establish health and equitable development in local, national, and international policy. These movements continue to campaign with a slogan reminiscent of the ideas of the 1930s: "Health for All."[160] However, with 30 percent of the population still under the poverty line and facing starvation and malnutrition, alongside an increasingly affluent and insulated middle class, these movements have struggled to establish the "right to health" as a fundamental right of the people of India.

The issue of personnel and politicization of recruitment has produced the most important and enduring legacy in India. Indian nationalism was a political struggle for state power, based often on demands for a greater share in jobs in technical and administrative posts, which had ultimately led to the crisis of imperial manpower. This politics found its manifestation in postindependence India in the socially repre-

[156] *Indian Council of Medical Research: A Review of the Activities during the Years 1950–57* (New Delhi, 1957), 33.

[157] Ibid., 34.

[158] Ibid., 35.

[159] "Mr MK Gandhi, in Opening the Tibb College," *Indian Medical Record*, Aug. 1921, 184–85, 185.

[160] Ravi Duggal, "Health and Development in India: Moving towards Right to Healthcare," mimeo (2003), Centre for Enquiry into Health and Allied Themes (CEHAT), Mumbai.

sentative reservation of government and academic jobs for the deprived communities for their own political and economic empowerment. Reservation has since become the most critical force and feature of Indian democracy and nationhood. Yet at the same time, the logic of efficiency and merit has remained fundamentally a rationale of privilege, and Indian nationhood has embraced it as much as it has generated its critique. In India today, the groups protesting against Reservation (as recently led by the students of AIIMS) loathe this very politicization of recruitment and manpower, using the same notions of "efficiency" and excellence.

MIT-Trained Swadeshis:
MIT and Indian Nationalism, 1880–1947

By Ross Bassett[*]

ABSTRACT

During the colonial period, roughly one hundred degrees were awarded by MIT to Indians. However their importance to India and to the historical understanding of India is disproportionate to their numbers. These men—and they were all men— often from elite families, formed a technological elite in the last days of colonial India. Their careers show a technological nationalism in India—several men came from families associated with Gandhi—and represent an important foreshadowing of the period after independence.

INTRODUCTION

As the introduction to this volume notes, Jawaharlal Nehru laid claim to science as one of the foundations on which independent India was to be built. The editors' focus on Nehru and his words emphasizes the historical resources that an appeal to a universal science offered a nation builder, enabling him to create a powerful forward-looking "imagined community." But as powerful as science and technology were as symbols, one should be highly skeptical about the ability of a single person to fashion a national identity out of whole cloth. The rhetorical connections to science would have meant little without resources on the ground in India to translate lofty aspirations into material form.

This essay looks at a small cadre of Indian engineers trained in the United States at the Massachusetts Institute of Technology during the colonial period. Its starting point is a database of every person claiming a hometown in India or South Asia who graduated from MIT in the twentieth century.[1] Between 1900 and 1947, MIT awarded roughly one hundred degrees to Indians, not a trivial number but one that made MIT graduates a tiny fraction of the Indian engineering community. However,

[*] Department of History, North Carolina State University, 478 Withers Hall, Campus Box 8108, Raleigh, NC 27695-8108; ross_bassett@ncsu.edu.

Research for this article was supported by the National Science Foundation under Grant No. SES-0450808. The author would also like to thank the following people for their assistance and encouragement: the staff of the Institute Archives at MIT, especially Nora Murphy; David Gilmartin; Rosalind Williams; the participants in the conference on Science, Technology, and National Identity, held at the University of South Carolina, September 20–22, 2007; and Ann Johnson and Carol Harrison.

[1] The author constructed this database, as part of a larger project on the history of Indians who went to the Massachusetts Institute of Technology in the twentieth century, by going through MIT commencement programs and picking out every graduate who listed a hometown in India. The programs are available at the MIT Institute Archives under the primary title Massachusetts Institute of Technology, *Graduation Exercises*.

© 2009 by The History of Science Society. All rights reserved. 0369-7827/09/2009-0011$10.00

their importance both to India and to the historical understanding of India is dispro-portionate to their numbers. These men—and they were all men—often from elite families, formed a technological elite in the last days of colonial India. Their careers show a technological nationalism in India and represent an important foreshadowing of the period after independence.

The examination of an Indian elite that was once content to operate within the colo-nial structure but in the twentieth century became increasingly disaffected under co-lonial rule has been a major theme in the historiography of Indian nationalism.[2] That nationalist elite is most closely associated with the legal profession and people such as Mahatma Gandhi, Mohammad Ali Jinnah, and Jawaharlal Nehru. Not completely surprisingly, the MIT-trained engineers also came from elite families in law, busi-ness, and government service. And by the early 1930s, western India had become an important center both for the nationalist movement and for MIT-trained engineers.

In fact, the connections between MIT-trained engineers and the nationalist move-ment run deeper than a few vague similarities between elites. In some cases, they were the same elites. Three families associated with Gandhi sent a total of nine sons to MIT during the nationalist movement. For these families following Gandhi and sending sons to MIT were not contradictory actions; both were part of the nationalist movement.

Writing on science and technology in colonial India has focused on the dominant role played by the colonial state. The major scientific and engineering actors were co-lonial institutions such as the Geological Survey of India, the Public Works Depart-ment, the Indian Medical Service, and Indian universities. Gyan Prakash observes that by the turn of the twentieth century, "colonial power" was about the "scientific and technological reconfiguration" of the colonies.[3] The colonial state was not the only actor though: Indians attended MIT largely based on individual private initia-tive, getting funding from outside the British Indian state. When the colonial govern-ment, in the last days of the raj, began looking for ways to produce more and better-trained engineers, its proposals were ones that had been in some ways anticipated by the lives of India's MIT-trained avant-garde.

---

[2] One can see this in almost any survey of modern Indian history, such as Barbara D. Metcalf and Thomas R. Metcalf, *A Concise History of India*, 2nd ed. (New York, 2006); Judith M. Brown, *Modern India: The Origins of an Asian Democracy*, 2nd ed. (Oxford, 1994); and Sugata Bose and Ayesha Jalal, *Modern South Asia: History, Culture, Political Economy*, 2nd ed. (New Delhi, 2004). The role of elites in the making and unmaking of colonial India has been a particular theme of the "Cambridge School," whose leading figure in recent years has been C. A. Bayly. His works include *Indian Society and the Making of the British Empire* (Cambridge, UK, 1988) and *Rulers, Townsmen and Bazaars: Northern Indian Society in the Age of British Expansion, 1770–1870* (Cambridge, UK, 1983). A summary and critique of this work is given in Nicholas B. Dirks, *Castes of Mind: Colonialism and the Making of Modern India* (Princeton, N.J., 2001), 303–13. Inevitably and rightly, recent Indian historiography has lessened the focus on elites. The thinking behind this article owes much to the work of Charles S. Maier, particularly his *Among Empires: American Ascendancy and Its Predecessors* (Cambridge, Mass., 2006), 19–77.

[3] Gyan Prakash, *Another Reason: Science and the Imagination of Modern India* (Princeton, N.J., 1999), 178. Other works that focus on the technology and the colonial state include the essays in *Technology and the Raj: Western Technology and Technical Transfers to India, 1700–1947*, ed. Roy McLeod and Deepak Kumar (New Delhi, 1995); David Gilmartin, "Scientific Empire and Imperial Science: Colonialism and Irrigation Technology in the Indus Basin," *Journal of Asian Studies* 53 (Nov. 1994): 1127–49; Daniel R. Headrick, *The Tentacles of Progress: Technology Transfer in the Age of Imperialism, 1850–1940* (New York, 1988).

An examination of Indians studying engineering at MIT might suggest to the reader's mind some American plan for "imposing modernity" on India. This is emphatically not what happened. Almost all the impetus for the activities described in this paper came from the Indian side and occurred before the work on modernization of MIT professor W. W. Rostow. And although none of these Indian engineers wrote treatises describing his view on the role of technology in India, the diversity of their careers suggests it would be wrong to subsume them into some one-dimensional modernization.[4] Instead, MIT-trained Indians stand within a long tradition of Indian interaction with other societies.

MIT-trained Indians in the colonial period represent an important foreshadowing. In the first three decades of the twentieth century, thousands of Indians studied in Great Britain to prepare for careers in India as barristers, civil servants, or engineers. Great Britain, as the colonial metropole, offered Indian students advantages they could never gain in the United States. But by 1947, the relative educational importance of these two countries to India switched: more Indians went to the United States for higher education than went to Great Britain. Foreign-trained Indian engineers would play a large role in building independent India, and those engineers were increasingly American, rather than British, trained.

In the early twenty-first century, a significant portion of India's identity as a high-technology nation, either implicitly or explicitly, comes through India's technological relationship with the United States. The gleaming IT (information technology) parks of Bangalore do most of their business with American firms. The achievements of NRIs (nonresident Indians) in American high technology companies have been a source of pride within India. Indians who went to MIT in the colonial period serve as a significant point of origin for a technological identity defined in relation to the United States.[5]

---

[4] Rostow's most complete formulation of his work was W. W. Rostow, *The Stages of Economic Growth, a Non-Communist Manifesto* (New York, 1960). The study of modernization theory is becoming alarmingly close to reaching the large-scale industry stage itself. Work includes Michael Adas, *Dominance by Design: Technological Imperatives and America's Civilizing Mission* (Cambridge, Mass., 2006); David C. Engerman, Nils Gilman, Mark H. Haefele, and Michael E. Latham, eds., *Staging Growth: Modernization, Development, and the Global Cold War* (Amherst, Mass., 2003); Nils Gilman, *Mandarins of the Future: Modernization Theory in Cold War America* (Baltimore, 2004); Nicole Sackley, "Passage to Modernity: American Social Scientists, India, and the Pursuit of Development, 1945–1961" (PhD diss., Princeton Univ., 2004). Ironically, Adas's chapter critiquing modernization theory, "Imposing Modernity," with his own reductionist single focus and lack of nuance, replicates some of the very features that Adas finds objectionable about modernization theory. The chapter is full of straw figures, generic "ideologues," "modernization theorists," and "development specialists." Adas's work practically grants modernization theory agency in and of itself, ignoring the complex mixture of motives and interests that drive most human activity.

[5] As of 2007, roughly two-thirds of India's IT business was done with the United States. Nasscom, *Strategic Review 2007*, Executive Summary, http://www.nasscom.in/upload/51054/Executive%20 Summary.pdf (accessed 5 May 2008). The importance of NRIs to India's identity can be seen by spending a few minutes with an Indian newspaper. Some representative articles that have appeared in the *Times of India* include: Chidanand Rajghatta, "Nine Indians in Elite US Science Academy," *Times of India*, 8 May 2003, http://timesofindia.indiatimes.com/cms.dll/html/uncomp/articleshow?msid=45781586 (accessed 23 June 2008); and Rajghatta, "IIT Madras Alumnus is MIT Dean," *Times of India*, 15 June 2007, http://timesofindia.indiatimes.com/World/The_United_States/IIT_Madras_alumnus_is_MIT_ dean/articleshow/2124324.cms (accessed 26 Sept. 2008).

THE FIRST INDIANS AT MIT AND
THE EARLY INDIAN NATIONALIST MOVEMENT, 1861–1920

Perhaps one reason Indians were drawn to MIT was that it was based on a notion of engineering and engineering education largely antithetical to that held by Indian engineering colleges. The first engineering college in India began operation in Roorkee in 1847. The British established the engineering college at Roorkee and the later engineering colleges in Sibpur, Poona, and Madras, as a way to produce intermediate-grade engineers for the British Public Works Department, which had control over the schools. As a consequence, these schools had a very limited curriculum, focused on civil engineering, the discipline most needed by the Public Works Department, and within civil engineering, focused on narrow vocational training in such areas as surveying and estimating. Given their restricted conception of engineering for India, the British in India argued that there was a limited need for engineering education and that expanding colleges beyond the needs of the Public Works Department would simply lead to unemployment. British mercantilist policies did not encourage the industrialization of India, as might have happened with a broader and wider technical education.[6] In contrast to its position on engineering, the British encouraged scientific training to improve agriculture, which would then lead to higher crop yields and higher tax revenues. In fact, the government of Bengal sent eight students to Cornell University to study agriculture between 1905 and 1909.[7]

In 1846, William Barton Rogers, then a professor at the University of Virginia, penned a prospectus for what would become MIT, titled "A Plan for a Polytechnic School in Boston," which showed his capacious vision of engineering. Rogers proposed to provide instruction in virtually all technical fields, combined with instruction in the sciences relating to those fields. Rogers expected the engineers trained at his school to be not merely competent at operating existing machinery but also inventors, men who would use their knowledge of scientific principles to improve on existing processes. Rogers planned to locate his school in Boston, to benefit from local industry, but aimed at national preeminence, claiming that the institution he envisioned would "soon overtop the universities of the land in the accuracy and extent of its teachings in all branches of positive knowledge."[8]

Rogers struggled to get support for his plan for many years; then the Morrill Act of 1862 provided critical funding and showed a country widely interested in technical and agricultural education. Although MIT was one of Massachusetts's land-grant colleges, and the great majority of students came from within the state, its distinctive approach to education attracted students from throughout the United States. The

---

[6] A typical case of an Indian educated in one of India's engineering colleges is described in Prakash Tandon's memoir, *Punjabi Century*. Tandon's father, Ram Das, went to Roorkee, then joined the Irrigation Department of the Punjab government. English engineers held the higher-level positions, while the senior Tandon was consigned to the lower ranks. Prakash Tandon, *Punjabi Century: 1857–1947* (Berkeley, Calif., 1968). See also Arun Kumar, "Colonial Requirements and Engineering Education: The Public Works Department, 1847–1947," in *Technology and the Raj: Western Technology and Technical Transfers to India, 1700–1947,* ed. Roy MacLeod and Deepak Kumar (New Delhi, 1995), 216–34; Headrick, *Tentacles of Progress* (cit. n. 3), 304–45.

[7] Ian C. Petrie, "Village Visions: Science and Technology in the Bengal Countryside, c. 1860–1947" (PhD diss., Univ. of Pennsylvania, 2004), 226–37.

[8] Quoted in Samuel C. Prescott, *When M.I.T. Was "Boston Tech"* (Cambridge, Mass., 1954), 332.

first graduate who can clearly be identified as a foreigner is Aechirau Hongma, from Tokyo, Japan, in the class of 1874, who was working in Tokyo as a government engineer one year after graduating. In 1880, the Chinese government sent nine students to MIT, making up 4 percent of the student body. (The Chinese government recalled them all the next year.)[9]

Although the first Indian attended MIT in 1880, only in the twentieth century is it possible to trace the lives of Indians who went to MIT with any specificity. The first MIT student to play a significant role in the technological development of India did so through a larger social movement. The early twentieth century marked the rise of the swadeshi movement in India, in which Indians developed indigenous industries as an act of resistance to British rule and dominance. In the late nineteenth century, the Poona area, on the western side of India, had been a seat of resistance to the British, led by pioneer nationalist Bal Gangadhar Tilak.[10]

In 1908, after two years of study at MIT, Ishwar Das Varshnei came to Poona to set up a glass factory under the umbrella of a nationalist organization, the Paisa Fund. Varshnei had grown up in the Aligarh region of northern India. Although little is known of his background, the fact that he came to MIT after receiving some training in Japan as well as the course of his later career suggests that the Indian technical education system did not suit him. The Paisa Fund, so called because it raised money by asking for donations of a paisa each (a paisa was a sixty-fourth of a rupee) from a broad spectrum of Indian society, was supported by Tilak and sought to develop indigenous Indian industry. After raising 10,000 rupees, the Central Committee of the fund decided to concentrate its efforts in glassmaking. The Central Committee convinced Varshnei to come to Poona to direct the glassmaking operation. Varshnei had apparently learned glassmaking during his time in Japan and at MIT. In 1908, Varshnei began glassmaking operations for the Paisa Fund, assisted by several Japanese apprentices. The founders of the Paisa Fund envisioned not just the opening of a factory but the training of a generation of people who could go out and run their own industrial enterprises, and so Paisa Fund combined education and production.[11]

Although by 1915, the Paisa Fund Glass Works had not succeeded as a business, it had successfully laid a foundation for the development of the glass industry in India. Under Varshnei's guidance, ten men had been well trained not only to work in the glass industry but also to start or run their own enterprises. A 1922 article on the state of the glass industry in India noted that Varshnei was running three or four factories in the Punjab, "trying to put his factories on a most up-to-date scale," and working

[9] Ibid., 3–127; Massachusetts Institute of Technology, *President's Report for the Year Ending September 30, 1875*, 203; *Class of '84, Forty-Fifth Anniversary Booklet* (Cambridge, Mass., 1929), 1; *Technology Review* 1 (April 1899): 245. Given the fact that over the long term Asia would be MIT's most important hinterland for foreign students, it was appropriate that the first foreign student and the first group of foreign students both came from Asia.

[10] Sumit Sarkar, *The Swadeshi Movement in Bengal: 1903–1908* (New Delhi, 1973); Metcalf and Metcalf, *Concise History of India* (cit. n. 2), 153–58; Manu Goswami, "From Swadeshi to Swaraj: Nation, Economy, Territory in Colonial South Asia, 1870–1907," *Comparative Studies in Society and History* 40 (1998): 609–36; Goswami, *Producing India: From Colonial Economy to National Space* (Chicago, 2004), 242–76. *Swadeshi* is a Sanskrit term meaning "of one's own country." Amartya Sen, "The Indian Identity," in *The Argumentative Indian: Writings on Indian History, Culture, and Identity* (New York, 2005), 337. On Tilak, see Stanley A. Wolpert, *Tilak and Gokhale: Revolution and Reform in the Making of Modern India* (Delhi, 1961).

[11] Paisa Fund, *Silver Jubilee Number* (Poona, India, 1935), 17–23; Dwijendra Tripathi, *The Oxford History of Indian Business* (New Delhi, 2004), 158.

to add the capability to manufacture window glass. The article further noted that students of the Paisa Fund operation had started a dozen other glass factories in India.[12]

Ishwar Das Varshnei did not earn a degree during his two years at MIT, and in a period when funding for Indian students came almost exclusively from India, and when the prestige of an MIT degree was very small there, most Indians who attended MIT during the early part of the twentieth century also failed to earn a degree. Although the historical records do not exist to allow for a detailed analysis of every case, one suspects that Varshnei and the students after him came to MIT to learn only specific skills, ones they could employ in India.

In the late 1890s and early 1900s, the British tried to maintain their rule by incorporating a few Indians into positions of responsibility and authority in the raj. Indians could compete to enter the Indian Civil Service, and enter, in a limited way, other government positions. In the early twentieth century, Krishna Gupta was one of two Indians who reached the highest levels within the raj. He passed the examination for the Indian Civil Service in 1871, and after a steady series of appointments, he became one of two Indians appointed to the India Council in 1907. One might see in Gupta the classic case of the co-opted elite that is necessary for the maintenance of empire. But from his own flesh came an elite with a different orientation—his son, Birendra Chandra Gupta, whom he sent to MIT, from which Birendra Chandra graduated in 1907. Although the exact circumstances under which the younger Gupta came to MIT are not clear, his father's positions would have given him knowledge of educational and technological developments worldwide that would not have been available to most educated Indians. (For example, the senior Gupta traveled to the United States.) The junior Gupta spent some time at General Electric's Lynn Works, and married an American woman before he returned to India. Gupta became a professor of electrical engineering at Bengal Engineering College. As might be expected from someone with an American wife, he kept ties with the United States—in 1922 the *Boston Globe* reported that Gupta was back for two years to do electrical research.[13]

INDIANS, MIT, AND THE AGE OF GANDHI, 1920–1940:
I. DEVCHAND PAREKH AND BHAVNAGAR

The Indian student experience at MIT in the 1920s and 1930s was different enough from the experience of the previous period to warrant being considered a new generation. The biggest change was that there were more Indian students and they were a constant presence. Before 1919, no more than two Indians had ever studied at MIT at a time, but between 1920 and 1939, in only one year did the number of Indians goes below five, and it rose as high as twelve.

---

[12] Paisa Fund, *Silver Jubilee Number* (cit. n. 11); G. P. Ogale, "Glass Industry in India," *Bulletin of the American Ceramic Society* 1 (Nov. 1922): 296.

[13] On efforts by the British to integrate Indians into official positions, see Brown, *Modern India: The Origins of an Asian Democracy* (cit. n. 2), 144–50. Information about Krishna Gupta is given in his obituary in *The Times of London*, 30 March 1926, 19. A visit of Gupta's to the United States is documented in "People Met in Hotel Lobbies," *Washington Post*, 10 June 1907, 6. The MIT Institute Archives considers Gupta the first Indian to have received a degree from MIT. (I did not find his name in the graduation program or in the corporation records of those awarded degrees.) Birendra Chandra Gupta, Pathfinder File, India, Institute Archives, Massachusetts Institute of Technology, Cambridge, Mass. (hereafter cited as MIT Institute Archives). His later visit to the United States is documented in "No Rebellion In India at Present, But—," *Boston Globe*, 22 Oct. 1922, E3.

The Kathiawar peninsula of what is today western Gujarat holds a special place in the history of Indian nationalism, being the home of Mahatma Gandhi and the ancestral home of Mohammed Ali Jinnah. During the 1930s, a small princely state in Kathiawar, Bhavnagar, was the leading source of Indian students at MIT. In fact, Bhavnagar, representing less than 2 percent of the population of India, produced almost half the Indians who earned degrees from MIT in the 1930s. This concentration in Bhavnagar was further concentrated in the family of a lifelong friend of Mahatma Gandhi, Devchand Parekh. Parekh had a vision of a technological India built around Indians trained at MIT and worked to realize this vision within his family.

Devchand Parekh was born in 1871 in the city of Jetpur on the Kathiawar peninsula, the son of a wealthy lawyer. In 1893 Parekh left to study at Cambridge, where he received a bachelor's degree in 1896 and his master's degree in 1899. While in Britain he studied for the bar exam and received his calling to the bar from the Middle Temple in 1897. Although Devchand Parekh returned to India in 1899 seemingly well placed to have a lucrative career as a barrister, according to Parekh's son, something happened in Cambridge that changed the course of his life.[14]

That something was an encounter with the economist Alfred Marshall. According to Parekh's son, Marshall counseled Devchand Parekh that Indians should not be coming to Britain to study liberal arts; instead they should go to America—specifically to MIT—to study engineering, and then return to India to set up industries that would improve the Indian standard of living. In response to this, Parekh went to the United States in 1893, visited MIT, and began a correspondence with MIT officials to receive catalogs.[15]

Although much of the testimony of Devchand Parekh's son (ninety-five years old in 2008, when interviewed by the author) cannot be verified directly, indirect evidence strongly supports the outline of his account. Students in Parekh's curriculum would have heard lectures from Marshall in political economy. Marshall was known for his openness—he set aside two afternoons a week in which any member of the university could call on him at home.[16]

Although no correspondence between Marshall and Parekh survives (and there may never have been any), a 1910 letter by Marshall is very instructive. In the letter, Marshall wrote apparently to a B. B. Mukherjee of Lucknow University:

> For twenty years I have been urging on Indians in Cambridge to say to others: "How few of us, when we go to the West, think of any other aim, save that of our individual culture? Does not the Japanese nearly always ask himself in what way he can strengthen himself to do good service to his country on his return?"[17]

Earlier in the letter, Marshall had written in praise of Jamsetji Tata, the great Indian entrepreneur, saying the country could use a "score or two" of men like him. But Marshall maintained a pessimism about India, writing, "[S]o long as an Indian who has

[14] M. D. Parekh, interview by author, 20 June 2008, Mumbai, India. Further information on Parekh is included in J. A. Venn, *Alumni Cantabrigienses*, Part 2, *1752–1900* (Cambridge, UK, 1953), 5:19.

[15] Parekh, interview (cit. n. 14).

[16] J. M. Keynes, "Alfred Marshall, 1842–1924," in *Memorials of Alfred Marshall*, ed. A.C. Pigou (London, 1925), 51. Marshall's lecturing to students in Parekh's curriculum is asserted by J. Cox, deputy keeper of University Archives, Cambridge University. Cox letter to author, 8 July 2008.

[17] Marshall to [B.B. Mukerji?], 22 Oct. 1910, in *The Correspondence of Alfred Marshall, Economist*, ed. John K. Whitaker (Cambridge, UK, 1996), 3:268–9.

received a high education generally spends his time in cultured ease: or seeks money in Indian law suits—which are as barren of good to the country as is the sand of the sea shore—nothing can do her much good." In an earlier letter, Marshall wrote:

> I do not believe that any device will make India a prosperous nation, until educated Indians are willing to take part in handling things, as educated people in the West do. The notion that it is more dignified to hold a pen and keep accounts than to work in a high grade engineering shop seems to me the root of India's difficulties.[18]

These all echo what Marshall is alleged to have said to Parekh. One might note multiple levels of irony in Marshall's advice. First, it represents complaints about a lack of industrial spirit in India coming from England at precisely the time when some historians have noted a decline in industrial spirit in England.[19] Second, an heir of Adam Smith and a colonizer is urging Indians to rise above individual self interest and think of the good of the country when an appeal to narrow self (or group) interests had been one of the main strategies of the British in colonizing India. Marshall was no Indian nationalist, and he showed traces of racism. If Marshall's statements had been made to a fellow Briton, they could sound like an apology for the status quo. But made to an Indian, who was in a position to act on the words, his counsel could be life changing. Of course Marshall's seeming hostility to lawyers carries an irony in that lawyers were one of the prime vehicles through which India was to receive its independence.

Marshall was familiar with MIT and its approach to technical education, and that approach was consistent with his way of thinking. In 1875, Marshall went to the United States, where he spent two weeks in Boston, hosted for part of the time by Charles Eliot, the president of Harvard and former professor of chemistry at MIT. Marshall was a regular correspondent with his fellow economist and the president of MIT Francis Amasa Walker. In 1886 Walker wrote Marshall telling him of the opening of MIT's school year and that he would send Marshall an MIT catalog so he could see "how unlike an English University is a Yankee School of Technology."[20] In Marshall's *Principles of Economics*, he described a failed effort of his at Bristol to introduce a technical education program of several years duration based on alternate six-month periods of studying science and six-month periods working on workshops, an approach consistent with MIT's. Marshall, whose own career had its origins in being the second wrangler in the mathematical tripos at Cambridge, and whose career was marked by the consistent, if sometimes surreptitious, use of mathematics in economics, can have reasonably been expected to have been enthusiastic about MIT, which introduced mathematical and scientific sophistication to technology.[21]

Although it is not possible to verify that Parekh visited MIT in 1893, strong indirect

[18] Marshall to [B.B. Mukerji?], 22 Oct. 1910 (cit. n. 17); Marshall to Manohar Lal, 28 Jan. 1909, Pigou, *Memorials of Alfred Marshall* (cit. n. 16), 457.

[19] Martin J. Wiener, *English Culture and the Decline of Industrial Spirit, 1850–1980* (Cambridge, UK, 1981).

[20] Quoted in James Phinney Munroe, *A Life of Francis Amasa Walker* (New York, 1923), 264. Marshall's visit to America is described in Peter Groenewegen, *A Soaring Eagle: Alfred Marshall, 1842–1924* (Aldershot, UK, 1995), 193–203.

[21] Alfred Marshall, *Principles of Economics,* 9th ed. (London, 1961), 209–10. On Marshall's mathematical background and use of mathematics, see Groenewegen, *A Soaring Eagle* (cit. n. 20), 91–94, 412–13.

evidence exists to support this claim. Passenger manifests show that in June 1893, Devchand Parekh disembarked in New York City from the *Paris*. In July 1893, the *Washington Post* ran a story describing the visit of Parekh and a companion to Washington, D.C. The report shows that the two had an interest in technology—Parekh's companion observed that the main difference he noted between America and India was the widespread use of machinery in America. The report suggests the two had a nationalist bent: they described conditions under British rule and then stated, "Now that is what you would call tyranny, would you not?" Parekh and his colleague had come to Washington from Chicago, where presumably they had been to the World's Columbian Exposition.[22]

Back in India, Parekh led a double life, practicing law but also pursuing technological development. With the help of his younger brothers, who had studied chemistry at the University of Bombay, he set up the Bhavnagar Chemical Works in 1910, which produced tinctures and liniments. By the late 1920s, the Bhavnagar Chemical Works was exporting papain, an extract of papaya, to Europe. A 1922 British publication called the company "satisfactorily worked."[23]

Parekh was also a close and lifelong friend of Mohandas Gandhi. Their times in Britain did not overlap, and apparently they came to know each other as young men in Kathiawar. In 1902, when Gandhi was a largely unknown lawyer, returning to South Africa after a brief stint working in Bombay, he wrote of the possibility that Parekh might join him in South Africa. Upon Gandhi's return from South Africa in 1915, at a time when Gandhi was more a regional than a national figure, one of his first stops was at Parekh's hometown of Jetpur, where Parekh and the other citizens of the town honored him.[24]

A picture of Gandhi and Parekh together in 1915 in Jetpur shows Gandhi wearing Kathiawar dress, while Parekh wears English clothes. In May 1921 Gandhi wrote to Parekh: "If you will, you can see that no home in Kathiawad is left without a spinning-wheel. But can a person ever rise to heaven except by giving up his life? Do you yourself spin? Do you use khadi exclusively, at home and outside?"[25]

The clear implications of Gandhi's questions were that he knew that Parekh did not spin and did not wear khadi. Family testimony is that in 1921 (apparently sometime after this letter), Parekh burned his English clothes and gave up his law practice to support Gandhi. Although that cannot be verified independently, what can be verified

---

[22] "Two Hindoos Here," *Washington Post*, 17 July 1893, 6. Documentation for Parekh's entry into the United States was found searching the ellisisland.org database using the search term "Parekh." See http://www.ellisisland.org/search/passRecord.asp?MID=06161316900145053280&LNM=PA REKH&PLNM=PAREKH&last_kind=0&TOWN=null&SHIP=null&RF=2&pID=103190080012 (accessed 23 Sept. 2008).

[23] Bhavnagar Chemical Works Vartej, *Golden Jubilee Souvenir, 1910–1960; Report on the Administration of the Bhavnagar State for the Year 1928–1929* (Bhavnagar, 1930), 44, India Office Records, British Library, London; Somerset Playne, *Indian States: A Biographical, Historical, and Administrative Survey* (London, 1922), 386. Parekh, interview (cit. n. 14).

[24] Gandhi's stay in Jetpur in January 1915, where he is said to have been "[p]ut up at Devchadbhai Parekh's," is documented in C. D. Dalal, *Gandhi: 1915–1948; A Detailed Chronology* (New Delhi, 1971), 1. The letter in which Gandhi talked about Parekh's joining him in South Africa is M. K. Gandhi to D. B. Shukla, 8 Nov. 1902, *The Works of Mahatma Gandhi*, http://www.gandhiserve.org/cwmg/ VOL003.PDF (accessed 28 Sept. 2008).

[25] Mohandas Gandhi to Devchand Parekh, 5 May 1921, *The Collected Works of Mahatma Gandhi*, http://www.gandhiserve.org/cwmg/VOL023.PDF (accessed 21 Sept. 2008). Khadi, homespun and home-woven cloth, became one of the foundations of Gandhi's movement in the 1920s. Lisa Trivedi, *Clothing Gandhi's Nation: Homespun and Modern India* (Bloomington, Ind., 2007).

is that Parekh took a leadership role in the khadi movement in Kathiawar. In 1925 Gandhi reported that Parekh had agreed to enlist one thousand volunteers who would always wear self-spun khadi.[26]

The imagery in Gandhi's phrase "can a person ever rise to heaven except by giving up his own life" suggested someone willing to sacrifice everything; Parekh sacrificed much, though he did not completely subordinate his will to Gandhi's. In 1925, Gandhi launched a public, semihumorous attack on Parekh. In February 1925, Gandhi came to Kathiawar. When Gandhi got to Parekh's hometown of Jetpur, Parekh promised Gandhi the use of his own spinning wheel for Gandhi's daily spinning session. When it arrived, Gandhi found a spinning wheel in very poor condition—obviously not being regularly used by Parekh. Gandhi reported that his arm started aching with just a half-hour's worth of spinning. Gandhi claimed Parekh was mocking the spinning wheel and threatened to remove him from his position if he did not set his spinning wheel right.[27]

On February 16, 1925, Devchand Parekh's daughter, Champabehn, married T. M. Shah in Jetpur. Gandhi was in attendance at the wedding. Shah was serving at the time as the registrar of the Gujarat Vidyapith, a nationalist school in Ahmedabad, founded by Gandhi. By 1927, apparently at the urging and with the financial support of his father-in-law, Shah was at MIT studying electrical engineering. In 1930 he earned both a bachelor's degree and a master's degree there as part of the cooperative program. Although he must have worn khadi at the Vidyapith, the 1929 *Technique*, the MIT yearbook, shows him in western clothes. The familial connection with Devchand Parekh would suggest that he was not repudiating Gandhi by going to MIT. If there were any doubt, it is refuted by a 1932 article in the *Tech*, the MIT student newspaper, reporting that Shah and a fellow MIT student were in prison in India as part of the nationalist movement. Shah endured eighteen months further imprisonment during World War II after he participated in a strike at Tata Iron and Steel in Jamshedpur as part of the "Quit India" movement. At first glance, going to MIT for advanced training and sitting in jail would seem to be incompatible—why would someone waste his skills that way? The reasonable answer is that for Shah, working for Gandhi, going to MIT, and going to jail were all nationalist acts.[28]

The evidence given thus far supporting family testimony about the link between Alfred Marshall, Devchand Parekh, MIT, and India is circumstantial. The strongest piece of evidence is also circumstantial. The Parekh family, between 1930 and 1940, received a total of eight degrees from MIT; the rest of India, not associated with the Parekh family, during that time received twenty-eight degrees from MIT. This extraordinary activity on the part of the Parekhs requires some explanation. After

[26] "To Kathiawaris," 12 April 1925, *The Collected Works of Mahatma Gandhi*, http://www.gandhi serve.org/cwmg/VOL031.PDF (accessed 21 Sept. 2008).

[27] "Reminiscences of Kathiawar-II," 8 March 1925, *The Collected Works of Mahatma Gandhi*, http://www.gandhiserve.org/cwmg/VOL030.PDF (accessed 28 Sept. 2008).

[28] Shah's wedding and his association with the Vidyapith is documented in Dalal, *Gandhi*, 54 (cit. n. 24). The history of the Vidyapith is given on its Web site, http://www.gujaratvidyapith.org/history.htm (accessed 28 Sept. 2008). Shah's imprisonment is documented in "Technology Graduates in Prison in India for Anti-British Activities," *The Tech*, 13 May 1932, 1, 5; and MIT, *Twenty-Fifth Reunion, Class of '29* (Cambridge, Mass., 1954), 60–61. Strikes at Tata Iron and Steel as part of the "Quit India" movement are documented in Vinay Bahl, *The Making of the Indian Working Class: The Case of the Tata Iron and Steel Co., 1880–1946* (New Delhi, 1995), 358–63. Shah's son, Anant Shah, asserted that his father had Gandhi's approval to go to MIT. Anant Shah, interview with author, 29 July 2008, by telephone.

Shah, Devchand Parekh sent three nephews and two sons to MIT. Almost all of them studied chemical engineering with the apparent idea that they would come back and work in the family chemical works. In 1940, Devchand's son, M. D. Parekh, earned a doctorate in chemical engineering from MIT, working under Warren Lewis, one of the founders of the modern discipline of chemical engineering.[29]

Although in some cases, families used MIT as part of a strategy to support the family business, this did not happen with the Parekhs. M. D. Parekh reports that his father did not insist that he work in the family business, and he did not. The 1948 MIT alumni directory shows none of the Parekhs working for the Bhavnagar Chemical Works.[30]

Devchand Parekh was unique, but in fact he was not the only father from Bhavnagar with connections to Gandhi to send multiple family members to MIT. Hiralal Shah was a wealthy cloth merchant who had transformed his business in line with Gandhi's movement from selling British clothes to selling swadeshi and eventually moved to Bombay. In the late 1920s, he corresponded with Gandhi about an idea for an improved charka. Shah had a keen interest in astronomy—his Bombay telex address was "Astronomy"—and in 1932, he sent Gandhi some books on the field. Shah's personal abilities in science and technology were strictly at the amateur level, but he was determined to develop greater capabilities in his family: two of Shah's sons went to MIT, where they earned master's degrees in engineering. Another son went first to Lowell Tech and then earned an MBA at Harvard.[31]

Both Gandhi and Parekh came from the Kathiawar peninsula, but Gandhi ultimately left Kathiawar. Its high density of princely states, where the British ruled indirectly through a local prince or maharaja, was not a favorable place for a nationalist movement. Any protests there would first have to be made against the local princes, confusing the issue. Although Gandhi left, first setting up his base of operations in Ahmedabad, Parekh stayed. One might see certain similarities between Parekh and Nehru, but whatever would have happened otherwise, the fact that Parekh stayed in Kathiawar almost guaranteed him historical obscurity. The very features that would ultimately make Kathiawar a backwater in the nationalist movement made it a favorable place for a technical movement based on sending students to MIT for education. In an environment without the constant protests, agitations, and arrests of British India, it was possible to think more for the long term. Rule by princely states allowed for the distribution of funds for students studying outside the empire in a way not possible in British India.

Outside of the Parekhs, eight more degrees were earned by residents of Bhavnagar in the decade of the 1930s, making Bhavnagar responsible for one half the MIT degrees earned by Indians. Whether the Parekhs played a role in the decision by others from Bhavnagar to attend MIT is not clear. What is clear is that Bhavnagar offered

[29] Parekh, interview (cit. n. 14).

[30] Devchand Parekh also sent two daughters to Boston University, where they earned master's degrees in the 1930s. Chanduben Parekh, "Acculturation in Marriage Institutions of India" (master's thesis, Boston Univ., 1938); Kamuben Valabhadas Parekh, "The Influence of Racial Prejudice in American Life Today" (master's thesis, Boston Univ., 1938).

[31] "Notes," from *Young India*, 5 Sept. 1929, reprinted in *The Collected Works of Mahatma Gandhi*, http://www.gandhiserve.org/cwmg/VOL047.PDF (accessed 28 Sept. 2008); Mohandas Gandhi to Hiralal Shah, 12 April 1932, *The Collected Works of Mahatma Gandhi*, http://www.gandhiserve.org/cwmg/VOL055.PDF (accessed 28 Sept. 2008). Further information has been provided by Shah's grandson Anand Pandya, in various emails and meetings.

a funding source not available elsewhere in India. Up until the mid-1950s, with the exception of doctoral students who might get a research or teaching assistantship, the money to fund an Indian student at MIT had to come from India. This money could come from several sources, such as families, private voluntary organizations, or philanthropies, such as the Tata Endowment. Indian princely states were another potential source.

The British often claimed that India did not need more Indian engineers, particularly engineers with advanced training, and that the training of engineers would just lead to more unemployment. However stingy the colonial government might have been in supporting technical education, the political organization of Britain's empire in India offered up a space to allow for government funding of engineers at MIT. During the colonial period, the British ruled 40 percent of India indirectly, through local princes. Although the British had a resident in many of the princely states to secure their interests, the local princes had a degree of autonomy.[32]

Most of the MIT students from Bhavnagar, with the exception of the Parekhs, appear to have been funded by the princely state. The grandson of the dewan of Bhavnagar entered MIT in 1936, and when the dewan himself came, he treated all the Indian students at MIT to a luncheon party at Boston's Ritz-Carlton Hotel. Later, the maharaja of Bhavnagar visited Boston and also treated the MIT Indian students.[33]

Bhavnagar also provided the most important Indian MIT graduate of the colonial period, Anant Pandya. Pandya was born in Bhavnagar in 1909 and largely raised by his grandfather. Pandya's father, a graduate of Cornell who had studied at Berkeley, served as an agriculturalist for several Indian princely states and spent large amounts of time away from the family home.[34]

In 1927, Anant Pandya entered NED Engineering College in Karachi, where he finished at the top of his class. Upon graduating, he had two options that typically would have been highly appealing to most recent engineering graduates: he could enter the Indian Engineering Service or take up a Prince of Wales Scholarship for higher studies. But Pandya did not want to work under the British, and the requirement that the Prince of Wales Scholarship be used only at British or Dominion institutions was unacceptable to him. Instead, he applied to and was accepted at MIT; a Bhavnagar state scholarship paid his tuition.[35]

Pandya earned a master's degree within a year and continued on for his doctorate, which he earned in 1933, in civil engineering. (After Pandya's first year at MIT, his cousin Upendra Bhatt joined him there.) Upon his graduation, Pandya returned to India and made a six-month tour of the country looking for an appropriate job. In the words of his cousin and closest friend, he did not meet a "proper response or

[32] Barbara N. Ramusack, *The Indian Princes and Their States* (Cambridge, UK, 2004); Manu Bhagavan, *Sovereign Spheres: Princes, Education and Empire in Colonial India* (Oxford, 2003).

[33] L. M. Krishnan, untitled memoir (epilogue, dated 10 Feb. 1997), in author's possession, 104–5. Independent evidence of students from Bhavnagar being funded by the state comes from B. V. Bhoota, unpublished memoir, 7 July 1997 (in author's possession); and Sriram Shastry, "Nautam Bhagwanlal Bhatt (1909–2005)," *Current Science* 89, 10 Sept. 2005, 895.

[34] The following account is based on R. S. Bhatt, "Anant Pandya: A Biographical Sketch," in *Dr. Anant Pandya: Commemoration Volume*, ed. Lily Pandya (np, 1955), 1–13; Upendra J. Bhatt, "Beloved Brother, Intimate Friend, Brilliant Contemporary," in ibid., 25–42; Kiran Bhatt, interview by author, 16 Feb. 2007, by telephone; and numerous discussions and emails with Anand Pandya, Anant Pandya's son.

[35] R. S. Bhatt, "Anant Pandya," 2–3; Upendra Bhatt, "Beloved Brother," 33–4. (Both cit. n. 34.)

appreciation" in India.[36] His degrees from an American institution seem not to have been fully valued, and his youth and lack of experience worked against him. One of his job offers was for a position paying a humiliatingly low 150 rupees a month. He finally accepted a position at McKenzies Limited in Bombay, but after a little more than a year, he became convinced that this company would not give him adequate scope for his skills. In 1935, he went to London, so desperate to find a position in which he could use his abilities that he considered working without pay. He eventually took a job at the Trussed Concrete Steel Company as a designer. Pandya had a very productive three-year stint with Trussed Concrete, serving as a consultant on making buildings more earthquake proof in India, writing an award-winning paper, lecturing throughout England, and designing an improved air-raid shelter.[37]

In 1939, the government of Bengal advertised the position of principal of the Bengal Engineering College in Sibpur (across the Hugli River from Calcutta). The posting of the advertisement in London was logical because all the previous principals had been English. Pandya applied and the Selection Committee judged him the most qualified candidate. When he officially took over as the principal of the Bengal Engineering College in September 1939, at the age of thirty, he became one of the Indian engineers holding the position of greatest responsibility in India, with many British professors reporting to him. His position as one of India's leading engineers was affirmed in 1941 by the Indian Science Congress, when it appointed him president of its engineering section.[38]

INDIANS, MIT, AND THE AGE OF GANDHI, 1920–1940:
II. BAL KALELKAR

In the 1930s, no young Indian had better nationalist credentials than Bal Kalelkar did. His father, Kaka Kalelkar, himself the son of a treasury officer for the raj, had developed nationalist and anti-British leanings in the early part of the twentieth century through reading the work of Tilak. The senior Kalelkar worked primarily in Indian schools, going in 1914 to Rabindranath Tagore's Santiniketan, where he was to meet Gandhi in February 1915. Shortly thereafter Kalelkar joined Gandhi at his newly established Satyagraha Ashram in Ahmedabad. Kalelkar became Gandhi's main educationalist, serving for a time as principal of the ashram school, and later professor at the Gujarat Vidyapith. Kalelkar spent time in jail with Gandhi, organized events, and took over some of Gandhi's publications when he was in jail.[39]

Kalelkar's younger son Bal grew up sharing his father's and Gandhi's work. At age eighteen, Bal became one of a select group chosen to participate with Gandhi on the Salt March. For Gandhi, this was a political, spiritual, and moral exercise, and he required each marcher to spin on a charka every day, to pray, and to keep a daily diary,

---

[36] Upendra Bhatt, "Beloved Brother" (cit. n. 34), 36–37.

[37] Anant Pandya to Frances Siegel, 25 Oct. 1935, folder 7, box 3, Frances Siegel Papers, SC 149, Radcliffe College Archives, Schlesinger Library, Radcliffe Institute, Harvard University, Cambridge, Mass.; R. S. Bhatt, "Anant Pandya" (cit. n. 34), 4–5.

[38] R. S. Bhatt, "Anant Pandya" (cit. n. 34), 6; Upendra Bhatt, "Beloved Brother" (cit. n. 34), 38; Anant Pandya, "Education for the Engineering Industry," *Proceedings of the Indian Science Congress* 29 (1942): 347–74.

[39] Information on Kaka Kalelkar is provided in Madho Prasad, *A Gandhian Patriarch: A Political and Spiritual Biography of Kaka Kalelkar* (Bombay, 1965), 355–64. Prasad gives evidence that the senior Kalelkar read Marshall's *Principles of Economics* (cit. n. 21).

which Gandhi would read. In 1944, Bal Kalelkar wrote the following synopsis of his life in the early part of the 1930s:

> [I]n the year 1930, the author found the country seething with political unrest, and though still in his teens, he decided to plunge into the social and political activities carried out by the Indian National Congress under the leadership of Mahatma Gandhi. In the years 1930–35 he devoted his entire time to organizing political activities in the villages of India and was imprisoned for the same. During this period and after his release from prisons, he also did extensive social and constructive work.[40]

In 1939, Bal was at Gandhi's side, at Rajkot, as a hired gang of thugs attempted to violently break up a prayer meeting. As the gang approached, Kalelkar joined Gandhi in his Hindu prayers.[41]

Then the next year, Bal Kalelkar was off to MIT to study mechanical engineering, funded by G. D. Birla, an Indian business magnate and close associate of Gandhi. He went with the blessing of Gandhi, who gave him the following letter to take to America:

> This is to introduce young Kalelkar to all my friends in America. He was brought up under my hands. He is one of the most promising among the boys brought up in Satyagraha Ashram. Any help rendered him will be appreciated.[42]

Kalelkar earned a master's degree in mechanical engineering from MIT in 1941 and then a PhD in mechanical engineering from Cornell in 1944. Just after Kalelkar finished his dissertation, which he dedicated to Gandhi as "that grand old man of India," Gandhi wrote to Kalelkar:

> I have your beautiful letter. I can understand that western music has claimed you. Does it not mean that you have such a sensitive ear as to appreciate this music? All I wish is that you should have all that is to be gained there and come here when your time is up and be worthy of your country.[43]

For Kalelkar, going to MIT was obviously the end point of a long process. His Cornell biographical sketch mentions that he studied at the NED Engineering College of Karachi between 1937 and 1940. At the point at which Gandhi and Kalelkar faced down a mob in Rajkot, Kalelkar had already set down the path to becoming an engineer.[44]

What are we to make of the fact that Bal Kalelkar, a young man at the very heart of the Indian nationalist movement and a trusted colleague of Gandhi's, who at the

[40] B. D. Kalelkar, "A Study of Intake Manifold Design, with Special Emphasis on the Distribution Characteristics of a Six-Cylinder Engine Equipped with a Twin-Carburetor Layout" (PhD diss., Cornell Univ., 1944), unpaginated front matter. Further details of the Salt March are provided in Rajmohan Gandhi, *Gandhi: The Man, His People, and the Empire* (Berkeley, Calif., 2008), 308–11.

[41] B. D. Kalelkar, "Potter: Through the Pot's Eyes," in *Incidents of Gandhiji's Life*, ed. Chandrashankar Shukla (Bombay, 1949), 99–100.

[42] M. K. Gandhi, 5 July 1940, *The Collected Works of Mahatma Gandhi,* http://www.gandhiserve .org/cwmg/VOL078.PDF (accessed 27 Sept. 2008). Support from Birla is asserted in Kalelkar, "A Study of Intake Manifold Design" (cit. n. 40).

[43] Gandhi to Bal D. Kalelkar, 3 Nov. 1944, *The Collected Works of Mahatma Gandhi,* http://www .gandhiserve.org/cwmg/VOL085.PDF (accessed 27 Sept. 2008).

[44] Kalelkar, "A Study of Intake Manifold Design" (cit. n. 40).

climactic moment when independence is within sight, decamps to the United States to study connecting rods and internal combustion engines? In the twenty-first century, scholars use terms "Gandhian" and "Nehruvian," as if they have a fixed set of meanings; however, these static terms do not correspond to lived experience. People came to Gandhi with their own interests. Gandhi drew significant support from elites, whether those elites were barristers such as the senior Parekh and Nehru or businessmen such as G. D. Birla. If the last twenty-five years of cultural history has taught us anything, it has been not to assume a simple linear relationship between texts and audiences. In this case, if Gandhi's life itself, combined with his spoken and written words form the text, each person would have his or her own experience of Gandhi and own definition of what "Gandhian" was. The only reasonable conclusion to be drawn from the case of the Parekhs, T. M. Shah, and Bal Kalelkar is that they saw no contradiction in their support for Gandhi and going to MIT: both were an integral part of building the Indian nation.[45]

ENGINEERING A NEW NATION:
INSTITUTIONALIZING MIT IN INDIA, 1944–50

The central role that Bhavnagar played in sending students to MIT demonstrates the idiosyncratic process by which Indians went to MIT for higher training; It was not part of a systematic countrywide process for developing talent. In 1943, the secretary of state for India asked the British Royal Society to send Nobel laureate A. V. Hill to India to provide advice on the organization of science, medicine, and technology in postwar India. Hill's mission to India lasted from November 1943 to April 1944. He saw himself as being helpful to India, but the overall tone of the report was patronizing, giving advice on how India could be brought up to British standards. In one area, Hill sounded a different note. In the section on technology, he spent a full paragraph lamenting Britain's lagging position in higher technical education, noting that the United Kingdom did not yet have an institution comparable in quality to MIT, although "responsible people" thought such institutions should be set up in the United Kingdom. Hill then went on to make an a fortiori argument with respect to India, stating that if the United Kingdom needed an MIT, India did even more so.[46]

Hill wrote as a wise father advising adolescents who were not quite as responsible as they should be. In the light of men such as the Parekhs, Pandya, and Kalelkar, who had recognized on their own the role for an MIT education in India, Hill's wisdom was not as great as it might have seemed to some. In closing his argument for high-level technical education in India, Hill wrote, "Nationalist fervor cannot replace first-class scientific ability and technical training."[47] Kalelkar, T. M. Shah, and the Parekhs would have agreed, although unlike Hill, they would have also asserted the converse.

In 1944, Ardeshir Dalal, formerly a high official working for the Tata business family, became a member of India's Executive Council for the Department of Planning and Development. In October of that year, he announced a plan to send 500 Indian

[45] Robert Darnton, *The Great Cat Massacre and Other Episodes in French Cultural History* (New York, 1984); Roger Chartier, "Texts, Printing, Readings," in *The New Cultural History*, ed. Lynn Hunt (Berkeley, Calif., 1989), 154–75.

[46] A. V. Hill, *Scientific Research in India* (London, 1945), 29–30.

[47] Ibid., 30.

students abroad in 1945 to institutes in the United Kingdom, Canada, and the United States to meet the demands for "urgent needs of post-war development." The government of India's plan was an acknowledgement that India did not have engineers with the advanced technical training it needed.[48]

Dalal's announcement produced a dramatic increase in the number of Indian students applying to MIT. Between 1920 and 1939, MIT had enrolled on average 7 Indian students per year. By the fall of 1944, that number had increased to 24. But in April 1945, with the Indian government's offering unprecedented funding for graduate training abroad, MIT had 271 applications on hand from Indian students, a number representing over half of the scholars the Indian government was planning to send to the United Kingdom, Canada, and the United States. Although MIT was able to admit 16 for the fall semester, it had placed 180 Indian students on its waiting list, implicitly stating that the students were well qualified for MIT but that there was no room for them. Although information does not exist to determine who was responsible for decisions about which schools prospective students applied to—the students themselves, their professors, or administrators—the increase in applications to MIT (particularly given that some of the 500 students would be working in fields, such as agriculture, where MIT offered no programs) suggests that for engineering students the preferred meaning of "studying engineering either in the UK, Canada, or the United States" was simply studying at MIT.[49]

In Dalal's first press conference, echoing Hill's report, he announced that the establishment of an institution on the order of MIT was being considered. Shortly thereafter, at Dalal's urging, the member of the Viceroy's Executive Committee with responsibility for the Department of Education, Health, and Agriculture constituted a committee to consider the development of higher technical institutions in India. This committee, called the Sarker Committee, after its chair, N. R. Sarker, has become well known in India for its role in laying out the framework of the Indian Institutes of Technology.[50]

This committee of twenty-two had nine Britons and thirteen Indians. Among the Indians were some of India's leading scientists, such as J. C. Ghosh and S. S. Bhatnagar, and representatives of India's leading industrial enterprises, such as A. D. Shroff, who worked for the Tatas. On the committee also, but unnoticed before by historians, were two young Indian engineers with doctorates from MIT: Anant Pandya and M. D. Parekh, with Pandya sitting on the working subcommittee. They came with a deep knowledge of what an MIT education was and what it might mean for India. The creation of the Indian Institutes of Technology was an act of the imagination, but

[48] "Progress in Reconstruction Plans for India," *Times of India*, 15 Sept. 1944, 1; "Report of the Selection Board Overseas Scholarships, 1945," 2 V/27/864/8, India Office Records, British Library, London.

[49] Committee of Stabilization of Enrollment, "Report on Foreign Students, to be presented to the Faculty on April 13, 1945," MIT Faculty Records, AC 001, MIT Institute Archives; Massachusetts Institute of Technology, *President's Report, October 1945*, 90. Out of fear that an increase in foreign students would squeeze out returning GIs and reshape the character of the institute, the faculty voted to place a quota of 300 on the number of international students the institute would accept.

[50] Kim Patrick Sebaly, "The Assistance of Four Nations in the Establishment of the Indian Institutes of Technology, 1945–1970" (PhD diss., Univ. of Michigan, 1972), 12–28; Central Bureau of Education, India, *Development of Higher Technical Institutions in India (Interim Report of the Sarker Committee)*, Feb. 1946.

it was not solely an act of the imagination: it institutionalized what Parekh and Pandya had done on their own, without the help of the government of India.[51]

In his report, Hill observed that one of the "most important needs today of Indian science, medicine and technology is of better facilities to send the ablest of their young people abroad, particularly to the United Kingdom."[52] After India won its independence in 1947, Hill proved to be half right. The government of India continued the late colonial policy of paying to send students abroad for higher studies, and in 1947–48, 900 Indian students studied in the United Kingdom. However, more than 1,200 studied in the United States. The preference for American universities over British universities among Indians studying abroad would grow more pronounced over the years. A variety of factors contributed to this. America had far greater capacity in its universities, and its schools had a greater orientation to engineering and agricultural education than those in Britain. Lingering resentment over colonialism may have caused some students to prefer the United States.[53]

Although by the late 1940s the numbers of Indian MIT alumni had grown too great to allow for a comprehensive examination of their careers, some sense can be gleaned from looking at the three men from Bhavnagar who earned doctorates from MIT between 1933 and 1940: M. D. Parekh, N. B. Bhatt, and Pandya. M. D. Parekh worked for Delhi Cloth Mills, where he designed plants for producing alcohol, vegetable ghee, and caustic soda. In 1949, he left to become the chief technical officer of the newly established National Rayon Corporation. Bhatt, who had earned his doctorate in physics, became the first head of the Department of Electrical Communications at the Indian Institute of Science and then in 1949 became one of the leading figures in India's Defence Science Organization. After World War II ended, Pandya took the position of director and chief consulting engineering for the contracting firm Hind Construction, where he sought to build up a firm with indigenous capacity to undertake large construction projects for newly independent India. In 1949, the Government of India asked him to take over the general manager position at Hindustan Aircraft Limited, a position he held for nine months before returning to contracting. In June 1951, Pandya died tragically in an automobile accident.[54]

MIT alumni saw themselves as having a corporate identity and a corporate responsibility to India. In 1945, a group of MIT alumni including M. D. Parekh and Pandya established an MIT Alumni Association based in Bombay. In independent India, the group saw itself as having an informal advisory function to the Indian government. In 1950, it published a report analyzing the Fischer-Tropsch process for producing liquid fuels from coal, arguing against its use in India. Between 1950 and 1968, the group published three other reports. The MIT Alumni Association also circulated

[51] *Development of Higher Technical Institutions in India* (cit. n. 50), 1–2; Parekh, interview (cit. n. 14).

[52] Hill, *Scientific Research in India* (cit. n. 46), 7.

[53] For data on Indian students in the United States, I have used Federal Security Agency, Office of Education, *Annual Report 1948*, 524; Robert C. Story, *Residence and Migration of College Students, 1949–50*, Office of Education, Misc. No. 14. For data from 1950 onward, I have used the annual reports of the Institute of International Education, *Open Doors* (New York). For Indian students in the United Kingdom in 1947, I have used *Yearbook of Commonwealth Universities, 1948* (London, 1948), 1002–3. Further data on Indian students in the United Kingdom comes from subsequent volumes of the same serial.

[54] Parekh, interview (cit. n. 14); Shastry, "Nautam Bhagwanlal Bhatt," 895 (cit. n. 33); R. S. Bhatt, "Anant Pandya" (cit. n. 34), 1, 6–13.

speeches of its members to the Planning Commission of India and government departments.[55]

For many years, India's MIT-trained engineers would have had a very limited public presence as "MIT-trained engineers." Even well-educated Indians would have had little reason to know of their existence. This first changed with the death of Anant Pandya in 1951. In 1952, the Gujarati upper-class youth magazine *Kumar* published a special issue memorializing Pandya. It featured pictures of his time at MIT and his later career and tributes from leading engineers in India and some of Pandya's professors in the United States. Pandya's education at MIT was a particularly prominent feature. Pandya received a level of acclamation rare among engineers anywhere.[56]

It is impossible to know exactly what this meant for young Gujarati men who read *Kumar*, but one person who claims the Pandya memorial changed his life was Kirit Parikh, the then seventeen-year-old son of a barrister in Ahmedabad. Parikh was considering the next step in his education after receiving his first college degree. After seeing the *Kumar* tribute to Pandya, Parikh, who was raised in a Gandhian school, which taught spinning on a charka, decided to become an engineer and attend MIT. He did both, earning a doctorate in civil engineering from MIT in 1962. The editors of *Kumar* would doubtless have been pleased with that result.[57]

CONCLUSION

The 1930s and 1940s were full of consequential events as Indians worked to create an independent nation, so it can hardly be considered surprising that in spite of all the writing on the period and on Gandhi, the connection between Gandhi's followers and MIT has not been noticed. The connection is subtle but important in understanding the India that a group of nationalists was fighting for—a technological India.

Indian MIT-trained engineers were an extraordinarily tiny group, utterly unrepresentative of the larger Indian population. In the 1930s, the Parekhs and the others from Bhavnagar were beneficiaries of a special set of circumstances that gave them access to MIT, in a way not available, even to elites, in other parts of India. However there is reason to believe that the Parekhs and other elites in Bhavnagar stood proxy for other elites in India: the latter would have done the same as the former if they had had the means and the opportunity to do so. In the decades after independence, they did have means and opportunity. The sons of Indian elites—lawyers, civil servants, educationalists, businessmen, and engineers—went to MIT in increasing numbers in the 1940s, 1950s, and 1960s.[58]

In the first forty years of independence, MIT graduates occupied an astounding

[55] Program, Eleventh Convention of MIT Alumni Association (1968), 7, folder "Bombay," box 54, Collection AC224, MIT Institute Archives; Parekh, interview (cit. n. 14).

[56] *Kumar*, Aug. 1952 (in Gujarati), in author's possession. I thank Anand Pandya, son of Anant Pandya, for providing me with a copy of this magazine. Many of the articles also appeared (in English) in *Dr. Anant Pandya: Commemoration Volume* (cit. n. 34). I thank Anand Pandya for providing me with a copy of this volume.

[57] Kirit Parikh, interview with author, 11 June 2008, New Delhi. Although I was aware of the *Kumar* tribute to Pandya, Parikh brought it up without prompting by me. As of 2008, Parikh is a member of the Indian Planning Commission.

[58] By the late 1950s, the Indian student population had grown to approximately 60 (out of approximately 6,000 total students). In the late 1960s, it was approximately 100 (out of approximately 8,000 total students). These data are from MIT's *President's Report* (available online at http://libraries.mit .edu/archives/mithistory/presidents-reports.html).

number of the highest-level positions in the Indian technical community—more than graduates of any other single school in the United States or the United Kingdom, and quite possibly more than the graduates of any single school in India. Although part of this disproportionate representation might be attributed to MIT and the training gained there, a greater part was due to the frequency with which people from the highest levels of Indian society sent their sons to MIT.[59]

The "dominance by design" narrative bespeaks American power and a lack of Indian agency in the face of that power. The lives of MIT-trained Indians suggests a confidence that they could use their MIT education to build India according to their interests and the interests of the nation. If policies pursued in India ultimately benefited elites more than they did the masses (which they certainly did), explanations should be sought in terms of Indian society and Indian politics, not in terms of theories made in America.[60]

To say that India's technological identity owed something to MIT is not to deny its legitimacy as a national technical identity, any more than to say the fact that the framers of the U.S. Constitution owed something to John Locke is to deny the legitimacy of an American identity. Rabindranath Tagore, the great Bengali poet (who sent his son Rathindranath to study agriculture at the University of Illinois), wrote to the British priest Charles Andrews: "Whatever we understand and enjoy in human products instantly becomes ours, wherever they might have their origin."[61]

In 1963, a memorial museum to Mahatma Gandhi opened adjacent to his former ashram in Ahmedabad. The museum was designed with the same types of materials used in the ashram buildings and attempted to give the feel of an Indian village. The museum, built without windows, used only wooden louvers and contained a water court at the center to provide cooling from the heat of Indian summers. The memorial's architect was Charles Correa, M. Arch. MIT 1955.[62]

The same year the Gandhi memorial opened, an Indian graduated with a master's degree in civil engineering from MIT, by this time hardly noteworthy, except for one fact: the graduate was Kanu Ramdas Gandhi, the grandson of Mahatma Gandhi.[63]

---

[59] Although proving this statement is beyond the scope of this paper, an example of how it happened can be seen in the case of Aditya Birla. Aditya Birla, born in 1943, was the grandson of G. D. Birla, the man who was the head of the Birla business empire and had funded Kalelkar to go to MIT. Aditya was seen by the Birla family as the likely heir to run the Birla business empire. When Birla went to college, the family sent him, not to an IIT, not to Cambridge, Oxford, or Harvard, but to MIT. Birla received a degree in chemical engineering and became the leader of the Birla empire and one of the most important business leaders in independent India. Minhaz Merchant, *Aditya Vikram Birla: A Biography* (New Delhi, 1997). My general statement about the social backgrounds of Indians graduates of MIT is based on dozens of interviews I have conducted with them.

[60] Adas, *Dominance by Design* (cit. n. 4). Daniel Headrick has made a similar critique. Daniel R. Headrick, "Dominance by Design," *Journal of World History* 18 (March 2007): 108–10.

[61] Rabindranath Tagore, *Letters to a Friend* (New York, 1929), 136. The work of Amartya Sen drew my attention to this quotation. Sen, *Development as Freedom* (New York, 1999), 242. For Rathindranath's time at the University of Illinois, see "History of the Tagore Festival," http://tagore.business .uiuc.edu/history.html (accessed 6 May 2008).

[62] Correa, an Indian of Goan ancestry, but not Portuguese, is by any standard the most important Indian architect and urban planner of the post-1947 period. Charles Correa, *Charles Correa* (Bombay, 1996); Jon Lang, *A Concise History of Modern Architecture in India* (Delhi, 2002).

[63] Kanu R. Gandhi, "Theoretical and Experimental Study of Kinematic Characteristics of Subtalar Joint in the Human Foot" (master's thesis, MIT, 1963). Kanu Gandhi is identified as the grandson of Mahatma Gandhi in "Grandson of Gandhi Weds," *New York Times,* 5 Oct. 1961, 34; "Cambridge Residents," *Washington Post,* 18 Oct. 1961, B15.

Taking to the Field:

Geological Fieldwork and National Identity in Republican China

By Grace Yen Shen[*]

ABSTRACT

This paper explores the significance of fieldwork for identity formation in three phases of modern Chinese geologic education: the prerevolutionary overseas phase (1903–12), the Geological School period (1916–18), and the early years of the Peking University Department of Geology (1918–28). This examination demonstrates the importance of the body in connecting the physicality of the individual to the dignity of the national collective and argues that Chinese identity was defined by the willingness to remake oneself for the sake of the nation, rather than by any intrinsic qualities of "Chineseness."

INTRODUCTION

In 1877, in the first installment of his five-volume treatise on Chinese geology, Ferdinand von Richthofen (1833–1905) opined that geology was the least likely of any modern science to take root in China:

> [T]he Chinese man of letters is sluggish and chronically loath to move rapidly; in most cases he simultaneously vexes one with his avarice and cannot free himself from native prejudices concerning decorum. In his view, to go on foot is demeaning, and the occupation of geologists a direct surrender of all dignity in the eyes of the world.[1]

Richthofen's researches became the standard work on Chinese geology as soon as they were published, and his assessment of the ignorance and indolence of Chinese elites helped foreigners justify both the exploration and exploitation of Chinese territory for several decades.

At the same time, Richthofen's comments spurred patriotic young Chinese caught up in the turmoil of the late Qing dynasty (1644–1911) to make the study of Chinese

---

[*] York University, Department of Humanities, 262 Vanier College, 4700 Keele Street, Toronto, Ontario, Canada M3J 1P3; gyshen@yorku.ca.

[1] Ferdinand von Richthofen, *China*, vol. 1, 38, as quoted in V. K. Ting, foreword to *Bulletin of the Geological Survey of China* 1 (July 1919): i. It is interesting to note that Richthofen himself traveled by horse with servants and, according to Jürgen Osterhammel, "did not lift a finger himself" as a sign of German racial superiority in China. (Osterhammel, "Forschungreise und Kolonialprogramm," *Ferdinand von Richthofen und die Erschließung Chinas im 19. Jahrhundert, Archiv für Kulturgeschichte* 69 [1987]: 179.) Bailey Willis, whose Carnegie expedition in 1903–4 was largely on foot, recalls being advised that to walk would be a loss of face, but he chose to do so anyway.

© 2009 by The History of Science Society. All rights reserved. 0369-7827/09/2009-0012$10.00

geology a new priority. Almost as soon as the republican revolution of 1911 had over-thrown the imperial order, "returned students" who had studied abroad, such as Ding Wenjiang (1887–1936) and Zhang Hongzhao (1877–1951), set up the Geological Section in the Bureau of Mines of the new provisional government. In 1913, Ding or-ganized the Geological Survey and Zhang established the Geological School to train native researchers. The Survey was up and running by 1916 and staffed with select graduates of the Geological School in 1918. A year later, the Survey began publica-tion of the *Bulletin of the Geological Survey of China*, and it followed this with a second series of lengthier *Memoirs* in 1920.[2] By the time the first issue of the *Bulletin* appeared, native geologic activity had progressed enough that Ding Wenjiang could confidently remind his readers of Richthofen's words and answer them with a senti-ment from Roman dramatist Terence: "*omnium rerum, heus, vicissitudo est*" (there are changes in all things).[3]

What is interesting here is that, while the original research that filled the pages of the *Bulletin* clearly dispelled the idea that Chinese could not excel at geology, Ding did not actually challenge Richthofen's basic characterization of Chinese flaws. In fact, his use of Terence suggests an arc that stretches from "the Chinese man of letters" to the modern Chinese geologist and links them in a narrative of national transformation that actually required the accuracy of Richthofen's assessment to be meaningful.

Although we see Ding in a moment of triumph and Richthofen as the disparag-ing outsider, the key element is the internalization of the injury: it is because Chi-nese saw themselves as flawed that they were compelled to act. As Jing Tsu notes in a recent study of twentieth-century Chinese literature, "Whatever suffering or hu-miliation may be inflicted by Westerners on the Chinese can never match the candor with which the Chinese relentlessly expose their own weaknesses."[4] In fact, self-criticism (and its attendant feeling of shame) did far more than simply inspire action and sweeten the pleasure of Chinese geologic successes. In the absence of clear po-litical or social leadership, self-criticism constituted an act of ownership that claimed the righting of wrongs (both external and internal) as a defining—and fundamentally personal—Chinese prerogative.

Throughout much of Chinese history, Confucian intellectuals had taken upon them-selves the burden of responsibility for social harmony and the well-being of the state. The failure of those intellectuals, whether in or out of political office, to respond to changing circumstances in the late Qing period was an indictment of their way of life and their ability to lead Chinese society. Unfortunately, the Republican period that emerged from the collapse of empire in 1911 was little more than a succession of

---

[2] The *Bulletin of the Geological Survey of China* was published continuously from 1919–45, and the *Memoirs of the Geological Survey of China* appeared in several series until 1947. Later, the Geological Survey also produced four series of its *Paleontologia Sinica* (1922–47), *Contributions from the Sin Yuan Fuel Laboratory*, *Contributions from the Cenozoic Research Laboratory* (in conjunction with the Peking Union Medical College), the *Seismological Bulletin*, and several pedological journals, of which the *Soil Bulletin* (1931–64) and the *Soils Quarterly* (1940–48) were most widely read. All of these journals published through the War of Resistance against Japan (1937–45), but few managed to survive the subsequent civil war between Nationalist (Guomindang) and Communist forces. Internal journals and working papers lasted throughout the period and were sometimes circulated informally.

[3] Terence, *Eunuchus*, act 2, sc. 2, 45 (276). Ding Wenjiang, *Bulletin of the Geological Survey of China* 1 (1919): 1.

[4] Jing Tsu, *Failure, Nationalism, and Literature: The Making of Modern Chinese Identity, 1895–1937* (Stanford, Calif., 2003), 112.

weak states, all torn by "internal strife and external threats" (*neiluan waihuan*), and none capable of sustaining a persuasive national consciousness to support its legitimate rule.

Amid this vacuum of convincing moral and political authority, self-criticism—or more precisely, criticism of China and the Chinese from a perspective of complicity—gave authentic voice to the chaotic times and offered an inverted vision for social cohesion. Identification of, and more important *with*, Chinese flaws allowed a new breed of modernizing intellectuals to reaffirm their Chinese identity by admitting individual fault for collective failings. Although definitions of "Chineseness" in terms of race, culture, history, or character remained divisively contentious in the Republican period, the very act of feeling China's shame as one's own assumed both the existence of the nation and the membership of the self in that unit. As a result, personal discipline could potentially redeem the group, and Chinese geologists, for instance, could look to strenuous fieldwork not only to reinvent themselves but also to identify with collective renewal.

Histories of geology have classically approached fieldwork as a manifestation of the epistemological and sociological ideals prevalent within specific eighteenth- and nineteenth-century national contexts. Most notably, in the case of Victorian England, fieldwork was an arena in which "gentlemen geologists" imprinted the earth sciences with the mores of public-school masculinity.[5] The great geologic controversies detailed by Martin Rudwick, James Secord, and David Oldroyd all hinged on fieldwork as an empirical practice that united preexisting "elements of romanticism and tacitly pantheistic religion with those of robust, manly Christianity and the gentleman's love of the countryside and its sporting pursuits."[6]

For Chinese eager to redress their nation's failures, however, the rigors of fieldwork were a bitter pill to take to cure what seemed explicitly wrong about traditional Chinese ideals. James Reardon-Anderson, in his history of modern Chinese chemistry, has characterized Chinese achievements in geology as playing "at the low easy end of the scientific spectrum" where simply "reporting on what they found in their own backyards" would allow Chinese to "quickly enter the game of world science."[7] But this view, which reflexively privileges experimental laboratory science, completely ignores the gulf between the Victorian vision of geologic fieldwork and Chinese elites' longstanding disdain for physical labor. As a result, it misjudges both the difficulty of geology for Chinese workers and the price of admission into international scientific circles.

The nature of working outdoors made geology far more of a challenge to traditional Chinese intellectual identity and masculinity than physics or chemistry ever were. It also afforded geologists unique opportunities for self-assertion in a period during which modernization often implied wholesale Westernization. When China's

[5] Roy Porter, "Gentlemen and Geology: The Emergence of a Scientific Career, 1660–1920," *Historical Journal* 21 (Dec. 1978): 809–36.

[6] Martin Rudwick, *The Great Devonian Controversy: The Shaping of Scientific Knowledge among Gentlemanly Specialists* (Chicago, 1985), 41. For more, see James Secord, *Controversy in Victorian Geology: The Cambrian-Silurian Dispute* (Princeton, N.J., 1986); David Oldroyd, *The Highlands Controversy: Constructing Geological Knowledge through Fieldwork in Nineteenth-Century Britain* (Chicago, 1990); and Bruce Hevly, "The Heroic Science of Glacier Motion," *Osiris* 11 (1996): 66–86.

[7] James Reardon-Anderson, *The Study of Change: Chemistry in China, 1840–1949* (Cambridge, UK, 1991), 5.

pioneering researchers began to fill in the geologic "blank" of their native country-side, they did so as much to reshape their bodies as they did to redefine their knowl-edge of the physical terrain, and this dedication of the individual body to the body of the nation gave Chinese a rare sense of possession over their own scientific labors. In their view, the geologic knowledge they produced fit into international theories and norms, but its foreignness was amended by the unmediated contact of local scientist and native land.

The aim of this essay, then, is to trace how fieldwork allowed Chinese geologists to identify with the nation and sublimate their feelings of shame over China's weakness into an empowering physical engagement with the landscape. To this end, I explore the significance of fieldwork for identity formation in three phases of Chinese geo-logic education: the prerevolutionary overseas phase, the Geological School period (1916–18), and the early years of the Peking University Department of Geology.

For the founding generation of Chinese geologists, fieldwork experiences abroad personalized China's failure to understand the national territory and framed it as the cumulative product of individual Chinese defects. Returned students took this lesson to heart and tried to stimulate a similar identification with nation in their students at the Geological School. The Geological School's empirical bent was designed to both inspire love of country through direct experience and instill a professional ethic that would discipline such feelings into credible scientific output. Fieldwork thus eased geologists' transition from empire to republic by rendering national identity and na-tional salvation a matter of pedagogy and training. The bookish academic culture of Peking University briefly threatened this training, but an examination of why field-work was reestablished as the centerpiece of geologic education demonstrates the importance of the body in connecting the physicality of the individual to the dignity of the national collective. Seen through the lens of national shame, Chinese identity was defined by the willingness to remake oneself for the sake of the nation, rather than by any intrinsic qualities of "Chineseness." In turn, hardships the body suffered for the nation both naturalized foreign science and transformed the ideals of Chinese modernity.

STUDY ABROAD AND RETURNED STUDENT CONSCIOUSNESS

In "The Nationless State: The Search for a Nation in Modern Chinese Nationalism," John Fitzgerald expresses the frustration of contemporary theorists with the idea of Chinese national identity. "Confucian reformers associated the collective self with a distinctive civilization; liberal republicans conceived of the nation as a body of citi-zens; Nationalist (Kuomintang) revolutionaries thought of a Chinese race; and Chi-na's Marxist-Leninists have qualified citizen and race by reference to social class." Given the bewildering variety of often contradictory forms that national identity claims have taken in modern China, the persistence and vehemence of Chinese na-tionalism has defied existing theories of the nation to the degree that Fitzgerald sees "China" as little more than a "floating referent of the state," and he expresses doubt that any "Chinese nation [exists] outside the state framework."[8]

---

[8] John Fitzgerald, "The Nationless State: The Search for a Nation in Modern Chinese Nationalism," in *Chinese Nationalism*, ed. Jonathan Unger (Armonk, N.Y., 1996), 56–85. Spellings and translitera-tions given as in the original, 57.

Fitzgerald's position, although extreme, is representative of the positions of other China analysts in at least two important ways.[9] His position is committed to constructing national identity as a set of inclusion criteria that should (however imperfectly) link nation and state, and because this linkage is particularly problematic in the Chinese case, it displays a tendency to accept the state and its manipulations as the only half of the Chinese nation-state dyad amenable to analysis. In contrast, many patriotic Chinese of the late Qing and early republic were also cognizant of the inconsistencies and ambiguities of Chineseness—they were, after all, deeply embroiled in contesting the ideal shape of Chinese modernity—but quite often they preferred the imprecision of the nation over the self-serving concreteness of the state.

When, for instance, Chinese geologic pioneer Zhang Hongzhao, at the age of thirty-two, traded his classical education for "new learning", he justified his choice to study geology abroad as follows:

> At this time [1909] I only knew that a great number of foreigners had investigated the geology of China, but I had never heard of any of my compatriots attending to such things. With so many [Chinese] nationals, not a one had explored the geology of a single portion of this divine land [*shenzhou*] in any detail. Instead they let outsiders extend deep into the hinterland without their knowledge; it was shameful enough.[10]

For Zhang, there was no need to define objective criteria for his "compatriots" or fellow "nationals," and he was not primarily concerned with China as a state formation. Rather, Zhang perceived unchecked foreign exploration as a violation or transgression of the nation, and he simply assumed the existence of a national cohort that should have preempted or responded to these incursions. More important, by expressing his own sense of shame at this situation, he identified himself with both the nation and its interests, and he rooted his national identity in an act of identification rather than a set of identifiable characteristics. Once begun, this act compelled him to "personally take on the responsibility" of being the first Chinese to study the geology of China, and in following through with his commitment, Zhang completed his assertion of and identification with the nation.[11]

Zhang was not, however, the first to accept this particular burden of national shame. Several years earlier, Gu Lang and Zhou Shuren (Zhou, who later became celebrated as the short story writer and social critic Lu Xun) also blamed Chinese themselves for allowing foreign encroachment. "If the owner is negligent," they wrote in 1906, "then brigands will have their way. Today we yield and offer, tomorrow we grant special privileges. Anyone with means at all can steal candy from an orphan."[12] Although

---

[9] A large body of work exists on Chinese nationalism and national identity. This has recently mushroomed as a result of mainland China's increasingly prominent political and economic profile, as well as its hosting of the 2008 Olympics. However, better starting points for understanding problems of nation theory in the Chinese context would be: Lowell Dittmer and Samuel Kim, eds., *China's Quest for National Identity* (Ithaca, N.Y., 1993); Unger, *Chinese Nationalism* (cit. n. 8); C. X. George Wei and Xiaoyuan Liu, *Exploring Nationalisms of China: Themes and Conflicts* (Westport, Conn., 2001); Zhao Suisheng, *A Nation-State by Construction: Dynamics of Modern Chinese Nationalism* (Palo Alto, Calif., 2004).

[10] 章鸿钊, 六六自述 [Zhang Hongzhao, *Liuliu zishu*] (Wuhan, 1987), 21.

[11] Ibid., 22.

[12] 顧琅 and 周樹人, 中國礦產志 [Gu Lang and Zhou Shuren, *Zhongguo kuangchan zhi*], quoted in 吴凤鸣, "關於顧琅及其地質礦產著作的評述," 中國科技史料 [Wu Fengming, "Guanyu Gu Lang ji qi dizhi kuangchan zhuzuo de pingshu," *Zhongguo kuxue shiliao*] 5, no. 3 (1984): 91.

their words were harsh, Gu and Zhou reminded Chinese that they had the power to redress past failings by acquiring personal knowledge of the geology and mineral resources of China's vast territory. Whether the state should have intervened or not, individuals could still make a difference by turning their shame into action.[13]

Larger historical circumstances made this goal increasingly attainable, as reforms enacted in 1905 by the ailing Qing dynasty eliminated the time-honored civil service examination system and encouraged the rapid expansion of overseas education. This outward movement created a class of young Chinese who could gaze at their home-land from a great remove while contemplating foreign ideas and peoples at close range. For Zhang Hongzhao, the comprehensiveness and advancement of Japanese geology not only exacerbated his sense of humiliation and urgency but also cemented his commitment to China. Zhang's professors at Imperial Tokyo University had "left their footprints" all over Japan, and after intensive field training in the Japanese coun-tryside, Zhang was determined to leave his own footprints on the area near his native home.

In the surrounds of the famous West Lake of Hangzhou, during the summer of 1910, Zhang completed the fieldwork for his final thesis. But his experience proved bittersweet. Despite Zhang's eagerness to represent China and Chinese geology, the trip was a poignant reminder of how backward China was. Whereas geologists in Ja-pan could find a clean and comfortable place to sleep in even the remotest regions in Japan, in China facilities were such that a researcher could get little rest after a hard day's work, and Zhang returned to Japan ill from the unclean conditions.[14] More im-portant, Zhang had to work without the assistance of even rudimentary modern maps of the region. The best maps available were in the traditional style, without longitude and latitude, elevation, or consistent scale.[15] Even as a student in Japan, Zhang could consult general maps of the country and check local reference materials before ever

---

[13] It may be instructive, as a contrast, to consult several recent studies on the narrative of "national humiliation" in the construction of Chinese nationalism, most notably: Luo Zhitian, "National Hu-miliation and National Assertion: The Chinese Response to the Twenty-one Demands," *Modern Asian Studies* 27, no. 2 (1993): 297–319; Paul A. Cohen, "Remembering and Forgetting: National Humilia-tion in Twentieth-Century China," *Twentieth-Century China* 27, no. 2 (2002): 1–39; William A. Calla-han, "National Insecurities: Humiliation, Salvation, and Chinese Nationalism" *Alternatives* 29 (2004): 199–218. The subject has even found its way into more popular outlets in Orville Schell, "China: Humiliation and the Olympics," *New York Review of Books*, 14 Aug. 2008, http://www.nybooks.com/articles/21715. These articles, however, focus on the pathology of humiliation and remembrance as, alternately, a sign of insecurity, a goad to xenophobic self-aggrandizement, and a narrative of victim-ization. Although several authors also highlight the multiplicity of ways that national humiliation is understood in China, their emphasis is on the state and its attempts to manipulate national psychology. For Chinese geologists, the mantle of shame was not put upon them, and as an act of self-identification that gave them an outlet for self-transformation—their shame was not about despising the self but about changing that which was shameful through, for their purposes, fieldwork. The critical link here is the construction of the self as directly (almost metonymically) linked to the larger collective. In this way, the narrative of humiliation discussed by the authors above is an important backdrop for under-standing Chinese geologists, but no more so than the traditional Confucian position articulated in *The Great Learning* that took moral action to be a conscious extension of the self to the society and state. More work is needed to understand what, if any, relationship exists between these Confucian ideas and modernizing Chinese geologists.

[14] 章鴻釗, 六六自述 (cit. n. 10), 24. Ding Wenjiang goes into great detail on the miserable state of both Chinese inns and local maps in the first few installments of his 1932 series of "漫遊散記" articles in 獨立評論 ["Manyou sanji," *Duli pinglun*].

[15] 章鴻釗, 中國地質學發展小史 [Zhang Hongzhao, *Zhongguo dizhixue fazhan xiaoshi*] (Shanghai, 1940), 12.

setting foot in the field.[16] In China, Zhang found no usable maps or books and had no other geologists to compare notes with. The study abroad experience had deepened his identification with China, but it had also trained him to see his own country with different expectations.

Although Ding Wenjiang followed a very different trajectory in his overseas education, he experienced a similar combination of reinforcement and distancing. In his two years (1902–4) in Japan, Ding witnessed the ways in which Chinese students were constantly led to compare their new surroundings with their homeland and how discrimination encouraged a strong sense of "common interest" among students from all over China.[17] His most formative intellectual experiences, however, occurred in England, where he completed his secondary education, and in Scotland, where he earned an undergraduate degree in biology and geology at Glasgow University. At Glasgow, the "broad-shouldered Chinese student of brilliant intellectual powers and of charming personality" was a favorite of geology chair John Walter Gregory, and in 1911, Ding graduated with first class honors and the Cowie Prize for fieldwork in geology.[18]

Like Zhang Hongzhao, Ding was eager to apply his newfound geologic skills in China, and rather than going directly home to Jiangsu Province after graduation, Ding took a circuitous route via Annam (now Vietnam) and then Yunnan, Guizhou, and Hunan provinces. Over the course of three months in the summer of 1911, Ding traveled by foot and by boat, observing the geology of southwestern China for himself.[19] His accounts of these travels (published much later) were aimed at a nonscientific audience, and he gave few details of his geologic investigations, instead emphasizing the ways that, after nine years abroad, fieldwork was a "revelation" that reintroduced him to the "real face" of the Chinese landscape.[20]

In part, this indicated his awe at the startling and unfamiliar karst topography of China's southwest, but it also reflected his realization of how little Chinese understood of their own territory. As he traveled through the southwestern provinces, he had expected to find lacunae in local geologic knowledge, but what shocked him was the degree to which provincial gazetteers and imperial route maps propagated patently inaccurate geographic information.[21] In his view, this Chinese preference for textual authority over empirical fact was a symptom of cultural decay and an embarrassing cause of national disunity that could be resolved only through fieldwork.

Both Zhang Hongzhao and Ding Wenjiang returned to China from study abroad in 1911, and they sat for the same returned-student examinations before watching the drama of the republican revolution unfold. Zhang had already published an article the previous summer suggesting that the Qing establish an official geologic survey, but he quickly redirected his argument to the citizens of the new Chinese republic, whom

16 章鴻釗, 六六自述 (cit. n. 10), 25.

17 V. K. Ting, "Chinese Students," *Westminster Review* 169 (Jan. 1908): 48–55.

18 G. W. Terrell, "Dr. V. K. Ting," *Bulletin of the Geological Society of China* 20 (1940): 369; *Glasgow University Calendar* (Glasgow, 1910–11), 547, 550; *Glasgow University Calendar* (Glasgow, 1911–12), 581.

19 See Hu Shi's biography of Ding for the most concise description of these travels. 胡适, 丁文江傳 (Haikou, 1993).

20 "漫遊散記," 獨立評論, 10 July 1932, 21–22. Ding mentions keeping a field notebook, collecting specimens, examining outcrops, sketching, and carrying equipment he brought from England.

21 "漫遊散記," 獨立評論, 26 June 1932, 12–17.

he felt would understand his sense of mission.[22] Zhang found it humiliating that foreigners monopolized scientific understanding of Chinese territory, so he wrote an essay promoting geologic investigation as part of owning the land or fulfilling the "proper duties of a landlord."[23] All civilized nations recognized the value of contributing to geologic knowledge; if China continued to shirk this duty, it would forfeit its claim to civilization.[24] As a Chinese, Zhang could not be satisfied with simply having the land surveyed; he wanted to lay the foundations for an independent native geologic community.

Unfortunately, the fledgling government, struggling as it was to legitimize itself, was less eager to accept Zhang's mantle of shame and showed little enthusiasm for homegrown geology when foreign exploration seemed a shortcut to practical resources. The provisional government established the Geological Section under the Bureau of Mines, but support proved so shaky that Zhang left his post as section chief in protest. He preferred to teach geology and keep up with fieldwork on his own rather than to bide his time as an ineffective bureaucrat. Ding Wenjiang, who had been teaching at the progressive Nanyang Middle School in Shanghai, was eager to try his hand at the project, but when Ding arrived at his new office, he found that

> the [Geological] Section had a porter and two junior functionaries, none of whom knew any geology. The "section" was an administrative unit, but we didn't have anything to administer. I repeatedly asked for travel [fieldwork] expenses, but the Ministry said there were no travel funds.[25]

Although Ding had partnered with the state to serve the nation, he had to work around the state to do so.

THE GEOLOGICAL SCHOOL AND PATRIOTIC PEDAGOGY

A year later, in 1913, state ministries were reorganized for the third time since the Republic's inception. Zhang Yiou, who had studied mining in Belgium before being appointed to head the Bureau of Mines, finally convinced the new minister and vice-minister of agriculture and commerce to create a geological survey from the inactive Geological Section.[26] The Survey's mission was to "plan and manage" all national surveying, and the Ministry of Agriculture and Commerce earmarked 50,000 yuan for start-up costs; but when Ding Wenjiang assumed directorship in late 1913, he was still the only technical staff member, and actual funding did not always materialize.[27]

The problem of personnel had been obvious to Ding since his frustrating days in

---

[22] 章鴻釗, 六六自述 (cit. n. 10), 30.

[23] "中華地質調查私議," 地學雜誌, ["Zhonghua dizhi diaocha siyi," *Dixue Zazhi*], 1912, part 1, 7.

[24] Ibid. Zhang also argued that China could not maintain its identity as a nation founded upon agriculture if it did not even understand its own soil.

[25] 胡适, 丁文江傳 (cit. n. 19), 54.

[26] According to Hu Shi, 59, this reorganization occurred in October, but 程裕淇 and 陈梦熊, eds., 前地质调查所 (1916–1950) 的历史回顾 [Cheng Yuqi and Chen Mengxiong, eds., *Qian dizhi diaochasuo (1916–1950) de lishi huigu*] (Beijing, 1996), 2, state that it was September 1913. I have adopted Zhang Hongzhao's June dating because the Ministry of Agriculture and Commerce was not established until 1914.

[27] Ibid., 2; 胡适, 丁文江傳 (cit. n. 19), 60. According to *Dizhi diaochasuo yange shilue* 地質調查所沿革事略 (n.p., 1922), 2, the Survey at this time had no fixed budget.

the Geological Section, and Ding capitalized on interest from Zhang Yiou to push for a geological school along lines originally proposed by Zhang Hongzhao in 1912. Although both returned students themselves, Ding and Zhang Hongzhao knew that China's development of its natural resources could not rely exclusively on "pilgrims heading west to procure the scriptures" like themselves. Not only was the cost of overseas education often prohibitive but "it could not be denied that for those who have had many difficult years of study at the remove of many hundreds and thousands of *li*, China's geology was nothing more than hearsay or a vision seen through a film."[28] China required a natively trained corps of geologists deeply familiar with the Chinese landscape and personally invested in the fate of the nation. Fieldwork was essential.

Unfortunately, government austerity left little money for a school. Coincidentally, budget problems at Peking University had shut down its entire Science College, and its flagging Geology Department, which had only graduated two students in four years, was facing permanent closure.[29] The timing was right, and the university offered to lend its equipment and facilities to the Ministry of Agriculture and Commerce, cutting its own losses and minimizing costs for the new government Geological School.[30]

In the absence of the civil examination system, Peking University became the preferred outlet for well-connected young Chinese with dreams of officialdom. In the early years of Republican China, the university was still viewed as a path to bureaucratic advancement rather than erudition, and students found it difficult to imagine how a degree in geology could further their political ambitions.[31] The Geological School, in contrast, lacked the high profile of Peking University, but it also dodged the air of decadence and intellectual indifference that the university had acquired. The Geological School's founders paid careful attention to students' practical needs and fought hard to eliminate tuition, guarantee housing, and cover fieldwork expenses so that its specialized curriculum would not be a financial burden.[32] As a government agency tied to the new national survey, the school also offered the tacit assurance of

[28] 翁文灝, "翁序," 地質研究所師弟修業記 [Weng Wenhao, "Weng xu," *Dizhi yanjiusuo shidi xiuye ji*] (Beijing, 1916), 2. One Chinese *li* is approximately a third of a mile.

[29] The Peking University Geology Department originated in the prerevolutionary Imperial University of Peking, established as part of the 1898 reforms. A Geology Department (地質門, *dizhi men*) was proposed immediately but was deferred until 1909, when the Science College (格致科, *gezhi ke*) was founded with chemistry and geology as subdivisions. Five students (王烈, 鄔友能, 裴傑, 陳祥翰, and 路晉繼 [Wang Lie, Wu Youneng, Qiu Jie, Chen Xianghan, and Lu Jinji]) entered the Geology Department in 1909 from the German language preparatory program, and no further students applied. Only two students graduated in May 1913. 于洸, 何国琦, 刘瑞珣, 李茂松, and 宋振清, "弘杨传统、把握机遇、再创辉煌—庆祝北京大学建立 100 周年、北大地质学系建立 89 周年," in 北京大学国际地质科学学术研讨会论文集, ed. 北京大学地质系 [Yu Guang, He Guoqi, Liu Ruixun, Li Maosong, and Song Zhenqing, "Hongyang chuantong bawo jiyu zaichuang huihuang—qingzhu Beijing daxue jianli 100 zhounian Beida dizhixue xi 89 zhounian," in *Beijing daxue guoji dizhi kexue xueshu yantaohui lunwen ji*] (Beijing, 1998), 1.

[30] 于洸 et al., "弘杨传统" (cit. n. 29), 1. Weng seems to have a lot of mistakes in his chronological "Biographical Note of V.K. Ting," *Bulletin of the Geological Survey of China* 16 (1936–37).

[31] The only student in the original Peking University Geology Department with a genuine commitment to the science, Wang Lie, left school just months before graduation to pursue more serious studies in Europe. Wang later became a professor at Peking University when the department reinvented itself in 1918, under the chancellorship of Cai Yuanpei.

[32] The school also covered classroom and laboratory materials, but board and personal expenses were the responsibility of the individual student. "本所章程," in 地質研究所, 農商部地質研究所一覽 ["Bensuo zhangcheng," in *Dizhi yanjiusuo nongshangbu dizhi yanjiusuo yilan*] (Beijing, 1916), 1.

future employment for high achievers.[33] In return, Ding and Zhang expected loyalty to the Geological School community, commitment to fieldwork, and a sense of duty to the nation, all of which they cultivated through mentorship and fieldwork.

The school began recruiting students in June 1913, and hundreds of youths disillusioned with the empty promises of Chinese educational reform turned out for the Geological School exams. Ding personally supervised entrance examinations in Beijing and Shanghai for middle school graduates with competence in mathematics, inorganic chemistry, English and Chinese language, and literature.[34] Thirty students were officially accepted, but Ding whittled this number down to the twenty-five whom he deemed physically and emotionally qualified to handle the school's intensive three-year program.[35]

Each school year was divided into trimesters, and the first year was devoted to general, or introductory, topics, such as mineralogy, petrology, paleontology, current ideas in geology, geography, chemistry, zoology, surveying, and German.[36] Several of these courses had associated laboratory components, and Saturdays were reserved for outdoor surveying practice. Summer vacations would include at least one long-distance field trip. The second school year added fundamental, or advanced, topics, including chemical analysis, structural geology, and historical geology, and Saturday excursions shifted to geologic fieldwork.[37] By the third year, students were expected to have a firm foundation in geologic principles, and they were directed toward practical applications and field research. Besides advanced topics in petrology, ore deposits, physical geography, mineralogy, historical geology, and paleontology, coursework included drafting, geologic report writing, and photography; but the overall emphasis was to be on independent field research.[38] Ding and Zhang wanted geology to take root in China through firsthand experience and not textbooks or memorization.

The Ministry of Agriculture and Commerce, however, took a far more instrumental view toward fieldwork. By October 1913, when classes began, Ding Wenjiang was occupied with research duties at the Survey, and Zhang Hongzhao had formally taken over as director of the Geological School. With Ding in the field and the Survey "active," ministry officials lost interest in geologic training, and Zhang had to fight to keep the Geological School open. He argued that every time programs for students with modern training were aborted, vital talent was wasted.[39] Moreover, it was shortsighted to think that Ding's isolated efforts could provide a sustainable foundation for Chinese geology. "The Geological School was created to cultivate survey personnel.

[33] This was buttressed by Ding Wenjiang's personal reputation, since several of the prospective Shanghai area students (including 李學清 [Li Xueqing], and brothers 徐厚甫 [Xu Houfu] and 徐韋曼 [Xu Weiman]) knew of him as a dynamic young teacher at the Nanyang Middle School, where he taught biology for a year in 1911–12. "Dizhi diaochasuo yigu keji renyuan xiaoshi" "地质调查所已故科技人员小传," in 程裕淇 and 陈梦熊, 前地质调查所 (1916–1950) (cit. n. 26), 242, 249.

[34] 李学通, 书生从政 (Lanzhou, 1996), 24.

[35] Ibid., 24; and "張軼歐序," in 農商部地質研究所一覽 (cit. n. 32), 2. Prospective students were given a battery of fitness tests to prove their readiness for field activities.

[36] "學科課程表," in 農商部地質研究所一覽 (cit. n. 32), 2–3; 農商部地質研究所一覽, 45. Students used both English and German textbooks, and language study continued throughout all three years.

[37] "學科課程表" in 農商部地質研究所一覽 (cit. n. 32), 3.

[38] Ibid., 3.

[39] 章鴻釗, "章序" [Zhang Hongzhao, "Zhang xu"] ,in 地質研究所師弟修業記 (cit. n. 28), 1.

If there is a Survey but no surveyors, its strength is not real strength, and without real strength its actions are not real action . . . the so-called Geological Survey is merely a name."[40] The school's goal was not to teach geology as a dead subject but to train skilled men who understood geology as a bond to the nation. Finally, after the Ministry of Education confirmed that it lacked qualified teachers for any comparable courses in geologic study, the Ministry of Agriculture and Commerce relented, but the school was barred from accepting any further students, and it had to close as soon as the first class was graduated.[41]

The second year of the Geological School brought with it a deep sense of mission. The school's existence was linked to the progress of one cohort, and students and teachers alike felt pressure to demonstrate the virtues of "Chinese students in a Chinese school under Chinese teachers studying Chinese geology."[42] During the school's administrative crisis, Zhang Hongzhao taught all first-year geologic courses himself. Beyond regular Saturday surveying, Zhang could find time to lead only one extended field trip in the summer of 1914, guiding his students through the Western Hills of Beijing as he had been guided through the Japanese landscape several years earlier.[43] By fall 1914, however, twenty-five-year-old Weng Wenhao had returned with a doctoral degree in geology from the University of Louvain, and Ding Wenjiang was back from his Yunnan survey.[44] Both men took up teaching responsibilities and taught specialty courses to round out Zhang's introductory curriculum.

Weng and Ding not only added breadth and depth to the school's teaching but also reinforced the school's commitment to fieldwork. Weng was the first Chinese researcher to make a significant contribution to the geology of another country. His 1913 thesis earned highest honors and was published in the University of Louvain's geologic journal, where it attracted a great deal of attention for its careful treatment of Belgium's previously overlooked igneous formations.[45] Having mastered the skills of independent field research in a well-studied area under the exacting standards of a mature geologic community, Weng was eager for the challenge of researching China's comparatively unexplored territory.

Fieldwork allowed Zhang, Ding, and Weng, despite differences in their overseas experiences, to fit China into a global frame of reference, and they agreed that this was a critical element upon which their efforts would stand or fall. Chinese students were struggling to master an unfamiliar subject using English or German textbooks based on inaccessible examples from Europe and America, and when they had questions, their teachers "did not dare confine themselves to the common references on the shelves or the received theories of foreigners."[46] All three men believed that "[i]f

[40] 章鸿钊, "章序," in 農商部地質研究所一覽 (cit. n. 32), 2

[41] 章鸿钊, 六六自述 (cit. n. 10), 33.

[42] "翁序," 地質研究所師弟修業記 (cit. n. 28), 2.

[43] 章鸿钊, 六六自述 (cit. n. 10), 33.

[44] Zhang Hongzhao originally hoped to hire Friedrich Solger, a German instructor in the old Geology Department at Peking University, but Solger accompanied Ding Wenjiang on fieldwork in Shanxi Province and soon afterward was captured by Japanese troops in Qingdao after the First World War began. Wong Wen-hao, "V. K. Ting: Biographical Note," ii–iii; 胡适, 丁文江傳 (cit. n. 19), 54–55.

[45] Wong Wen-hao, "Contribution de la porphyrite quartzifère de Lessines," *Mémoires de l'Institut géologique de l'Université de Louvain* 1 (1913): 298–325.

[46] "翁序," 地質研究所師弟修業記 (cit. n. 28), 2.

we want this discipline to move forward, the only choice is to shoulder [our] axes and go into the mountains, hacking through thorns and brambles and chopping down underbrush to further our knowledge of actual facts."[47]

What had to be done, had to be done firsthand, and students were encouraged to "climb Mt. Tai to study gneiss and ford the Yangzi River to observe alluvial deposits."[48] Fieldwork enabled them to fit the Chinese landscape into generalizable scientific categories without forgetting that "territories are not the same and many differences exist—only when one's shoes touch the ground can one see the reality beneath the surface."[49] Experiential learning thus bridged the gap between foreign concepts and natural phenomena, while heightening students' sensitivity to regional particularity and physical variation.[50] As this engagement became more intimate, students identified with their homeland through science and dedicated themselves to both interchangeably.

With its combination of geologic theory and physical practice, fieldwork inspired "boundless interest" in the Geological School's student body.[51] In part, the reason was simply the beauty of mountains and lakes and the opportunity to explore distant, storied scenery that others could never glimpse firsthand.[52] "Our country is now poor," Zhang claimed. "If we desire to remedy this, we must begin with industry, but if we want industry to flourish, we must in turn start from geological investigation."[53] Zhang hoped his students would be able find in fieldwork "joys of the soul to conquer the pains of the body," for unwillingness to endure the physical strains of geologic research had led to China's impoverished state.[54]

Foreign powers, unfortunately, were only too aware of geology's importance. Midway through the second school year, on January 18, 1915, Japan issued its Twenty-one Demands, including possession of resource-rich Manchuria and Shandong Province, control over China's main coal deposits, and acceptance of Japanese "advisers" on China's military, commercial, and financial affairs. Upon threat of war, the Chinese government under Yuan Shikai submitted to these terms in May, although Britain and the United States pressured Japan to eliminate the advisory clause. In the midst of this national emergency, Zhang Hongzhao immediately revised the Geological School's curriculum so that, by March 1915, students were studying mining, metallurgy, drawing, and mechanics instead of just advanced theory.[55]

Group fieldwork was accelerated during term time, and as a final project each student was assigned a district for independent research. China was in urgent need of

[47] Ibid., 1.

[48] Ibid., 2.

[49] Ibid.

[50] Later, in a speech to the Peking University Department of Geology, Ding would expand on this theme in a lecture titled "The Responsibilities of Chinese Geologists" (中國地質學者的責任, Zhongguo dizhi xuezhe de zeren).

[51] 章鴻釗, 中國地質學發展小史 (cit. n. 15), 16.

[52] 農商部地質研究所一覽 (cit. n. 32), 46–47.

[53] Ibid., 46.

[54] This sentiment is also reminiscent of his own feelings during his summer of fieldwork in 1910. When he encountered hardships, he found "joy within adversity" by reminding himself that "all that meets the eye is new knowledge and each beginning is an achievement." 章鴻釗, 六六自述 (cit. n. 10), 25.

[55] 農商部地質研究所一覽 (cit. n. 32), 4–7. Metallurgy was taught by Zhang Yiou himself, and Chinese mining engineers Zhu Kun (朱焜) and Li Bin (李彬) taught mining. Ibid., 13.

both a detailed geologic map of the entire country and a comprehensive catalog of the nation's mineral resources. Although the Geological School's students could not be expected to carry out such a monumental task, fieldwork guidelines were designed to maximize their contribution, for every student in the field was a pioneer. Students were graded on detailed field notes, to be taken on the spot and well illustrated with pencil sketches, photographs, planar and sectional figures, longitude and latitude measurements, and verbal descriptions. They were also responsible for collecting and carefully labeling specimens for later analysis and preservation, with particular attention to ore samples. Finally, all field reports had to include several maps for each surveyed area, including a 1:50,000 topographical map and a 1:10,000 geologic map (1:50,000 if detailed topographical maps were already available).[56] Instructors helped students process their raw materials in drafting and report-writing classes, simultaneously gaining intimate knowledge of student data and personally reinforcing good work habits.[57]

At the end of 1916, Weng Wenhao and Zhang Hongzhao published *Student-Teacher Studies of the Geological School,* based on group and individual field assignments over the course of the school's three-year program. Weng acknowledged that student work was "practice" rather than "directed research" but argued that "there was no reason why it could not augment that which our predecessors have not yet attained and guide the path of inquiry for our successors."[58] To this end, Zhang and Weng did everything they could to synthesize student data and existing (foreign) geologic research. They organized *Student-Teacher Studies* topically rather than geographically, covering systematics, igneous formations, structural geology, and mineral products across several parts of Zhili (now Hebei), Shandong, Shanxi, Jiangsu, Zhejiang, Anhui, and Jiangxi provinces, with supporting maps and sections. In conclusion, they reflected on the experience of teaching in the field and offered speculative thoughts on broad questions, such as China's north-south divide, its geologic history, and the relationship between geology and its natural resources.

Student-Teacher Studies was the first book-length original contribution to Chinese geology written in the Chinese language, and it was a model of how limited resources could make a significant contribution.[59] Like the Geological School it commemorated and the Geological Survey that followed, this book displayed a timely mix of practical results and pedagogy, framing fieldwork as both the product and producer of China's newly minted geologists. When the Geological School ended in July 1916, eighteen students graduated with an advanced degree, and three others received a basic diploma.[60] All of these advanced graduates entered the Geological

[56] Ibid., 7.

[57] Drafting was taught by Wang Shaoying (王紹瀛), while Zhang, Ding, and Weng all taught report writing. Each field group or individual was also assigned an adviser who either led students in the field or oversaw their final projects. Ibid., 7, 13.

[58] "翁序," 地質研究所師弟修業記 (cit. n. 28), 2.

[59] Because of the influence of the 1915 vernacular movement spearheaded by Hu Shi and Chen Duxiu, this book was also one of the last modern scientific texts written in the classical language.

[60] 農商部地質研究所一覽 (cit. n. 32), 45. Page 8 describes the introduction of the advanced degree and its qualifications, based on independent fieldwork and test results. Five of the eighteen advanced students completed more than one independent project (16–18). Secondary sources almost uniformly claim that twenty-one students graduated from the Geological School, and most references in the primary materials corroborate this number (e.g., graduation speeches). However, there are twenty-two students listed in the 農商部地質研究所一覽, 14–15. These seem to be in rank order, but there is

Survey, which was reorganized in October 1916.[61] The revamped Survey had three sections: a general affairs division headed by Director Ding Wenjiang, a geology division led by Zhang Hongzhao, and a minerals division under Weng Wenhao. In its early years, however, salaries and titles were irrelevant, and the Survey was essentially a continuation of the Geological School, with student-teacher relationships built into the ritual of fieldwork.[62]

FIELDWORK AS A HABIT OF MIND AND BODY

Despite the successes of the Geological School, its days had long been numbered. To guarantee future recruits and maintain the Survey's newfound momentum, Ding Wenjiang quickly began negotiations for a department of geology with newly appointed chancellor of Peking University Cai Yuanpei. Cai, who had been Minister of Education in 1912 before resigning in disillusionment, was a Hanlin scholar with several years' experience in Germany and France. He accepted the chancellorship of Peking University in late 1916 in the hope of rehabilitating the school's corrupt reputation and establishing a true center of learning that would inspire changes in the entire educational system. At the heart of Cai's educational philosophy was a belief that education should prepare the individual for a constructive role in society through a tripartite emphasis on utilitarian education, ethical education, and aesthetics. With a strong commitment to academic freedom, Cai hired faculty from across the political, cultural, and social spectrums, inviting, for instance, both the radical champions of the vernacular movement and the staunchest defenders of the classical literary style. A department of geology was, in Cai's view, a necessary component of utilitarian education, and he eagerly accepted Ding's proposal that the university resume geologic instruction. Although this was not completely in line with Ding's conception of geology's role in modernizing China, Ding hoped that shifting geologic training to an academic setting would put it on a more permanent footing and embed it in a broader program of cultivating modern youth.

In the fall of 1917, Peking University reclaimed the facilities and equipment that it had previously lent to the Geological School and established a new department of

---

no explanation of the status of the extra student, and in 1940, Zhang Hongzhao reports that twenty-two students graduated from the school (章鴻釗, 中國地質學發展小史 [cit. n. 15], 39). The eighteen advanced degree graduates are not contested.

[61] Most secondary sources list all eighteen advanced graduates as investigators in the Geological Survey, but the original constitution of the Survey (24 Oct 1916, 近史所檔案館 08-24 5 [1]) lists six technical experts (技師) and twelve investigators (調查員). The 地質調查所沿革事略 (cit. n. 27), 1, states that these eighteen graduates were hired as investigators, interns (學習員), "and other" positions, implying even that the basic graduates of the school might have taken up administrative and support positions (the entire survey had twenty-four members in 1916, which might comprise Zhang, Ding, Weng, and all twenty-one graduates), but there is no existing personnel roster for the survey in 1916. From January to October 1916, the Geological Survey was called the 地質調查局 and was directly under the executive branch, but it reverted to its original name (地質調查所) and administrative position in October when it was also granted a formal budget and constitution (see 地質調查所沿革事略 [cit. n. 27], 1; and 農商部公報, 15 Feb. 1916, 2:7–49); I have rendered both Chinese offices as the "Survey."

[62] 翁文灝, "對於丁在君先生的追憶," in 丁文江傳記資料, ed. 朱傳譽 [Weng Wenhao, "Duiyu Ding Wenjiang xiansheng de zhuiyi," in Ding Wenjiang zhuanji shiliao, ed. Zhu Chuanyu] (Taipei, 1979), 111.

geology,[63] with He Jie and Wang Lie as professors.[64] Now that the Geological Survey was active and Cai Yuanpei at the head of the university, the department had little difficulty attracting students; and from 1920, when its first eight students graduated, the Beida Department remained the Survey's most dependable source of new geologists.[65]

Even given the advantage of continuity, however, the department did not immediately match the Geological School in quality of education, especially with regard to fieldwork. The school was a tightly knit, closed community of students progressing through the same classes at the same pace and focusing on an urgent common goal. The department, by contrast, accepted new students annually, and until 1920, it had to maintain a full complement of courses for all levels with only two professors and insufficient facilities. By 1919, there were thirty-six students at three different stages of a geologic program that was transitioning from a three- to a four- year curriculum and trying to introduce electives in course selection.[66]

The workload was demanding enough "to make a person's head spin," yet geology students at Peking University were dissatisfied with their training.[67] They complained that resources were so scarce that they rarely had any personal contact with teachers,

[63] Some secondary sources claim that Beida's Geology Department began in 1918, but those sources most directly related to Beida itself uniformly place the department's founding in 1917, which conforms much more reasonably to the timing of Cai Yuanpei's chancellorship and the fact that the department's first eight students graduated in 1920. According to Amadeus Grabau, courses in geology were offered at Beida in 1916. ("A Decade of Research in Chinese Geology," 北京大學地質研究會年刊 [*Bulletin of the Geological Society of the National University of Peking*] 4 [April 1930]: 1). If so, these were not in any formal geology department, but it is not unlikely, since several of the 1920 graduating class entered the department in 1917 with credits from Beida's preparatory program, and He Jie was already teaching at Beida in the College of Engineering. Preparatory programs were common in much of the Republican period, since national standards of middle-school education were very uneven. According to Yu Guang, Beida students generally took three years of preparatory courses (heavily emphasizing fundamentals such as mathematics and foreign languages) and three years of university classwork until 1917, when this was changed to two years of preparatory work and four years of departmental courses. The preparatory program was closed in 1929. 于洸 et al., "弘杨传统" (cit. n. 29), 2.

[64] He Jie (何杰, 1888–1979) won a Boxer Indemnity scholarship in 1909 and received an undergraduate degree in mining engineering from the Colorado School of Mines. He then received a scholarship to Lehigh University and returned to China in 1914 with a master's degree in geology. He taught in the College of Engineering at Peking University until the founding of the Department of Geology, which he headed from 1917–24. Wang Lie (1887–1957) was one of three students in the original Department of Geology at Peking University (the Imperial University of Peking, before 1912). Wang completed his coursework but opted to go to Germany in February 1913 before graduating. He returned with a degree in geology in 1915 and taught both German and structural geology at the Geological School before joining the faculty at Peking University. He chaired the Department of Geology from 1924–27 and again from 1928–31.

[65] When the Geology Department was established in 1917, several students entered from Beida's own preparatory program. Others, such as the star of the 1920 graduating class, Sun Yunzhu (孫雲鑄), were enrolled in mining courses at other institutions and transferred to Beida to focus on geology proper. (江苏省政协文史资料委员会, ed., 一代宗师 [Yidai zongshi, ed., *Jiangsusheng zhengxie wenshi ziliao weiyuan hui*] [Nanjing, 1995].) Within a decade, several other geology departments were founded throughout China, often by graduates of the Beida program. Several of these, including the departments at Qinghua and National Central University (which merged with Jinling University to form what is now Nanjing University) were extremely successful, but Peking University still ran the preeminent department with the closest association to the Geological Survey.

[66] 于洸 et al, "弘杨传统" (cit. n. 29), 2.

[67] 趙國賓, "本會一年來的回顧和年來擬辦的事項," 國立北京大學地質研究會年刊 [Zhao Guobin, "Benhui yinianlai de huigu he nianlai niban de shixiang," *Beijing daxue dizhi yanjiuhui niankan*], 10 Oct. 1921, 1.

and they had to fight over laboratory materials.[68] Faculty had no time to take students into the field, and the university had no funds for geologic trips. The department did not maintain formal fieldwork requirements, and it focused so much on in-class teaching that students found it difficult to keep up with reading and memorization.[69] According to one student leader, Zhao Guobin:

> Routinely when we discuss various aspects of Chinese geology, we aren't even as familiar [with them] as certain foreigners; shame of this measure could not be greater! When classes are over and my colleagues get around to this subject, there are none that do not sigh.[70]

Like others before them, Peking University students felt a personal connection to the national plight and required direct interaction with the natural landscape to come to terms with Chinese realities.

Ding Wenjiang, though he was not officially connected to the Beida Department of Geology, was director of the Geological Survey and had a vested interest in maintaining the highest standards possible for geologic education. He understood that He Jie and Wang Lie were overtaxed and underfunded, but he felt that a reliance on textbooks and lectures was a giant step backward. As Ding had explained at the graduation ceremony of the Geological School in 1918, "[T]he greatest cause of our nation's weakness is that in general the upper classes do not have close contact with the [natural] world." Ding believed that once students learned to appreciate "the joys of nature," they could resist the temptations of wealth and status. Then, by climbing the great mountains of China's west and roaming where average men could not, young geologists would become "men of real abilities" who would take the nation's interests as their own and make China flourish.[71] In his view, which was shared by Zhang Hongzhao and Weng Wenhao, fieldwork was necessary for a solid understanding of geologic principles and scientific method as well as an enlightened training of patriotism and character.

The active lifestyle of field geologists challenged deeply entrenched prejudices against the assertion of physical culture among intellectuals—precisely those prejudices that Richthofen had disparaged. This negative attitude toward physicality is often traced to passages from the traditional canon, such as Mencius's claim that "[t]hose who labor with their minds govern men, and those who labor with their muscles are governed by men."[72] It is debatable whether Confucius himself was as disdainful of physical cultivation as later adherents were, since he taught students charioteering and archery as part of their training as *shi*, or minor nobility.[73] Cer-

---

[68] 王恭睦, "北大地質研究會夏季旅行團的提議," 國立北京大學地質研究會年刊 [Wang Gongmu, "Beidadizhi yanjiuhui xiaji luxingtuan de tiyi," *Beijing daxue dizhi yanjiuhui niankan*], 10 Oct. 1921, 1. (Journal was repaginated at each article.)

[69] Ibid., 1; "本會一年來得回顧和年來擬辦的事項," 國立北京大學地質研究會年刊, 10 Oct. 1921, 1; 于洸 et al., "弘杨传统" (cit. n. 29), 6.

[70] 趙國賓, "本會一年來的回顧和年來擬辦的事項" (cit. n. 67), 1.

[71] 農商部地質研究所一覽 (cit. n. 32), 46–47.

[72] Mengzi 孟子, 滕文公, 上. "勞心者治人, 勞力者治于人." [Mengzi, Qiwengong, "Laoxinzhe zhiren, laolizhe zhi yu ren"] (Classic texts are cited by chapter heading).

[73] The social meaning of *shi* has a long and involved history, which for simplicity's sake I have collapsed into "lesser nobility." At the time of Confucius, shi was a martial class, and it is significant that over time, the influence of Confucian *wen* (literary or civil) culture realigned shi toward the scholar-official.

tainly, though, these physical skills were not matters of pride the way mental and moral attainments were, and Han dynasty scholar Wang Chong noted that although "Confucius could lift the portcullis of the northern gate, he never flaunted his vigor, knowing that the strength of sinew and bone was not as honorable as the strength of benevolence and righteousness."[74]

It is perhaps less important whether China's sages and philosophers truly disapproved of physical culture than that their teachings were historically assimilated in this way. As a modified form of Confucianism became the state-sanctioned path to officialdom,[75] intellectuals used physical values to separate their civilized (*wen*) pursuits of literature and art from the violent habits of martial (*wu*) men,[76] and physical prowess was similarly employed to separate races by defining Chinese cultivation in opposition to barbarian brutality. The class implications of the denigration of physical skill and exertion are clear in Mencius's quotation, and they were reinforced by the concentration of wealth, social status, and intellectual skill in the hands of scholar-officials. Even among intellectuals themselves, passages referring to the sage-king Shun sitting "with hands folded in quietude and all under heaven in good order" could become pointed barbs suggesting that those who had to exert themselves to achieve a desired result were simply less competent.[77] It would have been shocking for traditional intellectuals to learn that Weng Wenhao considered it the highest praise to say that "even though there were several pack animals to transport luggage, . . . [Ding Wenjiang] walked for great distances, observing rocks and making sketches, hands and feet constantly at work."[78]

The privileging of the mental over the physical was also evident in art and literature. Although both *wen* (civilized, literary) and *wu* (martial, physical) heroes were popular in plays and fiction, *wu* figures were invariably subordinate to righteous *wen* characters and were often redeemed by a turn to scholarly pursuits. In China's famous romances, often referred to as *caizi-jiaren* (talented scholar–great beauty) stories, the protagonists were always fair, slender young men of literary skill who wooed women with poetry and brilliance rather than with bravery and physical prowess.[79] Once women were allowed in Chinese theater, these romantic protagonists were frequently played by female actors who could better capture their softness and mildness of manner.

In 1915, in an early volume of the magazine *New Youth*, cultural critic Chen Duxiu mirrored this image of upper-class masculinity but completely reversed its valuation. He wrote:

---

[74] 王充, 論衡, 效力, "孔子能舉北門之関, 而不以力自章, 知夫筋骨之力, 不如仁義之力榮也." [Wang Chong, Lunheng xiaoli, "Kongzi neng ju beimen zhi guan, er bu yili zizhang, zhi fu jingo zhi li, bu ru renyi zhi li rong ye."] Benevolence and righteousness are two of the Five Virtues of the *junzi*, or superior man.

[75] A Legalist-influenced brand of Confucianism was officially recognized as the governing philosophy of the Han dynasty, and later in the Sung dynasty a Buddhist and Taoist inflected form was canonized in the Four Books of the civil examination system.

[76] Several historical figures demonstrated both physical and intellectual achievements, but these men, like Wang Yangming, were exceptions that proved the rule.

[77] 尚書, 武成. "垂拱而天下治." [Shang shu, Wucheng. "Cuigonger tianxia zhi."]

[78] 翁文灝, "對於丁在君先生的追憶" (cit. n. 62), 112.

[79] See Kam Louie, *Theorising Chinese Masculinity: Society and Gender in China* (Cambridge, UK, 2002) for an excellent analysis of the wen-wu dyad and its many social and cultural manifestations.

> Whenever I look at our educated youth, I see that they have not the strength to catch a chicken, nor mentally the courage of an ordinary man. With pale faces and slender waists, seductive as young ladies, timorous of cold and chary of heat, weak as invalids—if people of our country are as feeble as this in body and mind how will they be able to shoulder burdens and go far?[80]

The educated youth Chen described might still make desirable romantic matches for some, but their atrophied physical traits boded ill for the nation.

The Opium Wars (1839–42, 1856–60) paved the way for this shift in consciousness. China's losses demonstrated that the nation was weak in comparison with Western powers,[81] sparking an interest among so-called self-strengtheners to adopt foreign technology on the basis of Zhang Zhidong's *ti-yong* formula: Chinese learning for substance (*ti*), Western learning for practical use (*yong*).[82] This movement resulted in several attempts at military modernization, but like the incorrigible old scholars that von Richthofen lampooned, even intellectuals who clamored for national "self-strengthening" were often weak and unwilling to engage in manual labor. In fact, it was only after China's unexpected defeat in the Sino-Japanese War (1894–95) that native thinkers began to attribute national weakness to the weakness of the Chinese people themselves.

One of the most influential of these thinkers was Yan Fu (1853–1921), who translated Huxley, Smith, Montesquieu, and Mill into classical Chinese and brought Spencerian Social Darwinism to China.[83] In "On Strength," one of four essays published in the aftermath of the Sino-Japanese War in 1895, Yan argued that the "weak become the prey of the strong, and the clumsy submit to the crafty. . . . This is true not only of animals and plants but also of people."[84] For Yan, the unit of natural selection was the group (alternately the society, nation, or race), and to win the "struggle for existence," China had to increase the population's "fitness" by developing its "bodily vigor, intelligence, and moral virtues."[85] Yan did not limit his endorsement of sound

[80] 陳獨秀, "近日的教育方針," 新青年 [Chen Duxie, "Jinri de jiaoyu fangzhen," *Xin qingnian*] [*New Youth*] 1, no. 2 (1915): 118.

[81] The Opium War commonly refers to the Anglo-Chinese War that ended in the Treaty of Nanjing and set the precedent for future "unequal treaties." The "second" Opium War is also known as the Arrow War or Anglo-French War and produced the treaties of Tianjin (1858) that established the contentious principle of extraterritoriality. Hostilities ended in 1860 after the sacking of the Summer Palace and the Qing dynasty's acceptance of the Peking Convention.

[82] The Self-Strengthening Movement (*ziqiang yundong*) and Foreign Affairs Movement (*yangwu yundong*) are largely synonymous and refer to the efforts of several prominent scholar-officials such as Zhang Zidong, Li Hongzhang, Zeng Guofan, and Zuo Zongtang to promote Western technology in answer to Chinese military weakness. Their efforts are associated with several military modernization as well as educational and commercial enterprises such as the Jiangnan Arsenal, the Fuzhou Shipyards, and the China Merchants Shipping Company.

[83] Yan Fu followed his classical training with several years at the Fuzhou Shipyard School studying navigation and two years of study in England, where he became interested in British social, political, and economic thought. Yan's essays and interpretive translations were "probably read by every eager student in China at the beginning of the [twentieth] century." (Y. C. Wang, *Chinese Intellectuals and the West* [Chapel Hill, N.C., 1966], 206.) For more information on Yan Fu, see Benjamin Schwartz, *In Search of Wealth and Power: Yen Fu and the West* (Cambridge, Mass., 1964).

[84] Translation follows Wang, *Chinese Intellectuals and the West* (cit. n. 83), 196.

[85] Bodily vigor, intelligence, and virtue were Yan's rendering of Herbert Spencer's three energies, the physical, intellectual, and moral. According to Yan, the same measures that had made the West strong (including the Self-Strengtheners' arsenals, railroads, and battle fleets) were ineffective in China because the people themselves were unfit. Yan Fu is discussed in most treatments of modern

physical culture to military drills and calisthenics; he also criticized foot binding and opium addiction, and he urged physical education for women because healthy mothers produced strong children.[86]

According to intellectual historian Benjamin Schwartz, this "affirmation of . . . physical virtues . . . [was] the most dramatic manifestation of the new transformation of values."[87] However, the humanists and educators who embraced this iconoclastic line were still quite instrumental in their motives.[88] Some, like the reformer and public thinker Liang Qichao, simply demanded: "With neither knowledge nor physical strength, how can the Chinese survive in a competitive world?"[89] In response, both military training and competitive sports filtered into modern educational reforms to build stronger bodies.[90] Others, like Cai Yuanpei, focused on the body in the service of mental and moral advancement, claiming that "perfection of the spirit requires perfection of the body; if the body is weak, then how can thoughts and spirit flourish?"[91] Strength was a goal, whatever the reasoning.

In contrast, Ding and other pioneering Chinese geologists viewed cultivation of the body and perfection of the spirit as an inevitable by-product of directly experiencing nature.[92] They shrugged off the traditional aversion to physical exertion as simply counterproductive to their intellectual aims. For them, fieldwork was inherently both manual and mental labor, and its physical activities—whether accessing, measuring, sensing, or collecting—transformed landscapes into a natural laboratory for practice and discovery. That these labors remade Chinese geologists in the image of an energetic modern manhood was a natural outcome of identifying oneself with the national cause.

Throughout his lifetime, Ding personally exemplified this ethos of muscular science in his own fieldwork and his pedagogical approach. Two of his mottos—"when climbing a mountain, always go to the top; when on the move, always go by foot" and

---

Chinese history, but for a comprehensive analysis of Darwinism's impact on China, see James Pusey, *China and Charles Darwin* (Cambridge, Mass., 1983).

[86] In 周振甫, 嚴復詩文選 [Zhou Zhenfu, *Yanyu shiwenxuan*] (Taipei, 1964), 14–32.

[87] Schwartz, *In Search of Wealth and Power* (cit. n. 83), 86.

[88] It is interesting to contrast the kinds of instrumentality suggested here to the student-driven and performance-oriented culture of athleticism discussed by Andrew Warwick in his study of the Cambridge Mathematical Tripos. Andrew Warwick, "Exercising the Student Body: Mathematics and Athleticism in Victorian Cambridge," in *Science Incarnate: Historical Embodiments of Natural Knowledge*, ed. Christopher Lawrence and Steven Shapin (Chicago, 1998), 288–326.

[89] 梁啟超, 飲冰室全集 [Liang Qichao, *Yinbingshi chuanj*], vol. 3, pt. 4, (Kunming, 1941), 109–18.

[90] The YMCA, which ran many government and Christian school athletic programs in China, noted that participation in competitions and practices was seriously hampered by traditional bias against physical activity. See chapter 1 of Jonathan Kolatch, *Sports, Politics, and Ideology in China* (New York, 1972).

[91] 蔡元培, "在南開學校全校歡迎會演說辭," in 蔡元培教育文選 (Taipei, 1956). These views were shared by the young Mao Zedong, who published "Study of Physical Education," in *New Youth* 3 (April 1917): 2. In this article, he claimed that "[i]t is the body that contains knowledge and houses virtue" and "[w]hen the body is strong, then one can advance speedily in knowledge and morality" (translation follows Stuart Schram, in *The Political Thought of Mao Tse-tung* [Harmondsworth, UK, 1969], 153–54). However, that article was not influential at the time of its publication.

[92] This did not mean that geologists were not concerned with the same nation-strengthening issues as their contemporaries. Much of their geologic research was aimed at providing the resources and materials for national development. My point here is simply that the physicality of fieldwork was seen as part of good geologic methodology, rather than as a form of self-cultivation. Ding's interest in strengthening the bodies of the people manifested itself in side projects, such as his promotion of eugenics, rather than in his geologic work.

"take the long path, not the short path; take the mountain trail, not the flat road"—influenced several generations of Chinese geologists.[93] Weng Wenhao described Ding as constantly active in the field, with one hand working with his hammer and the other taking measurements or making sketches. Ding transmitted this energetic style to his students.[94] He carried his own equipment and slept, ate, worked, and rested with his students, on the principle that fieldwork was always a collaborative endeavor.[95]As a teacher, Ding did not believe in leading students around and pointing out geologic phenomena; instead, he preferred to suggest problems and give students a chance to examine sites for themselves before influencing their observations.[96] Students could learn only by doing and by making their own mistakes. China's reliance on textual authority would have to give way to empiricism.

For this reason, Ding Wenjiang could not stand aside and allow Peking University's Department of Geology to neglect fieldwork and fall back on foreign textbooks. To free more time for fieldwork in the department, Ding informally began his own search for qualified teachers. While he toured Europe as part of Liang Qichao's unofficial Chinese delegation to the Versailles Peace Conference in 1919, Ding met with Li Siguang in England and, stressing that "cultivation of geological talent is a matter of vital urgency," asked Li to teach at Peking University.[97] On the American leg of the same trip, Ding consulted with the chief geologist of the U.S. Geological Survey, David White, who suggested that Amadeus Grabau might be interested in working in China. Li was a young revolutionary with a master's degree in geology from the University of Birmingham, and he was a protégé of William Savage Boulton, who published his work in British scientific journals. Grabau, whose pro-German sympathies led to his dismissal from the Geology Department of Columbia University in 1919, had been a beloved teacher to several Chinese overseas students, including mining engineer Wang Chongyou, and Ding was especially interested in gaining someone of Grabau's stature to teach paleontology and stratigraphy.

When the first graduating class from the Department of Geology at Peking University applied for positions in the Geological Survey in late spring of 1920, Ding seized on the chance to test the students' basic geologic skills.[98] The results proved to be worse than he feared, and he sought out his good friend Hu Shi, also a professor of history at the university, to help him approach Cai Yuanpei with the examination results.[99] After several hours with Ding and Hu, Chancellor Cai approved their proposal to bolster the geologic faculty and refocus on fieldwork, and both Amadeus Grabau and Li Siguang were hired immediately.[100] Additional funding for fieldwork

[93] 翁文灝, "對於丁在君先生的追憶" (cit. n. 62), 111; 胡适, 丁文江傳 (cit. n. 19), 85; 計榮森, transcriber, 講學會記事 [Ji Ji Rongsen, transcriber, *Jianxuehui jishi*], vol. 1 (handwritten), 26 Aug. 1930, meeting notes.

[94] 翁文灝, "對於丁在君先生的追憶" (cit. n. 62), 112.

[95] 高振西, "做教師德丁文江先生" [Gao Zhenxi, "Zuojiao shide Ding Wenjiang xiansheng"], in 丁文將傳記資料, 173.

[96] 翁文灝, "對於丁在君先生的追憶" (cit. n. 62), 111.

[97] 马胜云, 李四光年谱 [Ma Shengyun, *Li Siguang nianpu*] (Beijing, 1999), 36.

[98] 胡适, 丁文江傳 (cit. n. 19), 56.

[99] Hu Shi is considered the foremost liberal intellectual of the Republican period, one of the main promoters of the vernacular movement, and the leader of the so-called Chinese Renaissance. He also served as ambassador to the United States. Hu and Ding were lifelong friends and collaborators, and Hu is Ding's most famous biographer.

[100] Because Grabau came to China as both professor at Peking University and chief paleontologist of the Geological Survey, his generous salary of $1,600 each month was guaranteed by funds paid out

followed, and by the mid-1920s, the Peking University Department of Geology was as well known for its field training as it was for the caliber of its lectures.

By 1924, Ding, who had been so critical of the program a few years before, had come to feel that Peking University surpassed most Western institutions outside of the United States in its attention to fieldwork.[101] The 1927 course schedule listed a full day of geologic fieldwork each week for first- and second-year students, in addition to several hours of surveying practice and laboratory time. Upper-class students were required to participate in extended field trips during school vacations, and seniors researched their theses in small groups with the guidance of a faculty adviser.[102] The university's own student geologic association also sponsored fieldwork, and within a few years, Ding was convinced that China's homegrown geology graduates were the match of all but advanced-degree holders from overseas.[103] More important, they had made direct ties to their native land and would not forget their duty to "save China."[104]

CONCLUSION

According to Lucian Pye, in "How Chinese Nationalism was Shanghaied":

> Unlike in other countries, many Chinese intellectuals have at times adopted a totally hostile view toward their own great traditional culture, calling for a complete rejection of the past and a boundless adoption of Western culture.[105]

Science has generally been understood as one of these Western concepts that was brandished in opposition to the Chinese past, and statements such as revolutionary-anarchist Wu Zhihui's famous "All thread-bound [traditional] books should be dumped in the lavatory," and "all things in the universe can be explained by science" seem to affirm this.[106]

---

of the American Boxer Indemnity Fund rather than through the university itself. See chapter 9 of Allan Mazur, *A Romance in Natural History: The Lives and Works of Amadeus Grabau and Mary Antin* (Syracuse, N.Y., 2004), http://faculty.maxwell.syr.edu/amazur/Romancecover.htm.

[101] V. K. Ting, "Presidential Address," *BGSC* 3, nos. 1–4 (1924): 9–11.

[102] 地質系課程 [Dizhixi kecheng], 1927, BD 1927018, Peking University Archives, Peking.

[103] Ding Wenjiang, 6 Jan. 1924 speech, "The Training of a Geologist for Working in China [中国地质工作者之培养]," cited in 夏湘蓉 and 王根元, 中国地质学会史 (1922–1981) [Xia Xiangrong and Wang Genyuan, *Zhongguo dizhi xuehui shi (1922–1981)*] (Beijing, 1981), 61. See also 陶孟和, "追憶在君," 獨立評論 [Tao Menghe, "Zhuiyi Zaijun" *Duli pinglun*] 188, 16 Feb. 1936, 33–34. By the late 1920s, foreign students interested in the geology of China actually began applying to Beida to do postgraduate research under local professors. Because of the uncertain political and military conditions in Beiping, only two students (one of whom became a geologist for the Geological Survey of the Manchurian Railway and the other of whom earned a doctorate in Prague based on his research) actually studied in the Beida Department, although several applied. The Beida Department did not itself confer advanced degrees in the Republican period. Amadeus Grabau, "Contributions to Geologic Science by Graduates of the National University," 國立北京大學自然科學季刊 [*Science Quarterly of the National University of Peking*], 1 April 1930, 240.

[104] For more on this concept, see Zuoyue Wang, "Saving China through Science: The Science Society of China, Scientific Nationalism, and Civil Society in Republican China," *Osiris* 17 (2002), 291–322.

[105] Lucian Pye, "How China's Nationalism was Shanghaied," in Unger, *Chinese Nationalism* (cit. n. 8), 91.

[106] 吳稚暉, "一個新信仰的宇宙觀至人生觀" in 科學與人生觀, ed. 張君勱 [Wu Zhihui, "Yige xin xinyang de yuzhouguan zhi renshengguan," in *Kexue yu renshengguan*, ed. Zhang Junmai] (Taipei, 1977) 2:76, 137.

However, the loud saber rattling of public thinkers and social activists should not be allowed to drown out the perspective of China's early scientists themselves. Pioneering geologists, for instance, frequently came to their discipline carrying the cross of shame that Chinese had not been able to understand and therefore defend their own territory from foreign encroachment. This feeling was a cry against the dislocation of Chinese intellectuals from the realities of their time and place, not a call for another radical break. In fact, the humiliation that young geologists repeatedly expressed was an acceptance of continuity with the past and an act of identification with China's national plight. For practicing Chinese scientists, perhaps because they were already engaging with something unabashedly "modern," the Chinese past was less a threat than a spur to further action. Few of Ding Wenjiang's contemporaries would have been surprised that he reminded them of von Richthofen's disparaging words at the celebration of the Geological Survey's new *Bulletin*, for Chinese scientists understood that invoking past shame gave meaning to their self-transformation and connected it with the fate of the nation.

In an age when governments were unreliable and unconvincing, cultural values were in turmoil, and ethnic and regional tensions were being sharpened by new economic and military pressures, Chinese identity was understandably less tied to any specific definition or content than it was to the act of identification itself. Those who owned up to the mistakes of the past, who felt that Chinese had been wrong and not just wronged, were truly Chinese. Those who redressed these failings were truly modern.

By grounding itself in fieldwork and the ethos of physical engagement with the Chinese landscape, native geologic education cemented students' identification with their homeland even as it reshaped their values and aspirations. Science was a bridge between the China of memory and the China of future promise, and for early Republican geologists, it was not emotionally compelling without both points of attachment.

Justice, Geography, and Steel:

Technology and National Identity in Indonesian Industrialization

By Suzanne Moon*

ABSTRACT

This essay explores the connections between technology and the postcolonial project of national identity formation in Indonesia during the early years of industrialization, from 1950–75. Using a steel plant in Cilegon, West Java as a lens, this article examines the narratives of national identity that justified the building of this "impractical" plant from its beginnings as a Sukarno-era Soviet aid project, to its final realization in Suharto's New Order as a key piece of Indonesian oil giant Pertamina's "state-within-a-state" business empire. Advocates in both the Old and New Orders consistently drew on narratives of geography and justice to defend their visions of industrialization, but contrary to their hopes, the materiality of the technology did not necessarily make these moral tales of technology and nation more stable and enduring.

INTRODUCTION

Since Sukarno declared Indonesia's independence in 1945, the question of national identity has been at the center of the postcolonial experience. Defining Indonesia, both to the outside world and to the people of the archipelago, is an on-going project that traverses issues of ethnicity, language, geography, history, ethics, and political organization. In the developmentalist, postcolonial state, technology projects have been an important locus for the production and questioning of national identity. Public figures frequently frame technological works, large and small, as material markers of achievement (or lack thereof) that reflect the national character and, through their operation, make it possible to realize the aspirations that define the nation. This essay explores the connections between technology and the postcolonial project of national identity formation in Indonesia in the early period of industrialization, from 1950–75. Prominent political leaders struggled to articulate the identity of the new Indonesian nation, by asserting the shared values, history, and culture across the archipelago. These narratives of national identity were meant to serve two purposes: to unite the people of the islands, creating patriotic loyalty to the new state, and to

* Department of the History of Science, University of Oklahoma, 601 Elm St. PHSC 624, Norman, OK 73019; suzannemoon@ou.edu.

I would like to thank the University of Oklahoma College of Arts and Sciences for providing research funding to complete this article. I would also like to thank the editors and referees as well as the workshop participants for their valuable comments and feedback. I am particularly grateful to Katherine Pandora for practical and intellectual advice that helped bring this project to completion.

© 2009 by The History of Science Society. All rights reserved. 0369-7827/09/2009-0013$10.00

define Indonesia's place in the world, geographically and politically. Technology and national identity were made to play mutually reinforcing roles in these narratives. Technologies could produce materially compelling interpretations of the character of Indonesia, showing people concrete examples of Indonesia's new modernity, while simultaneously, particular interpretations of Indonesian identity justified to both internal and external audiences, the pursuit of technology projects that might otherwise seem impractical or unwise.

For the purposes of this article, I view national identity as a set of ideals and characteristics meant to define the essence of a nation and to provide the foundational logic for a coherent, enduring, and most of all, legitimate political order. Michael Hitchcock and Victor King argue that nationality "provides a significant, if not the significant, framework of personal allegiance and identity as well as collective solidarity and cohesion."[1] To define the reasons that a group of individuals should see themselves as part of a nation, proponents will cite characteristics that give reasons for social solidarity, such as shared geography, shared cultural practices and values, shared histories, and shared political aspirations. A functional national identity can explain not only why the nation exists but why it should continue to do so, as well as why citizens should exert themselves to participate in the collective goals of the nation, whether fighting wars, building industries, or engaging in public service. This process of national definition (which is at the same time justification) operates not only in reference to internal histories, as Jan Aart Scholte and Coen Holtzappel individually point out, but also in comparison with other nations.[2] A national identity may implicitly or explicitly delineate the place and significance of the nation with respect to the wider world, calling on narratives of similarity and difference in the process. When a nation identifies itself as "developing," for example, it presupposes a comparison with those countries that are "developed" and defines that trajectory toward development as a shared aspiration.

Despite their task to remind citizens of the enduring logic of the nation, national identities are by no means fixed. Because they are mutable, I find it helpful to think of national identities as narratives, or stories people tell about the nation, and to trace how those stories change over time. Some of the earliest assertions of regional identities in Indonesia came from nineteenth- and twentieth-century anthropologists who tried to define the elements of culture shared by the peoples of the region, who they dubbed "Malays," a set of definitions that anticolonial nationalists took up themselves as they argued that Indonesia should become a nation.[3] That creating a national identity is an ongoing process and not simply the assertion of a fixed set of ideals can be clearly seen in Benedict Anderson's *Imagined Communities*, in which he focuses on the processes that allow anticolonial political activists to reimagine themselves as nationalists and their colony as a nation. Coen Holtzappel adds to this analysis by exploring the interactions of European and indigenous concepts that informed defi-

---

[1] Michael Hitchcock and Victor King, introduction to *Images of Malay-Indonesian Identity*, ed. Michael Hitchcock and Victor King (Kuala Lumpur, Malaysia, 1997), 5.

[2] Jan Aart Scholte, "The International Construction of Indonesian Nationhood, 1930–1950," in *Imperial Policy and Southeast Asian Nationalism*, ed. Hans Antlov and Stein Tonnesson (Richmond, Surrey, UK, 1995), 191–226; and Scholte, "Identifying Indonesia," in Hitchcock and King, *Images of Malay-Indonesian Identity* (cit. n. 1); Coen Holtzappel, "Nationalism and Cultural Identity," in ibid., 63–108.

[3] Hitchcock and King, introduction (cit. n. 1), 3–4.

nitions of national identity in colonial and early postcolonial history, while Jan Aart Scholte emphasizes the importance of global ideas and trends on the project of nation building.[4]

Indonesia is a particularly compelling lens through which to study the process of national identity formation in a postcolonial nation. The struggle to define Indonesia to itself and to the world did not cease with Indonesia's declaration of independence in 1945; indeed it was then that leaders had to struggle hardest to establish the foundation of Indonesia's claim to nationhood. The process of defining a national identity for Indonesia has preoccupied politicians and intellectuals ever since.[5] One reason that this process has been so challenging is Indonesia's combination of geographic and social fragmentation. An archipelago consisting of approximately 7,000 inhabited islands (more than 13,000 total islands) and stretching roughly 3,200 miles from east to west, Indonesia has a profound cultural and ethnic diversity across and within the islands.[6] Governments (both postcolonial and colonial) have at times exploited and encouraged these divisions as a means of political management, leaving a legacy of social discord in its wake.[7] Although there is now an official language for the nation, Bahasa Indonesia, there are hundreds of other languages routinely spoken across the islands, which often take precedence in day-to-day life. Although Islam is the religion of the majority of the people, there are sizable minorities who are Hindu, Buddhist, Christian, and animist. Politicians and scholars, seeking to solidify the foundations of the Indonesian national identity must first define which aspects of the fragmentation represent a real problem for national integration and then propose a way to negotiate this fragmentation. Eka Darmaputera, in a thought-provoking study of Indonesian nationalism, highlights the fragmentation of culture and argues that an Indonesian national identity must "integrate the nation in spite of the centrifugal sub-national loyalties to racial, linguistic, ethnic, caste, or religious groups." One key strategy for doing this is to call on Enlightenment ideals, in which the will of the people, and the shared values emerging from a reasoned and agreed-on philosophy of rule, provide the sound and just basis for a nation. In this view, geographic and cultural fragmentation are reduced to superficialities, which can be overcome by a philosophy that harmonizes the most widely held and deeply cherished values of the

---

[4] Benedict O'G. Anderson, *Imagined Communities: Reflections on the Origin and Spread of Nationalism* (London, 1983); Holtzappel, "Nationalism and Cultural Identity" (cit. n. 2); and Scholte, "Identifying Indonesia" (cit. n. 2).

[5] See, e.g., the speech given by Sukarno, *The Birth of Pancasila*, 1 June 1945 (Jakarta, Indonesia, 1952); a copy of this speech is reprinted in D. R. SarDesai, *Southeast Asian History: Essential Readings* (Cambridge, Mass., 2006). See also Eka Darmaputera, *Pancasila and the Search for Identity and Modernity in Indonesian Society: A Cultural and Ethical Analysis* (Leiden, Netherlands, 1988); Pintamalem Singulingga, *Analisis Faktor-Faktor Yang Mempengaruhi Pemantapan Identitas Nasional Dalam PJPT II Untuk Meningkatkan Ketahanan Nasional Indonesia* (Jakarta, Indonesia, 1992); Abdullah Muchammad Ruslan, *Peranan Media Massa dalam Memlihara dan Memantapkan Identitas Bangsa Indonesia* (Jakarta, Indonesia, 1992).

[6] Geographic data was found in *Encyclopaedia Britannica Online*, s.v. "Indonesia," http://search.eb .com/eb/article-9106301 (accessed 24 April 2008). For general Indonesian history, see M. C. Ricklefs, *A History of Modern Indonesia since c. 1200*, 4th ed. (Stanford, Calif., 2008); Jean Gelman Taylor, *Indonesia: Peoples and Histories* (New Haven, Conn., 2003).

[7] This is a large and complex topic. A few recent works on the subject include Jemma Purdey, *Anti-Chinese Violence in Indonesia 1996–1999* (Honolulu, 2006); Leo Suryadinata, *Chinese Indonesians: State Policy, Monoculture, and Multiculturalism* (Singapore, 2004); and R. William Liddle, "Coercion, Co-optation, and the Management of Ethnic Relations in Indonesia," in *Government Policies and Ethnic Relations in Asia*, ed. Michael E. Brown and Sumit Ganguly (Cambridge, Mass., 1997).

people on the islands.[8] Once this intellectual assertion of national identity has succeeded, the problem is then simply one of education and of making sure that the idea of nation is communicated to one and all, so that through their use of reason, the citizens understand their place within, and their duties to, the state.[9] Other strategies emphasize the foundational importance of a geographic integrity justified on the historic boundaries of the colonial state. Here, as in Aceh (North Sumatra), where residents have called for an Islamic government, or in West Papua, where local peoples reject any cultural affiliation with Indonesia, Indonesia's military has brutally enforced the integrity of the nation.[10]

To understand the interactions of technology and national identity in Indonesia, this article explores the ways that technology is brought into these interacting narratives of heterogeneity and unity by those political elites who worked most assiduously to assert an identity for Indonesia.[11] At times, they deployed technologies to strengthen and disrupt narratives of identity; at other times, they deployed narratives of identity to gain support for technological projects. Among studies of technology in the formation of an Indonesian national identity Anderson's *Imagined Communities* has been especially influential.[12] Examining the colonial technologies of map, census, and museum that seemed to emphasize Indonesian cultural and geographic fragmentation, Anderson demonstrated the ways that anticolonial activists from diverse backgrounds reread these colonial symbols to create an idea of the colony as a nation in emergence.[13] Other scholars have taken up the question of geographic and cultural fragmentation in the postcolonial era, demonstrating the ways that technologies such as newspapers, radio, television, and satellites have been enlisted to breach both cultural difference and physical distance.[14] They offer insight into the ways that a

---

[8] Holtzappel, "Nationalism and Cultural Identity" (cit. n. 2); Darmaputera, *Pancasila and the Search for Identity* (cit. n. 5).

[9] This approach is reflected in Suharto's push for Pancasila education in schools. See R. E. Elson, *Suharto: A Political Biography* (Cambridge, UK, 2001).

[10] Elizabeth F. Drexler, *Aceh Indonesia: Securing the Insecure State* (Philadelphia, 2008); Anthony Reid, ed., *Verandah of Violence: The Background to the Aceh Problem* (Singapore, 2006); Jim Elmslie, *Irian Jaya under the Gun: Indonesian Economic Development versus West Papuan Nationalism* (Honolulu, 2002). For a 2006 statement from the Free Papua Movement, see http://www.eco-action.org/ssp/westpapua.html (accessed 23 Sept. 2008.). East Timor's battle for independence is another example of this process. See Richard Tanter, Gerry van Klinken, and Desmond Ball, eds., *Masters of Terror: Indonesia's Military and Violence in East Timor* (Lanham, Md., 2006).

[11] See for comparison studies that examine the significance of national identity for other groups in society, including Jean Gelman Taylor, "Identity, Nation, and Islam: A Dialogue of Men's and Women's Dress in Indonesia," in *The Politics of Dress in Asia and the Americas*, ed. Mina Roces and Louise Edwards (Eastbourne, Sussex, UK, 2007), 101–20; Leo Suryadinata, *Peranakan's Search for National Identity: Biographical Studies of Seven Indonesian Chinese* (Singapore, 1993); and Ariel Heryanto, "The Years of Living Luxuriously: Identity Politics of Indonesia's New Rich," in *Culture and Privilege in Capitalist Asia*, ed. Michael Pinches (London, 1999), 160–88.

[12] Anderson, *Imagined Communities* (cit. n. 4).

[13] For further attention to the role of mapping in the colonial period, see Eric Tagliacozzo, *Secret Trades, Porous Borders: Smuggling and States along a Southeast Asian Frontier, 1865–1915* (New Haven, Conn., 2005). The importance of manipulating maps is not limited to the colonial period. See John Pemberton's discussion of the Indonesian theme park Taman Mini Indonesia Indah (Beautiful Indonesia in Miniature) in *On the Subject of "Java"* (Ithaca, N.Y., 1994).

[14] For the colonial period, see Rudolf Mrázek, *Engineers of Happy Land: Technology and Nationalism in a Colony* (Princeton, N.J., 2002), especially for his discussion of roads and radio. Although not explicitly concerned with the technological side of the question, the role of newspapers in the colonial period is explored in Ahmat Adam, *The Vernacular Press and the Emergence of Modern Indonesian Consciousness (1855–1913)* (Ithaca, N.Y., 1995). For postcolonial history, see Joshua Barker, "Engineers and Political Dreams: Indonesia in the Satellite Age," *Current Anthropology* 46 (Dec. 2005):

national identity was both produced and deployed in the process of bringing technological projects to fruition. For example, Joshua Barker has shown that although the
Indonesian satellite project appealed in part because it made it possible for the same
television shows to be easily broadcast simultaneously nationwide, backers also used
its credentials as a national project to promote indigenous ownership of ground stations.[15] Sulfikar Amir's richly detailed study of the National Airplane project of the
1980s and 1990s shows how politically powerful it was for the backers to emphasize
the airplane's ability to easily traverse the distances of the nation. At the same time,
appeals to national pride made it possible to generate private financial support for this
project.[16]

This article moves away from the focus on communications and transport technologies to examine instead the interactions of technology and national identity in
Indonesia through the history of a steel factory in the town of Cilegon in West Java.[17]
Known initially as the Trikora steel plant, it later became the flagship plant for PT
Krakatau Steel, a major Southeast Asian steel producer. Superficially, this may seem
an unusual choice. It is not in any obvious way a medium that allows people and ideas
to circulate in the archipelago, overcoming geographic fragmentation. Although it
was a large and important project, it did not (so far as it is possible to gauge from
heavily censored newspapers) capture the public imagination more than other industrial projects did. Yet, there are reasons that this particular factory is useful for understanding national identity in Indonesia. The issue of industrialization was a crucial
one for Indonesia, as the ability to produce finished goods, and not merely raw materials, became a potent marker of postcoloniality for both domestic and foreign audiences. Defining the character of Indonesian industry and its role in the life of the nation was an ongoing project for industrialists and other advocates of industry as well
as politicians in the highest ranks of government. Because the steel mill in Cilegon
was begun (and halted) under Sukarno's leadership in the 1950s, revived in Suharto's
New Order (1965–98), and continues to exist as a key location of the Indonesian steel
industry today, it provides the opportunity to investigate changes and continuities in
the interplay of technology and identity between these two crucial political eras in
Indonesia's history.[18] Much current scholarship examines identity and technology either in the colonial era or the New Order, neglecting the early years of independence
and the period of Sukarno's leadership. It has been difficult therefore to discern the
longer-term contours of a technological politics of identity in Indonesia. This article

---

703–27; Barker and Bart Simon, "Imagining the New Order Nation: Materiality and Hyperreality in
Indonesia," *Culture, Theory, and Critique* 43, no. 2 (2002): 139–53; Barker, "Telephony at the Limits
of State Control: Discourse Networks in Indonesia," in *Local Cultures and the "New Asia": The State,
Culture, and Capitalism in Southeast Asia*, ed. C.J.W.-L. Wee (Singapore, 2002), 158–83; Sulfikar
Amir, "Power, Culture, and the Airplane: Technological Nationalism in New Order Indonesia" (PhD
diss., Rensselaer Polytechnic Institute, 2005), ProQuest Digital Dissertations, http://www.proquest
.com/ (publication number AAT 3183611) (accessed 24 April 2008); Phillip Kitley, "Pancasila in the
Minor Key: TVRI's *Si Unyil* Models the Child," *Indonesia* 68 (Oct. 1999): 129–52. For a discussion
of local identities and radio, see Jennifer Lindsay, "Making Waves: Private Radio and Local Identities
in Indonesia," *Indonesia* 64 (Oct. 1997): 105–23.

[15] Barker, "Engineers and Political Dreams" (cit. n. 14).

[16] Amir, "Power, Culture, and the Airplane" (cit. n. 14).

[17] Please note that during the course of the events described here, Indonesia reformed their system
of spelling. In some sources, the city is known as Cilegon, in others as Tjilegon. For simplicity, I am
using the contemporary spelling throughout this article.

[18] I have adopted the modern spelling of Suharto (used by R. E. Elson in his biography) rather than
the alternate spelling in the older orthographic system, Soeharto.

demonstrates that despite the significant political differences between the two eras, industrial technology was consistently enmeshed in the politics of national identity through narratives of geography and social justice.

SOCIAL JUSTICE AND TECHNOLOGY IN IDENTITY FORMATION

> "National identity discourse constructs a bridge between a mythologized past, and a coveted future."
>
> Gabrielle Hecht, *The Radiance of France*

> "[F]reedom, political independence was nothing more than a bridge. . . . [O]n the far side of that bridge we would rebuild our society."
>
> Sukarno, *The Birth of Pancasila*

When the Japanese occupiers convened a committee to prepare for Indonesia's transition to independence in 1945, nationalist leaders were deeply divided in their plans for the future of Indonesia.[19] The disagreements were not trivial; they centered on the very identity of the Indonesian nation. Some advocated an Islamic state, others secular. There were calls for a monarchy and for a military state. Many embraced some form of socialism, although the differences in their conception of what that would mean in practice varied dramatically, with some calling merely for a "family spirit" in the operation of the country, while others suggested that the government immediately take over all businesses.[20] On the last day of these meetings, June 1, 1945, in an effort to define a common ground for this divisive group, Sukarno elaborated his five principles, called the Pancasila, a clear statement of national identity that would constitute a philosophy of an independent Indonesian state. The Pancasila was intended to elaborate the common basis of belief and aspiration that would unite the people of Indonesia. The five elements of the Pancasila included: nationalism, or one state for all the peoples of Indonesia, asserting the claim to geographic unity for all the formerly colonized peoples of the archipelago; internationalism, or the promise to live in harmony with other countries of the world, asserting as a common value the desire for calm and steady relations between peoples; democracy through deliberation, underscoring the desirability of consensus for producing harmony in society; social justice for all the people of Indonesia, which was usually contrasted with the injustices of colonial rule; and belief in God, a compromise that would, not without some controversy, include all the nonatheistic religious beliefs of Indonesia.[21] The state that observed these principles was nothing less than a "gotong-royong" state, Sukarno declared. *Gotong-royong* is a Javanese term meaning mutual cooperation, of the kind that residents of villages engaged in to build irrigation works, celebrate festivals, and perform other community activities. Giving a new set of principles a veneer of tradition, Sukarno tied the past to his imagined future. Scholars have often noted the vagueness of the Pancasila and its susceptibility to multiple interpretations.

---

[19] George McTurnan Kahin, *Nationalism and Revolution in Indonesia* (1952; repr., Ithaca, N.Y., 2003), 147–212; John O. Sutter, *Indonesianisasi: Politics in a Changing Economy* (Ithaca, N.Y., 1959).

[20] Sutter, Indonesianisasi (cit. n. 19), 238–39.

[21] Sukarno, *Birth of Pancasila* (cit. n. 5). See also J. D. Legge's enlightening analysis, especially his discussion of Sukarno's early work to integrate disparate nationalist groups, in *Sukarno: A Political Biography* (Sydney, 1972), 184–88.

However, it served to assert a sense of unity despite (or perhaps because of) leaving crucial questions unanswered and remains a vital (if never uncontested) component of Indonesian political culture.[22]

Understanding the particular meaning of social justice in the Pancasila sheds light on one of the crucial ways that technology entered narratives of national identity in Indonesia. For Sukarno, "social justice" was formulated economically: a socially just nation was one in which all citizens, regardless of ethnic or religious background, had the same opportunities to prosper.[23] This may sound like the mundane promise of any politician, but in Indonesia it represented a profoundly postcolonial dream. Many nationalists asserted that Indonesian poverty had been primarily caused by the systematic exclusion of Indonesians from the most economically valuable professions and undertakings during the years of colonial rule, as well as the "drainage" of Indonesia's wealth out of the country and into foreign hands. During the 1930s, as part of a larger noncooperation movement, some nationalists argued against any dramatic efforts to reform the economy or the productive practices of Indonesians, because they claimed that the Dutch would always prevent Indonesians from prospering.[24] In 1947, when a Dutch blockade tried to prevent the export of products out of Indonesia, claiming that such goods were in fact Dutch property, Gani, the minister of prosperity, defended the rights of the new nation to export its own goods, saying that Indonesians would not go back to the colonial years when they were "a pauper people in a wealthy country."[25] Although the Japanese occupation (1942–45) at first seemed to promise something better for Indonesians, the Japanese quickly began to occupy the same economic niches as had the Dutch, offering few new opportunities for Indonesians. Calls by Sukarno and Muhammed Hatta to promote changes that would benefit Indonesians, like greater industrialization, were ignored.[26]

For nationalist leaders, the injustice of colonial rule was manifest in Indonesian poverty; the postcolonial state, an embodiment of justice by the very fact of its being postcolonial, would bring prosperity in train. As Gyan Prakash points out, such formulations play a crucial dual role for postcolonial states. The state gains legitimacy from its promise to provide a rightful distribution of resources. But defining justice this way reinforces the view that the territory and its people are resources to be managed. The definition of justice both legitimized the state and authorized it to exert power in fulfillment of its destiny.[27] Because justice was framed as an economic problem, the early debates about the state's political character also defined the state's

---

[22] For a detailed investigation of the Pancasila, see Darmaputera, *Pancasila and the Search for Identity* (cit. n. 5).

[23] In practice, however, this meant privileging the *pribumi*, or indigenous people of Malay heritage, over people of Chinese heritage, who were treated as Dutch collaborators and hence tarred with the same brush.

[24] On the loss of wealth to the Dutch, see Mohammad Hatta, "Drainage," in *Verspreide Gescriften van Mohammad Hatta* (Jakarta, Indonesia, 1952). Reprinted from the newspaper *Hindia Poetra*, issue 6, 1923. For a brief discussion of nationalist objections to technological and economic reform efforts, see Suzanne Moon, *Technology and Ethical Idealism: A History of Development in the Netherlands East Indies* (Leiden, Netherlands, 2007), 140–43. On Sukarno's advocacy of noncooperation, see Legge, *Sukarno* (cit. n. 21), 100–104.

[25] Sutter, *Indonesianisasi* (cit. n. 19), 494.

[26] Ibid., 174–75. For more on the Japanese occupation of Indonesia, see Saito Shigeru, *War, Nationalism, and Peasants: Java under the Japanese Occupation, 1942–1945* (Armonk, N.Y., 1994).

[27] Gyan Prakash, *Another Reason: Science and the Imagination of Modern India* (Princeton, N.J., 1999).

role as a technological actor. Many nationalists embraced some form of socialism as the path to justice, but the extent of their commitment to Marxism-Leninism varied dramatically. Radical socialists such as Tan Malaka agitated for complete state ownership of all business (including Dutch-owned businesses) and an immediate start to the development of state-operated heavy industry.[28] Sukarno and Hatta (soon to be the president and vice president, respectively) took a more moderate stance, advocating only a slow move toward socialism, rejecting the disruption and diplomatic difficulties of the more extreme position. The constitution of 1945 ultimately called for the promotion of cooperatives as the economic basis of society, with state ownership of those branches of production considered "important to the life of the state," which though ill-defined was meant to include any business in which a private individual could gain too much power and oppress the people through control of fundamental necessities.[29] The economic ideal (especially as Hatta formulated it) was one of mutual help, grounded in both village-based gotong-royong and a vaguely understood notion of *kekeluargaan*, or acting in the manner of a family.[30] These idealistic visions made the state seem less an authority dictating change than a facilitator for mutual help across the nation, reinforcing the idea that the state's economic and technological actions guaranteed that the postcolonial Republic of Indonesia would become a socially just nation.

The idea that the state promotes social justice is one that many Indonesians today would likely greet with polite incredulity. Whether or not the state has ever succeeded in producing the kind of justice described above is less important to this article than the connections these early debates established linking postcoloniality, social justice, and technology. In these early years, Indonesia's most salient national characteristic was its postcoloniality, but even this seemingly straightforward interpretation of political transition could be destabilized if old forms of exploitation reappeared. The postcolonial state had to be a just state, and justice required state action, some of which could be technological. It is through this logic that a state technology project could become a national project. When the state acted to build a factory or highway, it was also building a just nation.

JUSTICE, GEOGRAPHY, AND STEEL: INDUSTRIALIZATION IN THE SUKARNO ERA

From the early days of the republic, industrialization was a topic of considerable importance to Indonesian leaders. From the 1920s and into the 1930s, foreign-owned light industry had flourished on the otherwise largely agrarian islands of Java and Sumatra.[31] In the 1930s, M. H. Thamrin argued that industrial enterprises owned and operated by the indigenous people offered the best way out of the colony's economic problems. Citing the Japanese history of industrialization, Thamrin argued that in-

---

[28] Sutter, *Indonesianisasi* (cit. n. 19), 317–35. Tan Malaka was influential in both the Partai Kommunis Indonesia (PKI, Indonesian Communist Party) and the Partai Buruh Indonesia (PBI, Indonesian Workers' Party).

[29] Ibid., 275–77. Other businesses could be owned by individuals, so long as the public welfare was always paramount. According to Sutter, only Masjoemi, the political party made up of a coalition of Islamic organizations, advocated private ownership as the norm (ibid., 317).

[30] For Hatta's views on cooperatives, see Mohammad Hatta, *The Co-Operative Movement in Indonesia* (Ithaca, N.Y., 1957).

[31] Sutter, *Indonesianisasi* (cit. n. 19), 34–79. See also Mrázek, *Engineers of Happy Land* (cit. n. 14).

dustry would build national strength and help correct the wrongs of colonial rule.[32] As mentioned earlier, Sukarno and Hatta had tried unsuccessfully to convince the Japanese to support Indonesian industry during the occupation. In 1947, while the nation was still at war with the Dutch, the Ministry of Prosperity made a list of industries that Indonesia needed to build or improve during its first ten years. Number three, after new textile and aluminum plants, was a steel mill.[33]

One might assume that purely pragmatic reasons would be enough to explain the decision to create a steel industry in Indonesia. Producing steel locally would save the considerable foreign currency costs associated with importing it from Japan or elsewhere. Indeed one official suggested that the plant would save 15 million rupiah per year in imports.[34] New factories would provide employment for many people displaced by years of occupation and revolution. Although both of these reasons did matter, they do not convey the depth of meaning Sukarno and other advocates of industrialization in Indonesia would assign to this project using a potent mix of socialist idealism and postcolonial promises.

In 1956, during an extended state visit to the Soviet Union, Sukarno appealed to the Soviet government for technical and financial aid to construct a steel mill. Sukarno enjoyed cordial relations with the Soviets, although he maintained a neutral, nonaligned stance in international affairs. He had never embraced the hard-line approach of Indonesian communists such as Tan Malaka, but Sukarno did espouse a socialist framework for Indonesia's future, albeit one that looked significantly different from the atheistic variety the Soviets practiced. The Soviets proved receptive to Sukarno's request, signing an economic and technical agreement with Indonesia on September 15, 1956, which was followed in January of 1959 by a cooperative protocol.[35] This seemingly optimistic moment occurred during a time of intensifying political crisis in Indonesia, with the unity of the nation in serious jeopardy. Political divisions among leaders continued to be deep; party rivalries occupied politicians, while the spread of corruption, smuggling, and a decline in standards of living contributed to growing disillusionment in the results of the revolution. The military represented a difficult problem for Sukarno, as many military leaders objected to what they saw as Sukarno's procommunist leanings. They instead tended to align politically with Masjoemi, a Muslim political party that strongly opposed Sukarno's leadership. Certain army officers who, J. D. Legge argues, felt the need to set matters to rights in the divided and paralyzed government used the opportunity of Sukarno's visit to the Soviet Union to arrest his foreign minister on corruption charges (later dismissed). Two other military leaders played to the dissatisfaction with the central government in the Outer Islands (outside Java, the traditional center of power). One took power in Central Sumatra in December, while another tried (but failed) to do the same in North

---

[32] Moon, *Technology and Ethical Idealism* (cit. n. 24), 136–37. See also M. H. Thamrin, *De Koloniale Structuur als Oorzaak van de Crisis en de Massaverarming* (Batavia, Netherlands East Indies, 1932).

[33] Sutter, *Indonesianisasi* (cit. n. 19), 498. Other desirable projects include a fertilizer factory, a soda plant, a glass and porcelain factory, tile and brickworks, earthenware factories, a cement factory, and paper factories.

[34] Zakara Raib, a representative from the Ministry of Basic Industry and Mining, gave this figure in a press conference, as reported in "Trikora Steel Plant Project," *Indonesian Observer*, 22 May 1962.

[35] "Menteri Perdatam Chaerul Saleh: Industri dasar sjarat mutlak merobah ekonomi kolonial," *Warta Bhakti*, 5 April 1962.

Sumatra. Separatist threats appeared elsewhere in the archipelago as well.[36] National unity was fading from view.

Although the promise of a steel mill hardly registered in the midst of this crisis, and certainly had no power in itself to resolve conflicts or restore national unity, it is instructive to see how Sukarno defended his dealings with the Soviets, in light of this crisis. In short, he stressed anticolonialism, the one element that united all of Indonesia's diverse political players, rather than socialism, and argued that there existed a deep political harmony between Indonesia and the Soviet Union, a ploy that both invoked the Pancasila in its internationalism and subtly reinforced the *national* identity of Indonesia by playing up its role in the world. Sukarno claimed that his deal with the Soviets was not mere opportunism, because the histories of the Soviet Union and of Indonesia were similarly anticolonial and anti-imperial; both were engaged in building social justice.[37] The reporter for *Sin Po* who accompanied Sukarno's delegation (and the content of whose stories may have been controlled by Sukarno himself) prominently quoted Leningrad's Mayor Nikolay Ivanovich Smirnov, who declared that heavy industry was essential for Indonesia to "free itself of the remaining shackles of imperialism."[38] Sukarno's flattery of the Soviet Union was no doubt aimed at the Soviets themselves, but it also served to draw the attention of a divided and disillusioned public back to the ideals that defined the Indonesian nation: anticolonialism and social justice. By associating the Soviets with these same ideals, he positioned Soviet aid as a reasonable tool to build the Indonesian nation and himself as a unifying influence, pursuing constructive, rather than divisive, ends.

Over the next six years, Sukarno used the divisions within political leadership to take more power for himself, introducing his "guided democracy," which replaced an elected parliament with a nominated parliament, and eliminating the threat of weak coalition governments by creating a cabinet that included representatives from the main political parties, one that would operate (at least in theory) on the basis of deliberation and consensus. He called it the "gotong-royong" parliament.[39] By 1959, he had succeeded in removing the most rebellious elements in the military, although the military retained a great deal of power.[40] During these years, Sukarno stressed the incompleteness of the Indonesian revolution and the need to move beyond the colonial past and rebuild the nation, an image of Indonesia that would provide a framework for the public to interpret the value and necessity of technological development, particularly with respect to a steel industry.[41]

The steel project attracted little attention from the press between 1959 and 1962. During this time Soviet engineers and their Indonesian counterparts hammered out a design for the facility, which would include three open-hearth furnaces capable of

[36] Legge, *Sukarno* (cit. n. 21), 268–77. For more on Sukarno's relationship to the military in the late 1950s and early 1960s, see Herbert Feith, "Indonesia's Political Symbols and their Wielders," *World Politics* 16 (Oct. 1963): 79–97.

[37] "Djiwa Besar menghubungkan semua manusia: Hari-Depan Sovjet dan Indonesia gilang-gemilang," *Sin Po*, 3 Sept. 1956. See similar rhetoric in a speech given in 1962 by Subandrio, the minister for foreign affairs, "Pidato Wakil Menteri Pertama/Menlu Subandrio Menjambut Upatjara Pemberian Komando Pelaksanaan Permbangunan Pabrik Besi-Badja Tjilegon," *Kementerian Penerangan/Penerbitan Chusus, nos. 212–227* (Jakarta, Indonesia, 1962), 11–18.

[38] "Djiwa Besar menghubungkan semua manusia" (cit. n. 37).

[39] Feith, "Indonesia's Political Symbols" (cit. n. 36), 81–82.

[40] Ibid.

[41] Ibid., 82.

producing 100,000 tons of steel annually. In May 1962, the ceremony for the opening of construction brought the project back to the public eye.[42] Sukarno and his most trusted ministers consistently portrayed the project as a decisive step away from the colonial past. A critical characteristic of the steel industry (and other industrial projects) was that it produced not raw materials to be processed elsewhere, as colonial enterprises usually had, but finished goods for either domestic use or export, the economic role taken by the colonizers, not the colonized. Industry therefore carried connotations of domestic strength and international power, making it possible for a steel mill to become a potent symbol of national strength and a defining characteristic for a postcolonial nation. Language suggesting forward movement and passageways occurs repeatedly in newspaper articles and in the speeches given at the ceremony opening construction in May 1962. At a press conference held in April of that year, Chaerul Saleh, the minister of basic industry and mining, asserted that the fast development of a steel industry (and other industries) was the surest way to move from a colonial economy to a "healthy national economy."[43] At the groundbreaking ceremony in May, he enthused that the new steel mill would "open a door from the atmosphere and manner of agrarian life of centuries past to modern industrial times."[44] This was good not just for the national economy but also for individuals who had suffered under Dutch rule. Now they had the opportunity to fully develop their skills and talents, creating a new society with "a different form," embracing and enacting the aspirations of the Pancasila.[45] He underscored the moral obligations involved in building this project, not merely the economic opportunity. The "generation of '45" had the duty to "bequeath the conditions and basic requirements that make it possible for generations to come to continue to grow and finally reach the just and prosperous society we long for."[46] The steel mill would build steel and a nation imbued with the virtues of Pancasila.

Sukarno added another layer of nationalistic meaning to the project when he named the mill Trikora and thereby linked it to a major international crisis that occurred in 1962, the struggle for the "return" of West Irian (the western part of the islands of Papua New Guinea) to the Republic of Indonesia. When the Dutch had finally given up their claims to Indonesia in 1950, they had retained control of West Irian. Sukarno had long argued that West Irian should be given to Indonesia, making the Republic of Indonesia mirror the historical map of the Netherlands East Indies.[47] Sukarno used the situation with West Irian to justify the seizure of Dutch properties and the expulsion of Dutch citizens from Indonesia in 1957. In 1961, the Dutch announced plans for the creation of an independent state in West Irian. Sukarno immediately denounced the plan as a thinly veiled effort at neocolonialism and a violation of the

---

[42] There is little information in Indonesian or English about the details of this early period of design or building, nor is there much about the nature of the interactions between Soviet and Indonesian experts. Apart from difficulties of supply that slowed down production (detailed later), it is difficult to say whether the technology or design of the steel mill changed as a result of the process of transfer.

[43] "Industri dasar sjarat mutlak merobah ekonomi kolonial," *Warta Bhakti*, 5 April 1962.

[44] Chaerul Saleh, "Komando Pelaksanaan Pembangunan Pabrik Besi-Badja Di Tjilegon," speech delivered 21 May 1962. Reprinted in *Kementerian Penerangan/Penerbitan Chusus* (cit. n. 37).

[45] Ibid.

[46] Ibid.

[47] Legge, *Sukarno* (cit. n. 21), 247–51, 358–61.

rightful geographic territory of Indonesia.[48] By claiming that West Irian was a part of Indonesia, Sukarno framed his actions as a necessary anticolonial war, rather than an act of Indonesian imperialism, as the Dutch, many international observers, and more than a few Papuans interpreted it. In 1961, Sukarno issued what became known as the Trikora, or Trikomando Rakyat, the People's Triple Command, which called for the prevention of an independent state of West Irian, the raising of the Indonesian flag there, and the removal of the Dutch colonizers from control.[49] By calling the new steel mill Trikora (a name emphasized at the opening ceremony by three ceremonial explosions of dynamite), Sukarno symbolically tied this constructive, postcolonial technology project to narratives of geographic integrity in the Indonesian nation, a move that simultaneously upped the patriotic significance of the steel project and the postcolonial credentials of the West Irian movement.

The symbolic connection went deeper than simply the name of the new mill, however. The location chosen for the site by the Indonesian government, in Cilegon, West Java, wove narratives of geography and empire more tightly into the story of the steel mill. Cilegon was a small, rural, and impoverished town in the province of Banten, west of the capital in Jakarta. However, in precolonial times Cilegon had been a lively and wealthy part of the Majapahit empire, a Java-centered empire whose influence stretched all the way to Sumatra. Making Cilegon the site for the project highlighted the break from the colonial past (as represented by Cilegon's current poverty) and a revival of precolonial greatness. The Majapahit empire held particular appeal because unlike the later Mataram empire (whose infighting and deal making with the Dutch helped facilitate the spread of Dutch control of Java), it exerted influence into Sumatra, Borneo, and the Eastern islands of Indonesia, including Bali.[50] To many historians, the Majapahit empire represented not only a time of great military power but also an era in which a sophisticated high culture of music, dance, and design flourished. Evoking the greatness of Majapahit therefore conferred positive meanings of power and growth on the Trikora steel mill. It is not unusual for a state to portray the present-day nation as the logical continuation of a great or glorious past. This strategy has particular relevance in postcolonial settings.[51] In the postcolony, this use of history makes it possible to posit an underlying bedrock of identity, or primordial reality for the nation, that was only temporarily obscured by colonial rule, giving the nation a common tie that does not rely on the (usually despised) colonial past.[52]

In the case of Cilegon, the chosen location did two kinds of work for Sukarno and his supporters. As an area of past greatness and present-day poverty, it could stand as

---

[48] *Colonial Purposes in West Irian: An Exposé* (Jakarta, Indonesia, 1962); Sukarno, *Indonesia Wants Negotiations on the West Irian Problem, Based on Transfer of Administration from Netherlands to Indonesia: Address by President Sukarno, Commemorating the Revelation of the Qur'an, on February 21, 1962, at Istana Negara, Djakarta* (Jakarta, Indonesia, 1962). From the perspective of the present day, Sukarno's claims seem self-serving and ludicrous. However, his concern was not entirely unwarranted. The separatist generals operating in Sumatra in 1958 had received some encouragement from Western powers, including the United States, who were uneasy with Sukarno's communist leanings.

[49] Justus van der Kroef, "An Indonesian Ideological Lexicon," *Asian Survey* 2 (July 1962): 24–30.

[50] Ricklefs, *History of Modern Indonesia* (cit. n. 6).

[51] For a noncolonial example, see Gabrielle Hecht, *The Radiance of France: Nuclear Power and National Identity after World War II* (Cambridge, Mass., 1998). On the use of precolonial histories to create identities in postcolonial nations, see Anderson, *Imagined Communities* (cit. n. 4); and Prakash, *Another Reason* (cit. n. 27).

[52] Hitchcock and King, introduction (cit. n. 1).

both an example of the evils of colonialism and, with the new factory providing jobs and a modern economy based on the production of finished goods, the promise and power of postcolonial rule. Cilegon's imperial past was equally crucial to the story that Saleh told. In his speech, Saleh referred to the Banten area's "brilliant" past as a central link between east and west in an empire that stretched from Java to Sumatra. The steel mill, as a powerful Western technology, would in a slightly different manner, link the West to the East, reestablishing empire and restoring prosperity. Saleh makes the parallels with West Irian explicitly, even suggesting (disingenuously, given the time required to construct and make a factory operational) that the steel plant would help provide the resources needed to aid a war for West Irian. More effectively, he compared the "mental, moral, and physical" strength required to build the steel plant to that required to win back West Irian.[53] In another speech in the opening ceremony, the minister for the Department of Foreign Affairs, Subandrio, suggested the satisfactions that imperial expansion would bring, when he boasted that one year from that date the people of Cilegon would be able to travel to Papua for a vacation.[54] According to one newspaper report, the people of Banten who lined the streets for the opening ceremony chanted "With steel we are victorious."[55] The parallel between West Irian and the Trikora plant allowed this immobile technology, located in the heart of Java, to feed the narratives of geographic integrity that were (and are) crucial to Indonesian national identity.

That the site of Cilegon was purposely chosen to evoke a dream of greatness restored is suggested by both the practical shortcomings of the area and the comments made years later by Indonesian engineers. The harbor at Merak had to be significantly (and expensively) improved before it could be used to transport the products of steel production. There was no adequate source of fresh water nearby, requiring the construction of a forty-kilometer pipeline to feed the plant. Access to raw iron resources was equally limited. Early on, although they expected to get a small percentage of raw iron from Lampung, Sumatra, not far away, they had always planned to use as a major resource scrap iron, which had to be transported from more urban areas to the relatively remote town of Cilegon. Practically speaking, the site in Cilegon had little to recommend it, while the alternative site in East Java would have had none of the problems and several practical advantages. It is difficult to know exactly what factors may have influenced Sukarno's insistence on the Cilegon site, and it is certainly possible that some other political machinations or personal favoritism influenced his choice. However, engineers writing years later insisted that the government had chosen this site for purely "political and social" reasons.[56] East Java did lack a convincing history of interconnection between east and west, and speaking speculatively, its location seemed to reinforce the view of power as centered on Java, rather than pointing outward, toward other islands. Whatever the symbolic shortcomings of East Java, however, it seems likely that the West Java location did gain favor in large

[53] Saleh, "Komando Pelaksanaan Pembangunan Pabrik Besi-Badja Di Tjilegon" (cit. n. 44), 6–7.

[54] Subandrio, "Pidato Wakil Menteri Pertama/Menlu Subandrio" (cit. n. 37), 15.

[55] "Pembangunan Besi Badja Tjilegon Djawaban Terhadap Tututan Djaman," *Nasional*, 23 May 1962.

[56] J. L. Rombe, T. Ariwibowo, Manurung et al., *Feasibility Study for Tjilegon Steel* (Jakarta, Indonesia, 1970), 12.

measure because of its symbolic and historic geography, rather than its instrumental usefulness for steel production.[57]

A final contextual element may help explain the heavy emphasis on duty in the ceremonial speeches. By 1962, Indonesia's economy had gone into a tailspin, with out-of-control inflation, rampant corruption, and illegal trade. One analyst estimated inflation to be running at 182 percent for 1962.[58] Sukarno was (and still is) roundly criticized in the foreign press for his lack of sound economic management. His large-scale projects, such as the stadiums built for the 1962 Asian games, were at the time, and have since been portrayed more as irresponsible monuments to his ego than sensible elements in a plan for national development.[59] Assigning Trikora a national significance that drew on intertwined narratives of geography and justice, Sukarno made the project an urgent work of patriotic, national transformation that could not wait for mere economic considerations. By placing Trikora at a currently impoverished site with a "brilliant" past as a meeting place of east and west, Sukarno positioned it as both a modern development project and a return to the greatness of precolonial Java, a return whose glory was leavened by renewed imperial power.

FROM TRIKORA TO KRAKATAU STEEL

The path from Trikora's opening ceremony to a functioning steel plant would prove to be a long one. Progress was slow between 1962 and 1965 as Indonesian and Soviet engineers struggled to construct the water pipeline, improve the port at Merak, and build housing for 1,000 workers and their families, in addition to the work on the plant itself.[60] Construction lags became the norm. Any changes to the original plans had to be approved in Moscow, making it difficult for project engineers to respond to changing circumstances.[61] The Indonesian government frequently lacked the rupiah (related to Indonesia's hyperinflation) to pay laborers and Indonesian engineers. (Soviet experts were paid by the Soviet government.) By late 1965, the government had reportedly lost interest in the project due to the political chaos surrounding Suharto's coup and the removal of Sukarno from effective leadership.[62]

In late 1965, power changed hands in Indonesia during a coup whose full contours are still not entirely clear. The official version of events was that leaders of the Indonesian Communist Party, using dissident members of the Indonesian Army and Indonesian Air Force, schemed to kidnap and kill six generals as part of an attempted coup to take over the government. Suharto, then a general in the Indonesian military, assumed control of central Jakarta during the ensuing chaos and, over the

---

[57] My interpretation here conflicts with that of H. W. Arndt, whose description of the project assumed that the site choice was simply a matter of technological wishful thinking (with respect to Lampung's resources) and incompetence. Arndt, "PT Krakatau Steel," *Bulletin of Indonesian Economic Studies*, 11, 1 July 1975, 120–26.

[58] Don D. Humphrey, "Indonesia's National Plan for Economic Development," *Asian Survey* 2 (Dec. 1962): 17–18.

[59] For a typical discussion from academia, see Guy J. Pauker, "Indonesia: The Year of Transition," *Asian Survey* 7 (Feb. 1967): 138–50. For an example of a portrayal of Sukarno from the popular press, see "Djago, the Rooster," *TIME*, 10 March 1958; and Norman Sklarewitz, "Trouble for Sukarno: Indonesia's Economy Falters as Output Slips and Inflation Worsens; Parts Pinch Slows Socialized Industries; President Buys Stadiums and Soviet Arms," *Wall Street Journal*, 5 April 1963.

[60] Rombe et al., *Feasibility Study for Tjilegon Steel* (cit. n. 56).

[61] Ibid., 13.

[62] Ibid., 2–4, 14.

course of the next several months, effectively took control of the nation out of Sukarno's hands.[63] A letter of authority, signed by Sukarno on March 11, 1966, formally
transferred executive power to Suharto. The transition to power was bloody: some
historians estimate that 500,000 or more Indonesians were killed in a massacre of
(suspected) communists over the next year.[64] The fate of the Trikora plant was not
the most urgent problem for the leaders of the nation. Symbolically, the name Trikora
would for some time be tainted by the memory of Sukarno's overreaching ambitions
and the social chaos that closed his years of rule. When Suharto took power, his first
order of business was to restore the fundamental basis of Indonesia's economy, primarily by encouraging foreign investment. As Suharto focused on fundamentals, the
steel mill languished. In the ensuing political chaos and poor economic conditions,
construction came to a halt, and the remaining Soviet engineers left Cilegon in 1967.
They had supplied 80 percent of the equipment; only 25 percent of the construction
was complete.[65] After 1967, all work except basic maintenance had ceased.

Suharto initiated changes to the economic organization of the country, in consul-
tation with the Intergovernmental Group on Indonesia, an economic assistance organization made up of representatives from Australia, Japan, the United States, and
Western Europe, and a group of American-trained Indonesian economists (known
as the technocrats, or "Berkeley Mafia"). Decreasing inflation, and work on basic
infrastructure, as well as a quieter political situation enforced by repressive social
policies, gave Suharto the means to lure foreign investors. He cultivated an air of
cool pragmatism, an image that many Indonesians and foreign observers readily accepted.[66] Yet this turn away from the bombast and instability of the Sukarno years did
not mean that Suharto engaged less in identity work than had Sukarno. Suharto eschewed the anticolonial stance that Sukarno had taken, stressing instead Indonesia's
identity as a developing country, creating the image of a quietly modernizing country,
both to attract foreign investors and to create a climate in which they would be willing to work. A boom in oil production, increasing oil prices, and deals with foreign
oil companies favorable to Indonesia substantially improved the economic situation
through the early 1970s.

During this period of economic recovery, the Indonesian government began to consider what to do with the incomplete works at Cilegon. They went to the United
Nations, the United States, and the Soviet Union for feasibility studies, ultimately
producing one themselves as none of the foreign studies was encouraging.[67] The Indonesian authors maintained that the project ought to be completed, not for the "political" reasons that it had been started in the first place, which they openly criticized,
but because of the investment already made and the capital equipment already available. In the pragmatic spirit of Suharto's leadership, they argued simply that it would
be wasteful not to complete the project and that a domestic steel industry would ultimately help the nation grow.[68] Although between 1968 and 1970, the government

[63] For a detailed discussion of the coup, and the multiple interpretations of events, as well as a discussion of the transition of power from Sukarno to Suharto, see Elson, *Suharto* (cit. n. 9), 99–166.
[64] Ibid., 123–27.
[65] Rombe et al., *Feasibility Study for Tjilegon Steel* (cit. n. 56), 4.
[66] See, e.g., Guy Pauker, "Indonesia: The Age of Reason?" *Asian Survey* 8 (Feb. 1968): 133–47; and
Theodore Friend, *Indonesian Destinies* (Cambridge, Mass., 2003).
[67] Rombe et al., *Feasibility Study for Tjilegon Steel* (cit. n. 56).
[68] Ibid., 34.

was able to provide enough support to finish a minor portion of the plant, a cold wire drawing shop, the budget could not support completion of the project as planned. The government found the ideal partner to complete it in the state-owned oil company Pertamina. Pertamina formed a subsidiary company to run the mill, Krakatau Steel.

Krakatau Steel was just one of many businesses, both related and unrelated to oil, that Pertamina started in the early 1970s. Pertamina itself had been formed from a number of smaller military-operated state oil companies established to take over Dutch oil interests after the transfer of West Irian to Indonesia. During these years of economic hardship, the military had used profits from these (and other) enterprises to purchase weapons, arms, and uniforms and otherwise supplement the unpredictable and inadequate national budget. This practice continued under Suharto, because he did not wish it to appear to foreign donors, or to his own economic advisers, as if he were pouring too much money into defense.[69] Pertamina, run by a former general, and Suharto's handpicked executive, Ibnu Sutowo, continued these practices as well as performing various kinds of development projects with Suharto's blessings. There was virtually no oversight from the tightfisted technocrats, allowing Pertamina to act as an "off-budget development agency," as William Ascher names it.[70]

Using this strategy, Suharto crafted two images of Indonesia, aimed at two audiences. Both images would become crucial to the new interpretation of Indonesian identity that Suharto used to justify his own place at the apex of national power. The first, primarily for foreign consumption, was of a rationally and responsibly managed economy, in which discipline and austerity were steadily producing dividends and a favorable environment for foreign investment. This aspect of Indonesia's identity established Indonesia's place in the world as an open "developing country," eager to pursue development in the ways approved by international agencies such as the World Bank and align and integrate itself with developed, capitalist nations. The second image, for domestic consumption, was that of a rapidly developing and modernizing country, the reality of which was attested to by new schools, mosques, hospitals, roadways, and jobs in new industries. A scandal involving Pertamina and the construction of Krakatau Steel would show the difficulty of maintaining these narratives of Indonesian identity simultaneously.

The problems started when Ibnu decided to expand the size of the Krakatau project considerably, first to 500,000 tons per year rather than the original specification of 100,000, and then to 2 million.[71] Ibnu believed that the increasing demand for steel in areas such as shipbuilding and container manufacturing warranted this dramatic ramp up in size. From the standpoint of the early 1970s, the original project plans were small by international standards, and Ibnu, with faith in the profits to come, saw no problem in increasing the size to something more in keeping with his plans for Indonesian industry.[72] He partnered with several German companies, including Siemen's Kloeckner and most prominently Ferrostaal, to construct the new facili-

---

[69] William Ascher, "From Oil to Timber: The Political Economy of Off-Budget Development Financing in Indonesia," *Indonesia* 65 (April 1998): 37–61; Harold Crouch, "Generals and Business in Indonesia," *Pacific Affairs* 48 (Winter 1975–76): 519–40.

[70] Ascher, "From Oil to Timber" (cit. n. 69).

[71] Arndt, "PT Krakatau Steel" (cit. n. 57), 122. Please note that in Indonesia, first and last names function a bit differently than is common in other parts of the world. It is considered both proper and respectful to shorten Ibnu Sutowo's name to simply Ibnu. Indeed, he was usually known in the press as General Ibnu. Many Indonesians go by one name as a matter of course.

[72] "Pertamina's Loan Problems," *New Standard*, 26 April 1975.

ties. Although Ibnu drove this expansion, it seems he did not make Suharto aware of his plans, at least not right away. In early 1975, trouble became apparent as Pertamina experienced serious financial problems resulting from the overuse of short-term credit to finance long-term projects such as Krakatau Steel.[73]

Foreign journalists and scholars roundly criticized Ibnu and Pertamina for what were seen, at best, to be fiscal misjudgments and, at worst, examples of corruption and otherwise shady financial dealings. They blasted Krakatau Steel as one more inappropriate venture, citing Pertamina's involvement in a seemingly incoherent range of projects including rice farms, hotels, and telecommunications as nothing more than empire building: "a state within a state."[74] A second criticism concerned the cost of mill construction. H. W. Arndt reported that the Indonesians had paid Ferrostaal three times the average price for the plant (the comparison figure was obtained from a similar plant built in Taiwan), which raised the suspicion of kickbacks. Ferrostaal, which took the brunt of the criticism, responded that the high prices were due entirely to Indonesia's insistence that the plant be built to the highest quality and safety standards met in the rest of the world.[75] A separate, although related, critique questioned whether such an expensive project, built in such a luxurious manner, was appropriate in an overwhelmingly poor country. As of 1975, it appeared more effort had been put into building the "German colony" (the "Steel City," or *Kota Baja*), including facilities such as grocery stores, golf courses, and cinemas, than into constructing the plant itself.[76] Critics noted with disapproval that executives commuted from Jakarta in private helicopters and lived lavishly, while construction seemed to move at a snail's pace.[77] The image of Indonesia that emerged was one in which rationally and carefully managed development was a mirage obscuring the reality of corruption and incompetence.

Suharto's response, in the early days of the crisis at least, emphasized Indonesia's identity as a modern nation with a not atypical modern problem of misbehaving corporations. He compared Pertamina's difficulties with those being experienced by the U.S. and West German governments with their own corporations.[78] The difference, he argued, lay not in competence or honesty but in Pertamina's enthusiasm for development: "The problem with Pertamina lies in the fact that Pertamina has been extending a helping hand to accelerate Indonesia's development efforts; the reason was that only through such development can we improve the living standards and general well-being of the Indonesian people."[79]

---

[73] Arndt, "PT Krakatau Steel" (cit. n. 57); Bruce Glassburner, "In the Wake of General Ibnu: Crisis in the Indonesian Oil Industry," *Asian Survey* 16 (Dec. 1976): 1099–112.

[74] Glassburner, "In the Wake of General Ibnu" (cit. n. 73), 1099. See also Arndt, "PT Krakatau Steel" (cit. n. 57); Dan Coggin and Seth Lipisky, "The High Price of Pertamina's Big Dreams," *Far Eastern Economic Review*, 30 May 1975; Crouch, "Generals and Business in Indonesia" (cit. n. 69). See also Harold Crouch's book-length study, *The Army and Politics in Indonesia* (Ithaca, N.Y., 1988).

[75] Ibid.; Arndt, "PT Krakatau Steel" (cit. n. 57), 123.

[76] Ibid.; Glassburner, "In the Wake of General Ibnu" (cit. n. 73); Coggins and Lipisky, "High Price of Pertamina's Big Dreams" (cit. n. 74).

[77] See, e.g., Arndt, "PT Krakatau Steel" (cit. n. 57); and "The Army Has it All," *TIME*, 12 Jan. 1970. For a response to these sorts of criticisms, see Clement Masenas, "The Tiger in Indonesia's Tank: Pertamina Plays a Vital Role," *New Standard*, 21 Dec. 1974; and Feris Yuarsa, *Perintis Krakatau Steel: 70 Tahun Dr. Ir. H. Marjoeni Warganegara* (Jakarta, Indonesia, 2004), 65–68.

[78] Heinz Möller, "Soeharto Warns of New S.E. Asia Communist Threat," *New Standard*, 26 July 1975. Suharto's party line was echoed by General Ali Moertopo, who compared Pertamina's problems to those experienced by Pan Am, Ford, and Chrysler in the *New Standard* article.

[79] Ibid.

The Indonesian press tended to defend Ibnu's actions as understandable, if not entirely wise, by emphasizing the particular circumstances of a developing nation. Of course, newspapers at the time were censored, and given that Ibnu was a close associate of Suharto's, it would be difficult for any press outlet to promote a more critical view. Without necessarily interpreting these stories as fully representative of Indonesian views, we can find them illuminating when it comes to understanding the ways that a narrative of Indonesia's identity as a technologically developing country was deployed to explain and excuse Ibnu's actions. Instead of criticizing the multitude of businesses operated by Pertamina, newspaper accounts extolled Ibnu as an energetic and successful entrepreneur (foreign reporters acknowledged his skill as well), implicitly a model for Indonesians to follow.[80] As reported in the article "A Tiger in Indonesia's Tank," Ibnu emphasized that as a national company Pertamina was fully owned and operated by Indonesians, and that the company ran a hospital, an air service, radio and telex facilities, and a tanker fleet, employing thousands of Indonesians. Those employees enjoyed access to training schools in Indonesia, Japan, and the United States and the opportunity to learn modern technologies.[81] Pertamina was not just any company, it was a development company: "We always give top priority to the development of the community, whenever we carry out our operations."[82] Projects included construction of roads, housing, schools, mosques, hospitals, and TV stations.

Ibnu's emphasis on the Indonesian ownership of Pertamina is reminiscent of the Sukarno-era narratives of postcoloniality and social justice, a victory for Indonesians as they took control over a highly lucrative sector of the economy. Although Suharto encouraged foreign investment and rejected Sukarno's framing of foreigners as threats to Indonesian sovereignty, many Indonesians continued to perceive foreign countries and foreign businesses as potentially threatening. In 1971, Franklin Weinstein, an international relations scholar, interviewed sixty-four Indonesian politicians to understand how they viewed Indonesia's place in the global political order. Weinstein reported that a popular analogy compared Indonesia to a pretty girl, who is always threatened by the advances of men and therefore in need of protection.[83] His findings indicated that the majority of Indonesian leaders, even those young leaders who had earned their credentials as anti-Sukarnoists, continued to find this analogy compelling. Indonesians' anxiety over the intentions of foreigners however, was focused less on the unlikely prospect of political incorporation than on the threat of economic colonization through the foreign investors that Suharto had used to help bring economic stability to the country. The idea that Indonesia was technologically underdeveloped increased this sense of vulnerability, particularly with respect to Japan. One author lamented that although Japanese businessmen could (and did) behave unethically, Indonesians still relied on their machinery and equipment, because Japanese equipment was smaller, lighter, and cheaper than American or European

---

[80] Masenas, "Tiger in Indonesia's Tank" (cit. n. 77); Glassburner confirms Ibnu's reputation in Indonesia, "In the Wake of General Ibnu" (cit. n. 73). For a discussion of the problem of entrepreneurship in Indonesia, see Arifin Siregar, "Indonesian Entrepreneurs," *Asian Survey* 9 (May 1969): 343–58.

[81] Masenas, "Tiger in Indonesia's Tank" (cit. n. 77).

[82] Ibid.

[83] Franklin Weinstein, "The Indonesian Elite's View of the World and the Foreign Policy of Development," *Indonesia* 12 (Oct. 1971): 97–131.

equivalents.[84] This criticism was widespread; in January 1974 student riots took place on the occasion of the Japanese ambassador's visit.[85] Yet however much it seemed that these riots were aimed at the Japanese, the real targets were Suharto's government and the highly placed officials who facilitated—and were assumed to profit the most from—foreign business ventures. The similarity of this situation to that of the colonial period, when profits accrued primarily to foreigners and a few elites, raised the specter of neocolonialism, even as that term had become unfashionable.[86] Ibnu's emphasis on the Indonesianness of Pertamina (and its subsidiaries, such as Krakatau Steel) from the ownership to the jobs and career opportunities, to the projects it pursued for the good of Indonesia, would stand in powerful contrast to foreign companies suspected of continuing a long tradition of exploitation. Indeed Ibnu's stress on the training opportunities he offered to countless Pertamina employees made the strongest case for the value of his enterprise from a social standpoint. A more technologically educated workforce would eventually bring Indonesians into positions of power and allow them to be equal players in a competitive global economy. Technology was the best answer to vulnerability.

Yet it is important to note the significant differences between Ibnu's development projects and the social justice ideals of the early 1950s. The original notions of social justice, in which opportunities abounded for ordinary Indonesians, were often framed as the elimination of an unbalanced distribution of wealth that benefited the few, in favor of a more widely shared prosperity. Ibnu's luxurious lifestyle and the great wealth of his compatriots certainly did not fit into the egalitarian dreams of the revolutionary era. His work on useful infrastructure projects, while assuredly welcome, was charity (a concept with great value in a predominantly Muslim country), not socialism. The economist Daoed Joesoef explained, that with more widely available technical training: "the principle of social justice, which is one of the aims of our development efforts, will be better assured. Justice in this context will be achieved not through the classical policy toward more equal distribution of income, but through increased utilization in the process of development of the sources of energy possessed by virtually every citizen, namely, human labour."[87]

Ibnu positioned himself as a moral exemplar for Indonesians both by providing training and by being a living example of what an Indonesian could achieve. Such a claim would seem incredible to Western observers (and probably to more than a few Indonesians), but within the logic of the development state, his ideas were reasonable. He argued that it was "not enough to build factories only, because the development of the Indonesian people is more important."[88] And further: "But if we are to build a modern nation we must discard this notion of poverty being normal, or in any case inevitable, and see that in the modern world, poverty is abnormal. If we show

[84] G. A. Utama, "Time for a Redirection," *Jakarta Times*, 14 Jan. 1974. A less critical, but still instructive, discussion of the problems can be found in "Proses Indonesianisasi Tergantung Buruh Indonesia Sendiri," *Kompass*, 11 Feb. 1972.

[85] Gary E. Hansen, "Indonesia 1974: A Momentous Year," *Asian Survey* 15 (Feb. 1975): 148–56.

[86] On the reproduction of colonial practices under Suharto's leadership, see Benedict O'G. Anderson, "Old State, New Society: Indonesia's New Order in Comparative Historical Perspective," *Journal of Asian Studies* 42 (May 1983): 477–96.

[87] Daoed Joesoef, "Indonesian Development in Global Perspective," *New Standard*, 22 March 1975. See also "Proses Indonesianisasi Tergantung Buruh Indonesia Sendiri" (cit. n. 84).

[88] "Pertamina—A Role to Play," *New Standard*, 31 May 1975.

people that there is another way, and that way can be followed by Indonesians, then we will accomplish a great deal."[89] Shared prosperity would come not through socialist control but through the benefits of technology and techniques (such as roads and education) and through the power of example. For foreigners, who saw "development" as exclusively emergent from careful adherence to the five-year plans constructed by the "Berkeley Mafia," such arguments on behalf of a company whose operations seemed to enrich a small sector of Indonesian society, would seem absurd. Indeed, journalists used words such as "gaudy" and "shaky" to describe Pertamina, adjectives not dissimilar to the "bombastic" and "impractical" labels attributed to Sukarno and his projects. Yet in Indonesia itself, Pertamina's flaws could be excused, in part, because they brought this new vision of development in tow.

In the media frenzy over Pertamina, Krakatau Steel got relatively little attention, beyond its portrayal as the very expensive straw that broke the camel's back. When it was mentioned in the Indonesian press, however, its significance was framed with respect to geographic narratives of Indonesian identity, although the specific nature of these narratives did change from Sukarno's time. Defending Pertamina, Ibnu pointed out that it was one of the few companies that pursued development across the archipelago, not just in Java, playing on a sore point of long standing: the government's Java-centric development efforts. Pertamina's lists of subsidiaries, which seem so illogical from a business standpoint, actually attain greater coherence when considered in the context of a fragmented archipelagic nation. Ships, containers, radio, roads, telex, television, and even hotels were all technologies that facilitated communication and circulation around the islands. Krakatau Steel would literally lay the groundwork for Indonesia's networks of transport and communications technologies: "The Indonesian archipelago is almost as vast as Europe. Indonesia finds it has to maintain and build as many harbors, bridges, railways, ships, telecommunications and electrical networks, as a number of countries lumped together." Domestic steel could build rails, ships, bridges, cars, and motorcycles, uniting the nation and keeping oil-boom wealth in the country.[90]

Despite his favored position, Ibnu finally went too far in his financial overcommitments, and Suharto, with little fanfare, but apparently a profound sense of betrayal, removed him from his position at Pertamina in March 1976.[91] But Krakatau Steel survived Ibnu's departure, finally going into full operation in 1977. As William Ascher has argued, it seems clear that Suharto purposely used Pertamina and other military-operated companies to do development projects of which his economists would disapprove, as monuments to his leadership. Giving himself the title of "Father of National Development," he was no less ambitious and single-minded than Sukarno in getting what he believed was needed to materialize the Indonesia of his imagination, and he was equally as heedless of the consequences. His monuments, if less grandiose than those symbols of postcoloniality constructed by Sukarno, were just as important as material statements of one vision of Indonesia, an identity that reflected the greatness of the leader.[92]

[89] Ibid.

[90] Wasidesa, "Indonesia's Appetite for Steel Begins to Spiral," *New Standard*, 1 March 1975.

[91] Elson, *Suharto* (cit. n. 9).

[92] Ascher, "From Oil to Timber" (cit. n. 69). On Suharto's title "Father of National Development," see *Bapak Pembangunan Nasional* (Jakarta, Indonesia, 1988).

KRAKATAU STEEL: A VIEW FROM CILEGON

Because social critique could be a dangerous practice under Suharto, it can be difficult to discern from written sources how Indonesians, educated or otherwise, responded to claims about the justice of projects such as Krakatau Steel, or whether such projects shaped their views of national identity. A hint that this project produced rather different notions of national identity in the town and surrounding area of Cilegon is suggested in a Ford Foundation report published in August of 1975. The Ford Foundation and the Institute for Social and Economic Research, Education, and Information (Lembaga Penelitian, Pendidikan, dan Penerangan Ekonomi dan Sosial, or LP3ES) held a seminar on entrepreneurship (also called "auto-activity" in Indonesian) in Cilegon that year. LP3ES, a nongovernmental organization formed by students and other activists in 1971, aimed to find new approaches to development that would be alternative or complementary to those pursued by the government, approaches that would spread the benefits of development more widely. In this case, their emphasis on entrepreneurship in small business was offered as a complement to the employment opportunities becoming available at Krakatau Steel, which helped sponsor the seminar. The seminar participants, working as trainees at LP3ES, focused their attention on young people in the region. Using interview data, the trainees explored (and reported on) ways to encourage the growth of private business initiatives that could take advantage of the presence of Krakatau Steel, but operate independently.[93] It was an effort to create a trickle-down economy of independent businesses, which ideally would also increase the prevalence of small-scale skilled enterprises (carpentry firms, for example.) As part of their investigation, student trainees at LP3ES surveyed residents in the region concerning their attitudes toward Krakatau Steel and their interest in small business.[94]

One striking finding of the survey is the invisibility, in some respects, of Krakatau Steel in the local area. The authors found that many people in the surrounding villages knew that there was a factory but not what business the factory was in. One group of people guessed that it was a local coconut fiber company, because there was a warehouse for that company nearby. Another group of people interviewed believed it was a water processing plant.[95] This latter guess may be explained by the substantial water pipeline that had been built for the original Trikora site. Although it is true that the plant was not fully operational at the time of the interviews (though it had been under construction for several years), the lack of knowledge of the plant in surrounding areas suggests that it could not have functioned particularly well as a symbol of Indonesia's industrial accomplishments, or as proof of its international power, at least not locally. Closer to the city of Cilegon itself, people were more aware of the company, but the response to it was mixed. Although many responded positively to the possibilities for employment, and to the idea that the factory was producing

---

[93] *Seminar Pengembangan Usaha Swakarya dan Swadaya Masyarakat Merak, Anyer, dan Cilegon, 29, 30, dan 31 Agustus 1975,* transcript, Serang, 1975, sec. 1, 1–4. (Sections are individually numbered.)

[94] Ibid., sec. 2, "Kesimpulan Pokok Hasil Penelitian Latihan Mahasiswa Angkatan ke VI LP3ES Tentang Potensi Dan Perkembangan Swakarya Masyarakat di Daerah Industri Baja PT Krakatau Steel Merak, Anyer, dan Cilegon," 1–6.

[95] Ibid., sec. 3, "Kesimpulan Pokok Hasil Penelitian Latihan Mahasiswa Angkatan Ke VII LP3ES tentang Aspek-2 Sosial Budaya dalam Rangka Pengembangan Usaha Swadaya dan Swakarnya di Daerah Sekitar Industri Baja PT Krakatau Steel Merak, Anyer, dan Cilegon," 5.

development in the area, there were some who were unhappy with the plant. The trainees conducting the survey observed that those who disliked the plant saw it as something that benefited outsiders and hurt the many local people who lost their land or livelihood with the arrival of the steel plant and its nonlocal employees.[96] Because this was a national project, the authors explain, the resources for the project (raw materials, expertise, and to some extent, even labor) largely came from elsewhere, and the product was also marketed elsewhere. They obliquely suggest the existence of ethnic conflicts arising from the influx of "strangers," unfortunately without providing details.[97]

Seeing a "national" project as the exclusive province of outsiders echoes the exclusions of the colonial experience that earlier revolutionaries had fought to overcome.[98] The findings of these researchers suggest that national projects could easily be read as "foreign" projects, whose organization favored the central government and those in positions of high power, at the expense of local communities. As such, national projects did not necessarily create a sense of unity and belonging, nor did they unproblematically support the idea of technology as a vehicle to produce social justice. Instead, the socially just nation remained a distant goal, while the present reproduced old forms of inequity. Cilegon sits a mere fifty-five miles from Jakarta, the capital of the country, but for some residents, it existed at an enormous distance as measured by power and benefit. The existence of even a massive material representation of the nation within the region itself could not counteract for every resident the sense of being disconnected from their rightful participation in the nation.

CONCLUSION

The history of Krakatau Steel offers a view into the effort to turn a state technology project into a national project in the era of early industrialization in Indonesia. The changes in interpretive frameworks that assigned social value to this project map to the well-understood political transition from the anticolonial socialism of Sukarno to the crony-capitalist developmentalism of Suharto. In both eras, however, official actors, advocates, and apologists create these divergent interpretive frameworks from narratives of social justice and geography that were central to Indonesian identity. These narratives were flexible: the social justice of the Suharto era, which embraced the power and wealth of a few wealthy individuals such as General Ibnu as a positive example for society, differs significantly from the narratives of Sukarno's years, which emphasized the belief that eliminating all colonial-like relationships through the agency of state control would produce a society of equitable prosperity. Sukarno used the Trikora plant to ground in reality a new imperial geography for Indonesia, while advocates for Krakatau Steel more mundanely used the story of a far-flung state in constant need of more solid connection to justify the huge expense of Ibnu's expanded project.

---

[96] Ibid., 6–7.

[97] Ibid., sec. 1, 3. I have not been able to confirm the existence of ethnic tensions from other sources, in part because of the inaccessibility of newspapers from the region of Cilegon for the era. Additionally, all newspapers were heavily censored in this period.

[98] Local complaints of outsider influence, unfair distribution of opportunities, and lack of local involvement are not limited to postcolonial industrial projects. See Hecht, *Radiance of France* (cit. n. 51).

Today, in the post-Suharto period, Krakatau Steel still exists and is at the time of this writing still a state-owned corporation. Its contribution to Indonesian identity today would take far more analysis than can be offered here. Nevertheless, it is clear that many people see it as a relic of Indonesia's past, a symbol of the corruption that ran rampant at state-owned corporations during the Suharto era. Recently, the Indonesian government announced an IPO (initial public offering) of up to 30 percent of Krakatau Steel.[99] Rumors have circulated for years that it will be bought by Tata Steel or some other international steel power. It has become, in some respects, a negative symbol in modern Indonesia, out of step with the slimmed-down, globally integrated companies of the present. Its continuity as a state-owned company runs counter to the move toward decentralized power that has characterized recent history in Indonesia.

Is Indonesia's postcolonial experience of embedding technology in narratives of national identity unique? At first glance, it is the similarities with other postcolonial nations that stand out. Most postcolonial nations have pursued industry with an eye to building national strength. India is perhaps the iconic example of this, with Nehru's vision of a powerful industrial India winning out over Gandhi's image of an India of self-sufficient villages. Indonesia's claims to be building the gotong-royong state, based on cooperation, harmony, and the model of the family find echoes elsewhere in Asia and Africa as a philosophical antidote to the hyperindividualism of European and American culture on the one hand, and the hyperrational subordination of the individual in Soviet culture on the other.[100] As Eka Darmaputera points out, it is a common problem for postcolonial nations to cope with profound ethnic, cultural, and linguistic diversity, requiring them to build unity to inspire people to work together while embracing enough diversity to secure the necessary emotional commitment of individuals to the nation.[101] The interest in social justice, usually defined in direct contrast to the colonial past, is another common theme—indeed, how many nations in the contemporary world would deny the maintenance of a just society as a central part of their mission?

There are some distinctions, however, when we look deeper. Indonesia's geographic fragmentation is profound, reinforced in every iconic or cartographic image of the nation. This gives discussions about technology and national identity a particularly enduring emphasis on geography in the Indonesian context. Technologies that work successfully to build national identity are those that explicitly or symbolically negotiate Indonesia's geography. Another difference is the value of indigenous technology. For Indonesia, social justice comes less through technical mastery of industrial skills than through Indonesian ownership and control of industry. Indonesia has had advocates of "fully Indonesian" designed technologies, primarily Suharto's protégé and former president of the Republic of Indonesia, Jusuf Habibie, but it has been at best a sporadic interest.[102] In the history of Krakatau Steel, it is the Indonesian

---

[99] "RI May Delay Krakatau Offering Due to Market Turbulence," from the Krakatau Steel Web site news page, http://www.krakatausteel.com/home.php?page=news&action=view&id=1211&PHPES SID=20a6cdf90c3d3b3197328b599b983ef8 (accessed 23 Sept. 2008).

[100] Leaders as disparate as Julius Nyerere of Tanzania and Ngo Dinh Diem of Vietnam aimed to reinvent community life on "traditional" models. See James Scott, *Seeing Like a State: How Certain Schemes to Improve the Human Condition Have Failed* (New Haven, Conn., 1998), on Julius Nyerere. For Diem, see Phillip Catton, *Diem's Final Failure: Prelude to America's War in Vietnam* (Lawrence, Kans., 2002).

[101] Darmaputera, *Pancasila and the Search for Identity* (cit. n. 5).

[102] Amir, "Power, Culture, and the Airplane" (cit. n. 14).

entrepreneurs and (to a lesser extent) Indonesian employees who made this an enterprise in which national pride could be invested, not the creation of specifically Indonesian expertise or technologies.[103] This contrasts strongly with the value of indigenous technology in Indian nationalist rhetoric, or in the Soviet Union for that matter, and with both of those countries' need to display a profound intellectual self-sufficiency. And while Sukarno worked closely with Soviet authorities to build Trikora, and made claims himself about the similarity of the two countries, his vision of the justice that Trikora would produce, through Indonesian ownership, and anticolonial power does not resemble in any of its particulars the Soviet promise of the workers' paradise.

The question of whether Indonesia is unique may actually be a red herring, pulling us away from the more productive question of why the similarities between nations are so profound. Do our stories of national identity, when carefully and truthfully told, resemble a collection of dolls, each decked out in colorful and distinctive national dress, while those with insufficiently unique clothing are banished to the back of the closet as uninteresting? Or should we look beyond the trappings and ask why so many of those dolls look so much alike? The often-repeated themes in national identity are as worthy of investigation as the entirely unique, because the process of creating an identity is as much outward looking as it is inward looking. National identity not only defines how a nation is unique but also defines how a nation is like other nations. The desire for industrial power, in addition to its pragmatic utility, is also spurred by the understanding that industrial, developed nations are powerful actors in the world. It should be no surprise that so many nations link industry and the image of a strong nation, although the specific ways this is done vary. Likewise, anticolonial nationalists were frequently as cosmopolitan as their postcolonial successors, traveling to metropoles, colonies, and as decolonization took place, other emerging nations, trading ideas and plans along the way. We should not therefore be surprised, or disappointed, to see that many postcolonial countries have adopted similar framings of the problems of social justice and diversity, and that they have adopted similar answers. Rather we should take these similarities as an opportunity to understand more clearly the international dimensions of national identity formation, particularly the ways that the desire to establish a recognizable place in the world can make similarity as powerful a goal as uniqueness in constructing a national identity.

Regardless of which particular technologies are of interest, one of the reasons we see technology and national identity paired so often is the mutual reinforcement they offer each other. For a state such as Indonesia, whose identity as a unified nation is not easily maintained, a technology project can give a material dimension to a national imagining. Interpreting a visible steel mill in an economically depressed region through the lens of postcoloniality made the construction of the factory equivalent to the achievement of postcoloniality, reinforcing Indonesia's march to modernity as a key characteristic of its national identity. However, it is important to realize that the materiality of the technology does not necessarily stabilize a particular view of national identity. I spoke with several Indonesian scholars who were quite knowledgeable about Indonesian industry, and Indonesian history, and none of them knew that Krakatau Steel's Cilegon mill had once been called Trikora. Sukarno's carefully

---

[103] The same can be said of the Indonesian satellite project discussed by Joshua Barker in "Engineers and Political Dreams"; and Barker and Bart Simon in "Imagining the New Order Nation." (Both cit. n. 14.)

designed symbolic role for Trikora was lost when he lost power. Interpretive flexibility also militates against the stability of particular technology-identity packages. It is not clear that Krakatau Steel functioned in the areas around Cilegon as it did in Ibnu's imagination, as a shining example to inspire increased national self-confidence and the belief that any Indonesian could achieve great things.

Just as particular views of national identity might benefit from being tied to a technology, so can a technological enterprise benefit from being linked to national identity. Deploying ideas of national identity created a powerfully patriotic, even moral, urgency to proceed with this steel mill, although in both the Old and New Orders, it seemed overly risky or even unwise. Foreign economic analysts and Suharto's Indonesian technocrats generally regarded this undertaking as too expensive for a poor country, unlikely to ever be competitive with the steelmaking powerhouses of the world, and therefore a foolish and wasteful use of limited resources, a symbol of postcolonial incompetence. This interpretation seemed to restrict Indonesia to a weak, subordinate role as raw materials supplier, a role that could never fulfill the postcolonial promises of the Indonesian revolution and the promises of successive Indonesian leaders. For the government, what was at stake was not just the pragmatic need for a steel mill but also their own credibility in national and international communities. These political needs provided powerful motivations when pragmatism alone could not justify the work. In the face of economic chaos, Sukarno's defense of Trikora framed it as part of a larger plan to reestablish Indonesia's power and prosperity. Suharto justified Ibnu's improprieties because of what Ibnu's projects, including Krakatau Steel, did for the nation. Sukarno and Suharto used narratives of national identity to prove that steel production was necessary to produce a powerful and independent Indonesian nation and, not incidentally, to demonstrate that the state had exercised its power to achieve worthy national ends. Krakatau Steel's Web site (May 2008) attests to the ongoing relevance of this interpretation. On opening the Web page, the headline encourages us to view "Steel as a National Power."[104]

[104] See http://www.krakatausteel.com (accessed 23 Sept. 2008).

European Experiments

By Alfred Nordmann[*]

ABSTRACT

European science policy, as well as the creation of research agendas for converging technologies, functions as a testing ground for a transnational European identity. In light of competing conceptions of European identity formation, the European Commission's Sixth Framework Program (2002–2007) adopted an experimental mode. Given that the quest for European identity is already an open-ended experiment, the policy process invites experiments in governance, for example, by developing participatory schemes. Moreover, as European science studies scholars advance the notion of "real experiments" in the laboratory of society, one basis for identity formation among Europeans is to be the very fact that they all partake in "collective experiments" with new technologies. This analysis draws for its central cases on the creation in 2003/2004 of a "European vision" for converging technologies and a 2007 report on innovation processes in the "European knowledge society." It thereby highlights also the contribution of historians, sociologists, and philosophers of science and technology to the European quest for identity.

TRANSNATIONAL IDENTITY

In 1957, the Treaties of Rome laid some of the cornerstones for what is today known as the European Union. Fifty years later, European leaders gathered in Berlin to commemorate this founding moment and to sign another declaration. Two aspects are interesting about this brief, perhaps incidental, Berlin declaration.[1] In the shadow of the so-called Lisbon Agenda of 2000, with its ambitious economic goals, the Berlin Declaration advanced an idea of Europe that is not primarily a union defined by economics but one defined by shared values and "a unique way of living and working together." At the same time, however, a key phrase of the document got lost in translation. The German document proposed that "We, the citizens of the European Union, are united for our happiness," but the English version renders this "We, the citizens of the European Union, have united for the better."[2] The incantation of unity is thus

[*] Institut für Philosophie, Technische Universität Darmstadt, 64283 Darmstadt, Germany, nordmann@phil.tu-darmstadt.de.

I would like to thank Christopher Coenen, Kristine Bruland, and the editors and referees of this volume for their thoughtful suggestions.

[1] Council of the European Union, "Declaration on the Occasion of the Fiftieth Anniversary of the Signature of the Treaties of Rome," Brussels, 24–25 March 2007, http://www.consilium.europa.eu/cms3_applications/applications/newsroom/LoadDocument.asp?directory=en/misc/&filename =93282.pdf (English), and http://www.consilium.europa.eu/cms3_applications/applications/news room/LoadDocument.asp?directory=de/misc/&filename=93284.pdf (German).

[2] "Wir Bürgerinnen und Bürger der Europäischen Union sind zu unserem Glück vereint." Since this is an ambiguous sentence, it invites an alternative translation: "We citizens of the European Union are fortunate to be united." The fact that the official translations do not offer either of these readings has

© 2009 by The History of Science Society. All rights reserved. 0369-7827/09/2009-0014$10.00

deflated and has become a reminder of cultural and linguistic as well as political differences.

There is no mention in the Berlin Declaration of science and technology, but an implicit reference may be found in the following: "We are facing major challenges which do not stop at national borders. The European Union is our response to these challenges." This formulation harkens back to the October 2005 Hampton Court Summit, which produced a rather more substantive document on "European values in the globalised world." It defined Europe as "25 countries with shared values and strong institutions acting together" and declared that "[t]oday's policies are challenged by new technologies, ageing and globalisation."[3]

All of these apparently superficial political statements add a new dimension to the familiar ways in which science and technology are related to questions of identity. Science and technology participate in national aspirations, they belong to a specific cultural heritage, and they can be subservient to territorial and economic interests. But in the context of the European quest for transnational identity, they also become strategic sites for the formation of a uniquely European response. Accordingly, Maurizio Salvi of the Bureau of European Policy Advisers argues that the Berlin Declaration and the Hampton Court Summit demand the integration of ethics into European policy on research and development.[4] It is not enough, along this line of reasoning, for the European Commission to attend to economic goals that might belong to a classical national agenda that is merely scaled up to the European level. Instead, the Commission has to take on global challenges that call upon shared values of a community of nations. One such challenge might be the "responsible development of nanotechnology," and Salvi presented his views at a conference dedicated to the Commission's recommendation of a code of conduct for responsible research in the area of nanotechnology.[5]

But is it really possible to understand the European Commission's high-level interest in the low-level activity of laboratory research as a contribution to the formation of transnational European identity and citizenship? The question raises a host of methodological problems, if only in that it attributes to the Commission a certain understanding of scientific knowledge production and the social construction of technology. The very idea that shared European values can and ought to be inscribed into research practice presupposes notions of social shaping or of the co-construction of technology and society. And this presupposition, in turn, draws attention to the complicated relation between the realms of science studies and politics. As will become abundantly clear in the following pages, the language of analysis is inextricably bound up with political programs, and vice versa. Even an apparently analytic

---

been taken to be a matter of politics; see Helena Spongenberg, "Berlin Declaration's 'Fortune' Is Lost in Translation," eu*observer*.com, 27 March 2007, http://euobserver.com/9/23786.

[3] Commission of the European Communities, "European Values in the Globalised World," Communication from the Commission to the European Parliament, the Council, the European Economic and Social Committee, and the Committee of the Regions, Brussels, 20 Oct. 2005, 11–12, 5, http://ec.europa.eu/growthandjobs/pdf/COM2005_525_en.pdf.

[4] Bureau of European Policy Advisers (BEPA) European Commission, presentation at the Conference on Governance and Ethics of Nanotechnology, Brussels, 7–8 May 2008, http://ec.europa.eu/research/science-society/document_library/pdf_06/salvi-m-presentation-nano_en.pdf. On the role of the BEPA, see http://ec.europa.eu/dgs/policy_advisers/.

[5] Commission of the European Communities, "Commission Recommendation of 07/02/2008 on a Code of Conduct for Responsible Nanosciences and Nanotechnologies Research," Brussels, 2 Feb. 2008; see ftp://ftp.cordis.europa.eu/pub/fp7/docs/nanocode-recommendation.pdf.

term such as "experiment" serves not only to describe the multiplicity of tentative and open-ended approaches to research policy and identity formation but also to valorize a certain attitude and understanding that might be shared by all Europeans. A similar difficulty holds for texts and their authors: Commission Recommendations, the reports by various expert groups, proceedings from Commission-funded conferences and research projects, and finally seemingly independent academic scholarship such as the present essay all inform each other.[6] These inextricable relations are part of the phenomenon under investigation, and one cannot completely extricate oneself merely by acknowledging the situation or reflecting upon it.[7]

This is most evident in the multitude of studies on the question of European identity itself. Some of these are informed by specific European traditions in a way that allows them to mobilize the elusive idea of a transnational European citizenship and thereby to motivate a quest for European identity that takes the form of an experiment: "It mobilizes . . . the myth of a community of citizens brought into existence through the practice of citizens."[8] Accordingly, a substantive notion of identity might be giving way to multiple occasions and practices of identification, and national conceptions of citizenship might become reoriented toward a transnational order, for example, by focusing civic participation on the development of new technologies.

THE FRAMEWORK

The temporal frame for the following case study was set by the Sixth Framework Program (FP6) of the European Union. This program provided research funding during the five-year period beginning January 1, 2003. Its first year overlapped with FP5, and its last year, 2007, with FP7. Administered by the Directorate General Research of the European Commission, FP6 was introduced in the following terms: "Past FPs have helped to develop a culture of scientific and technological co-operation between different EU countries and they have been instrumental in achieving good research results. They have not, however, had a lasting impact on greater coherence at the European level."[9] With FP6, European research policy thus set out to go beyond its traditional brief to strengthen Europe as a civilian power that simultaneously competes and partners on an equal footing with U.S. research.[10] What may have been implicit before now became an explicit goal of FP6: European research policy takes on the task of helping to create an integrated European knowledge society and thus Europe itself. Three novel features, in particular, were to advance this goal.

The main aim of FP6 was to prepare and support the emergence of the European

---

[6] These remarks apply to the present contribution in that I served as rapporteur of a European expert group on converging technologies and will draw on this background for my main case study.

[7] For a proposal of how to deal with this difficulty, see Alfred Nordmann, "Knots and Strands: An Argument for Productive Disillusionment," *Journal of Medicine and Philosophy* 32 (2007): 217–36.

[8] Klaus Eder and Bernhard Giesen, "Citizenship and the Making of a European Society," in *European Citizenship between National Legacies and Postnational Projects*, ed. Klaus Eder and Bernhard Giesen (Oxford, 2001), 245–69, 263f.

[9] "Frequently Asked Questions" on the homepage of the Sixth Framework Programme, http://ec.europa.eu/research/fp6/.

[10] John Krige, "The Politics of European Scientific Collaboration," in *Science in the Twentieth Century*, ed. John Krige and Dominique Pestre (Amsterdam, 1997), 897–918; Krige, *American Hegemony and the Postwar Reconstruction of Science in Europe* (Cambridge, Mass., 2006); cf. Kalypso Nicolaidis and Robert Howse, "'This is my EUtopia . . .': Narrative as Power," *Journal of Common Market Studies* 40 (2002): 767–92.

Research Area (ERA), which included the creation of a European Research Council (ERC) and, subsequently, a European Institute of Technology (EIT). Since the framework programs are dedicated to research that supports industrial development, broadly conceived, they were set up to stimulate research above and beyond the various national research councils and funding agencies. The creation of the ERA, ERC, and EIT also allows for EC funding of so-called basic research, and it challenges the national agencies to coordinate on a European level and thereby to cede some of their autonomy. An April 2007 green paper provides an assessment of the ERA and adopts as its title the promising slogan "inventing our future together." It employs a programmatic singular by declaring that the "ERA is essential to making Europe a leading knowledge society."[11]

The integration of research is explicitly conceived of as a vanguard for the integration and expansion of Europe. A second main objective of FP6 was therefore to "use the scientific potential of candidate countries to prepare and assist their accession to the EU for the benefit of European science at large."[12] The research area is thus to provide a stage of sorts on which member and accession states can become European.

Finally, the program for integrating European research opened a major funding line for investigations of "social cohesion in the knowledge-based society" and of "citizenship, democracy and new forms of governance." In particular, this led to the creation of the Science and Society Directorate and the support not of self-reflective but of self-exemplifying projects. Research on ethical issues of nanotechnology, for example, can involve philosophers and social scientists from various European countries, whose methodologies often include public engagement exercises or deliberative forums that create experimental situations for the interaction between citizens, scientists, and policy makers. Aside from discovering how Europeans think about nanotechnology and what they consider to be relevant ethical issues, such research projects institute occasions for coming together as Europeans, and they testify to the Commission's commitment to involve citizens in science policy decisions.[13] Thus, research on "implications of European integration and enlargement for governance and the citizen" is often self-exemplifying in that it integrates researchers and citizens from the enlarged Europe and in that it tries out novel forms of interaction to represent the views of citizens. This accords with the recommendation that the integration of European researchers involves experiments in governance.[14] Accordingly, a fair number of FP6 projects directly took on the idea of European citizenship

[11] European Commission, *The European Research Area: New Perspectives,* Green Paper, 4 April 2007 (Brussels, 2007), 6, http://ec.europa.eu/research/era/consultation-era_en.html. See also Didier Buysse, "The Debate on Relaunching ERA," *Research EU: The Magazine of the European Research Area* 53 (2007): 18–19. The notion of the European knowledge society was prepared by the so-called Lisbon Strategy, which set out to transform Europe into "the most competitive and most dynamic knowledge-based economy" and referred repeatedly to life in "the knowledge society." See Lisbon European Council, "Presidency Conclusions [The Lisbon Strategy]," Lisbon, 23–24 March 2000, http://ue.eu.int/ueDocs/cms_Data/docs/pressData/en/ec/00100-r1.en0.htm. See also Council of the European Union, "Follow-up of Lisbon European Council Conclusions," Brussels, 19 April 2000, http://register.consilium.europa.eu/pdf/en/00/st07/07953en0.pdf.

[12] "Frequently Asked Questions" (cit. n. 9).

[13] The author of this essay contributes to two such projects: Deepening Ethical Engagement and Participation in Emerging Nanotechnologies (DEEPEN) (http://www.geography.dur.ac.uk/projects/deepen/), and Nanotechnology Capacity Building NGOs (NanoCap) (http://www.NanoCap.eu). Both projects began in fall 2006 and are to be completed by fall 2009.

[14] European Commission, *The Sixth Framework Programme in Brief* (Brussels, 2002), 10, http://ec.europa.eu/research/era/pdf/era-greenpaper_en.pdf.

and European identity—empirical studies of national attitudes toward the European Union complementing explorations of the kinds of narratives that might foster identification with "Europe."[15]

The time frame of FP6 coincided with the failed ambitions of integrating the European Union by giving it a political constitution and of acting as one of the superpowers to prevent the war in Iraq. Just as several books celebrated the emerging Europe as an alternative or corrective to the hegemony of the United States,[16] Europeans were reminded, once again, of the difficulties of constructing a meaningful political union that goes beyond economic expediencies. This highlights the poignancy of the FP6 goals: At a time when the "European Union" is still a political experiment that might succeed or fail,[17] research and research policy become arenas for advancing this political experiment by conducting experiments on European governance and identity—and, indeed, by suggesting that the European knowledge society is rooted in collective experimentation with emerging technologies. This suggestion may prove untenable in that it may just be too much to ask of European citizens to identify with Europe on the grounds that they are all part of and subject to somewhat risky experiments. However, this suggestion makes explicit what was already implicit in the FP6 goals. And finally, it aids the formation of a European identity by evoking a forceful contrast between the political values associated with Europe and those of the United States.

Accordingly, the following case study does not show how a notion of European identity is expressed by or imposed upon policies for the development of science and technology. This is not a story of European science heralding and propagating European ingenuity, virtues, and values. Instead, it shows how the organization of research is a testing ground on which the notion of European identity is articulated, even invented. It is a testing ground, quite literally: the implicit notion that Europeans are together engaged in a collective experiment to technologically shape and reshape their world is thought to be a means also of ascertaining the existence of this fragile collectivity.[18]

THE PARADOX

If research policy is a testing ground, laboratory, or arena for the invention of "Europe," it will reflect the ambivalence that attends all efforts to define the European Union during a time of European expansion. At one extreme, the Union can be seen as an ideal community that commands patriotic allegiance and displaces the affective bond to any particular member state ("I am European first, and German second"). At

---

[15] The "Eurobarometer"-projects are particularly visible. These investigations of public opinion across the EU include scientific and technological issues as well as the question of European citizenship. Gallup Organization Hungary, *European Union Citizenship*, Flash Eurobarometer no. 213 (European Commission, Brussels, 2008), http://ec.europa.eu/public_opinion/index_en.htm.

[16] Sheila Jasanoff, "Citizens at Risk: Cultures of Modernity in the US and the EU," *Science as Culture* 11 (2002): 363–80; Jasanoff, *Designs on Nature: Science and Democracy in Europe and the United States* (Princeton, N.J., 2005); Jeremy Rifkin, *The European Dream: How Europe's Vision of the Future is Quietly Eclipsing the American Dream* (New York, 2004); T. R. Reid, *The United States of Europe: The New Superpower and the End of American Supremacy* (New York, 2004).

[17] Stefan Aust, ed., *Experiment Europa: Ein Kontinent macht Geschichte* (Hamburg, 2002); Soledad Garcia, "Europe's Fragmented Identities and the Frontiers of Citizenship," in *European identity and the Search for Legitimacy,* ed. Soledad Garcia (New York, 1993), 18.

[18] This is not to claim, of course, that research policy is the only such testing ground.

the other extreme, the Union is constituted procedurally and can draw nothing but trust if it learns to involve European citizens in a transparent and effective manner.[19] The two extremes leave considerable room to maneuver and allow for considerable opportunism, as research policy refers, on the one hand, to specifically European traditions, virtues, and values and, on the other hand, to the job of strengthening competitiveness at a European level. It is in this space that the so-called European paradox arises, which prefigures the formulation of a "European vision" for the convergence of enabling technologies and prefigures also the suggestion that collective experimentation can lead us to take the singular European knowledge society seriously.

In 1994, the first *European Report on Science and Technology Indicators* identified the "European paradox." In a subtle but effective manner, this established research policy as the arena in which the pragmatic goal of economic competitiveness needs to be related to questions of identity. Overtly, the paradox is said to consist in the "gap between Europe's strong science base and its poor performance in terms of technological and industrial competitiveness"[20] and "refers to the fact that Europe plays a leading world role in terms of scientific excellence and the provision of highly skilled human capital. But it largely fails to convert science-based findings and inventions into wealth-generating innovations."[21] This is a failure, to be sure, but not one to be ashamed of. The rhetoric of the paradox suggests, after all, that Europe is perhaps not very good at capitalizing commercially on the ideas generated in Europe, but all the same it has been and continues to be the most fertile ground for the creative development of ideas. Even as it urgently identifies a desperate economic need, the paradox flatters Europe. And thus, the paradox reappears on another level as praising oneself by way of identifying shortcomings—it affords a process of identification. In light of the broadly accepted assumption that Europe is the cradle of science as well as capitalism and that Europe is therefore the place in which the modern world originated

---

[19] To be sure, these two views are not mutually exclusive. (See Jürgen Habermas's conception of a "constitutional patriotism" that suggests the possibility of strong ties to the procedural norms of democratically constituted society.) Moreover, these two views do not arise as merely theoretical possibilities. Each member state brings its own more or less troubled relation to its own national identity into the European Union and thus colors the debate about European identity. The present essay, for instance, brings German experiences and struggles to bear on the question (which explains the reference just now to Jürgen Habermas). Not surprisingly, a survey of research on European citizenship (from the FP6 time frame and partly financed through FP6) stresses multiple processes of identification with Europe over the articulation or discovery of a given identity and thereby stresses also the efficacy of numerous weak ties to Europe over few strong ties. Michael Keating, *Plurinational Democracy: Stateless Nations in a Post-Sovereignty Era* (New York, 2001), 142; Garcia, *European Identity* (cit. n. 17); Jürgen Habermas, "Ist die Herausbildung einer europäischen Identität nötig, und ist sie möglich?" in *Der gespaltene Westen* (Frankfurt, 2004), 68–82; Ireneusz Pawel Karolewski, "Citizenship and Collective Identity in Europe," in *European Identity: Theoretical Perspectives and Empirical Insights,* ed. Ireneusz Pawel Karolewski and Viktoria Kaina (Münster, Germany, 2006), 23–58; Yannis Stavrakakis, "Passions of Identification: Discourse, Enjoyment, and European Identity," *Discourse Theory in European Politics: Identity, Policy, and Governance,* ed. David Howarth and Jacob Torfing (New York, 2005), 68–92.

[20] European Commission, *Towards a European Research Area: Indicators on Science, Technology and Innovation; Key Figures 2003–2004* (Luxembourg, 2003), 422, http://cordis.europa.eu/indicators/publications.htm. The first extended discussion of the European paradox can be found in the European Commission's *Green Paper on Innovation,* Brussels, 20 Dec. 1995, http://europa.eu/documents/comm/green_papers/pdf/com95_688_en.pdf.

[21] European Commission, *Towards a European Research Area: Indicators on Science, Technology and Innovation; Snapshots, "Key Figures 2003–2004"—From "European Paradox" to Declining Competitiveness?* (2003), 1; ftp://ftp.cordis.europa.eu/pub/indicators/docs/pckfbd_snap4.pdf.

and subsequently spread to other continents, the paradox celebrates Europe as a producer of ideas, reminds Europe that its economic success depends on this tradition,[22] and challenges Europe to accept just one further challenge. Accordingly, the presentation of the European paradox has been prefaced by a lofty motto: "The ultimate limits to growth may lie not as much in our capacity to generate new ideas, so much as in our ability to process an abundance of potentially new seed ideas into usable forms."[23]

To be sure, the European paradox is a fragile occasion for identification with Europe in that it affords unambiguous identification only with Europe's past and thus with the "old Europe." To accept the challenge and turn one's attention to wealth-generating innovations might amount to an invitation to join the "new Europe"—in which the new Europe is identified with the aspirations of those member states that joined the European Union only recently.[24] After decades of communism, these member states are said to exhibit an unfettered enthusiasm for capitalism and therefore to be less encumbered by Western traditions. Thus, negotiations of the European paradox always involve the question of who the Europeans are and who they might wish to become, ranging from emphatic affirmations of the old Europe, with its tradition of free and creative inquiry, to calls for a radically reformed new Europe that enters a scientific and technological race with the United States and especially with countries in the Far East.[25]

A EUROPEAN VISION

Moving from the European paradox to specific developments within the FP6 time frame, two particular documents come to the center of attention.[26] Both carry the "European Knowledge Society" on their sleeves, and the first especially suggests a transition from many diverse member states to a single European knowledge society.[27] The report's title uses the plural and thus acknowledges the plurality of states:

---

[22] It is readily apparent that the formulation of the paradox presupposes the "linear model," according to which technological innovation and economic growth descend from basic research. This assumption will be challenged by the FP6 documents on converging technologies and collective experimentation.

[23] European Commission, *Towards a European Research Area* (cit. n. 21), 1.

[24] Jan Ifversen, "It's About Time: Is Europe Old or New?" in *Discursive Constructions of Identity in European Politics*, ed. Richard C. M. Mole (New York, 2007), 170–89.

[25] This is especially true for the FP6 time frame, which included, on January 21, 2003, Donald Rumsfeld's infamous contrast of old and new Europe, considerably facilitating identification with old Europe. Anne Applebaum, "'Old Europe' versus 'New Europe,'" in *Beyond Paradise and Power: Europe, America, and the Future of a Troubled Partnership*, ed. Tod Lindberg (New York, 2005), 39–59; and Ifverson, "It's About Time" (cit. n. 24).

[26] HLEG (High Level Expert Group) on Foresighting the New Technology Wave, *Converging Technologies: Shaping the Future of European Societies*, rapporteur: Alfred Nordmann (Luxemburg, 2004), ftp://ftp.cordis.europa.eu/pub/foresight/docs/ntw_report_nordmann_final_en.pdf; Ulrike Felt (as rapporteur for the Expert Group on Science and Governance), *Taking European Knowledge Society Seriously* (Brussels, 2007).

[27] Since I drafted the report and served as its advocate on numerous occasions, the remarks about the HLEG *Converging Technologies* report (cit. n. 26) are those of a participant-observer even more so than the rest of this contribution. Accordingly, I do not pretend to deliver here a perfectly neutral account. However, since the entanglement of analysis and advocacy are part of the phenomenon under consideration, these obstacles to neutrality may prove to be heuristically useful and serve as a magnifying glass that brings to light the European dimension. In addition, the temporal remove of just a few years affords a more distanced view. One should not assume that all the features of the report

Converging Technologies—Shaping the Future of European Societies. However, the report's key concept suggests a convergence upon a singular common goal. CTEKS might be shorthand for *converging technologies* ("c-teks"), but it is also an acronym for Converging Technologies for the European Knowledge Society. This acronym expresses an implied contrast, namely that of converging technologies in Europe as opposed to converging technologies in the United States, with its so-called NBIC program. But the CTEKS designation also includes an ambiguous "for." It suggests, on the one hand, the existence of a single European knowledge society as a beneficiary of the convergence. Yet it suggests, on the other hand, that converging technologies might have as one of their goals to bring about the European knowledge society.

NBIC—CONVERGING TECHNOLOGIES FOR IMPROVING HUMAN PERFORMANCE

The history of CTEKS began with a volume of proceedings published in July 2002 by the U.S. National Science Foundation (NSF) and the U.S. Department of Commerce: *Converging Technologies for Improving Human Performance: Nanotechnology, Biotechnology, Information Technology, and Cognitive Science.*[28] Editors of the NBIC report were Mihail Roco and William Bainbridge, both from the NSF. Roco already had a high degree of visibility worldwide as architect and promoter of the National Nanotechnology Initiative. As such, he was perceived as a powerful actor who could set and fund a research agenda.[29] The report gathered approximately eighty

highlighted here were actually subject to explicit deliberation within the expert group or the staff of the European Commission. The telling plural in the title of the report and the teleological singular of the CTEKS conception are products of many forces (of commission and omission) at work simultaneously.

[28] This is not to say that the notion of converging technologies originates with this report. This notion has a long and varied history. The specific use of the term in the U.S. document has been traced to a 1999 report on research directions of nanotechnology in which James Canton wrote about a "convergence of nanotechnology with the other three power tools of the twenty-first century (computers, networks, and biotechnology)": James Canton, "The Social Impact of Nanotechnology: A Vision to the Future," in *Nanotechnology Research Directions,* ed. NSTC-IWGN, Workshop Report, The National Science and Technology Council's Interagency Working Group on Nano Science, Engineering, and Technology workshop, 27–29 January 1999, 178–80, http://www.wtec .org/loyola/nano/IWGN.Research.Directions/IWGN_d.pdf. See also Christopher Coenen, Torsten Fleischer, and Michael Rader, "Of Visions, Dreams, and Nightmares: The Debate on Converging Technologies," *Technikfolgenabschätzung—Theorie und Praxis* 13, no. 3 (2004): 118–25, http:// www.itas.fzk.de/tatup/043/coua04a.pdf. Canton was later one of the contributors to the converging technologies workshops. In addition, William Sims Bainbridge, the coeditor of the U.S. report, has on various occasions referred to Manuel Castells's notion of technological convergence. Castells had argued that "technological convergence" is a characteristic of the information technology revolution. In Castells's view, the "ongoing convergence between different technological fields in the information paradigm" results from "their shared logic of information generation" and "increasingly extends to growing interdependence between the biological and microelectronics revolutions." Manuel Castells, *The Rise of the Network Society,* vol. 1 of *The Information Age: Economy, Society, and Culture* (Malden, Mass., 1997), 63f.

[29] William Bainbridge has gradually emerged as the driving force behind the reflections on converging technologies: As a certain understanding of the meaning of the report emerged, so did an image of its editor. See, e.g., William Bainbridge, "Converging Technologies and Human Destiny," *J. Med. & Phil.* 32 (2007): 197–216; Bainbridge, *Nanoconvergence: The Unity of Nanoscience, Biotechnology, Information Technology, and Cognitive Science* (Englewood Cliff, N.J., 2007); and George Khushf, "The Ethics of NBIC-Convergence," *J. Med. & Phil.* 32 (2007): 185–96. The difference between the two editors of the NBIC report corresponds to different interpretations of the historical self-positioning of NBIC-convergence as a "new renaissance." Mihail Roco appears to use this term as a historical reference to a transdisciplinary culture of knowledge production, whereas William Bainbridge refers to the birth of the new human being.

contributors from a wide variety of agencies, companies, and academic disciplines, including the social sciences and humanities. Academic and popular discussions of NBIC convergence focus mostly on 26 of the report's 396 pages, namely the executive summary and overview that appear to speak for the entire group of authors.[30]

The scientific content of the report concerns the complementarity of nanotech, biotech, information technology, and cognitive science and was summed up by a "statement of workshop participant W. A. Wallace":

If the *Cognitive Scientists* can think it
the *Nano* people can build it
the *Bio* people can implement it, and
the *IT* people can monitor and control it.[31]

The major benefit foreseen from the convergence of the four fields is that of improving human performance. This program has since been discussed under the heading of human enhancement, lending visibility and credibility to a debate that had originated in the field of biomedical ethics.[32] Most striking in this program are visions of mind-machine and mind-mind communication without the cumbersome detour through the human body or language.[33] There are also programs for expanded physical strength. One article in the volume adds an "S" for "socio" to NBIC. Rather than bring in a social science perspective, it proposes a GULP (Giant UpLoad Process) sense: "the most valuable sixth sense for our species would be a sense that would allow us to quickly understand, in one big sensory gulp, vast quantities of written information (or even better, information encoded in other people's neural nets)."[34] Although most papers are concerned with individual humans, there is also a focus in the report on enhancing group performance and interpersonal communication, including a paper that focuses on the benefits of NBIC convergence for the environment.[35] The discussion of explicitly military applications is rather limited,[36] yet it would appear that many suggestions make sense primarily in a military context. Finally, the report envisions, especially in the overview, a reorganization of research, a "holistic" approach that

[30] The size and diversity of the group, the boldness of the claims and their unqualified presentation raises the question whether this introduction could have been jointly authored or an outcome of a genuine process of consensus formation. For example, the report's assuredly predictive voice is not easily reconciled with the presence in the group of careful intellectuals such as Sherry Turkle and Mike Gorman. Daniel Sarewitz published a brief memoir of his short-lived participation in the process: Daniel Sarewitz, "Will Enhancement Make Us Better?" *Los Angeles Times*, 9 Aug. 2005.

[31] Mihael Roco and William Bainbridge, eds., *Converging Technologies for Improving Human Performance: Nanotechnology, Biotechnology, Information Technology, and Cognitive Science*, NSF/DOC-sponsored report, Arlington, 2002, 11.

[32] See, e.g., Erik Parens, *Enhancing Human Traits: Ethical and Social Implications* (Washington, D.C., 1998).

[33] The third canonical human enhancement theme, life extension and immortality, plays only a marginal role in the report, but see Roco and Bainbridge, *Converging Technologies for Improving Human Performance* (cit. n. 31), 14, 162–69.

[34] Jim Spohrer, "NBICS (Nano-Bio-Info-Cogno-Socio) Convergence to Improve Human Performance: Opportunities and Challenges," in Roco and Bainbridge, *Converging Technologies for Improving Human Performance* (cit. n. 31), 95; cf. Gregor Wolbring, "Improving Quality of Life of Disabled People Using Converging Technologies," in ibid., 240–42.

[35] Roco and Bainbridge, *Converging Technologies for Improving Human Performance* (cit. n. 31), 242–86.

[36] Ibid., 287–320.

harkens back to the Renaissance.[37] There are also in the overview some indications that systems theory may provide a paradigm for this new way of thinking, but only one paper in the volume is explicitly committed to this paradigm.[38] Many readers were more impressed by a kind of reductionism that is especially directed at the social sciences and humanities:

> Some partisans for independence of biology, psychology, and the social sciences have argued against "reductionism," asserting that their fields had discovered autonomous truths that should not be reduced to the laws of other sciences. But such a discipline-centric outlook is self-defeating, because as this report makes clear, through recognizing their connections with each other, all the sciences can progress more effectively. A trend towards unifying knowledge by combining natural sciences, social sciences, and humanities using cause-and-effect explanation has already begun, and it should be reflected in the coherence of science and engineering trends and in the integration of R&D funding programs.[39]

CTEKS—CONVERGING TECHNOLOGIES: SHAPING THE FUTURE OF EUROPEAN SOCIETIES

One of the first readers of the NBIC report was Mike Rogers, at the time program officer at the Foresight Unit of the Directorate General Research of the European Commission. In various memoranda, he made the case that the Commission should consider the technical issues raised by the report and frame them yet more comprehensively. Although a bit cryptic, Rogers suggests two ways in which the Commission ought to go beyond the U.S. report. The first way is to place a greater emphasis on the social sciences and humanities and take a more comprehensive approach to the cognitive sciences. The second way is to integrate this convergence within European values to allow for the acceptance of the emerging technologies.[40] This is motivated by the identification of specific deficits in the U.S. report. One of the texts that was eventually produced by the European expert group first presents the NBIC report and then confronts it with Rogers's comments:

> If the visions presented in this report are striking, the values underlying much of this work are strongly positivistic and individualistic. Science and Technology are out to be harnessed for our good, and given the right incitements, a "new Eden can be created on Earth". However, the emphasis is not so much, as one might have expected in view of the new Eden, on increasing the quality of life, social cohesion or on solving humankind's main challenges of access to safe water, sustainable development, peace etc. It was commented that "*it says nothing about the rest of the world, the issues of poverty and deprivation, of sharing, of any benefits to the global challenges facing the 95% of the world's population who are not US.*" (Svanfeldt & Rogers 2003) The emphasis is among others on "*accelerating advancement of mental, physical and overall human performance,*" that is to say to increase human efficiency and productivity.[41]

---

[37] Ibid., x, 1–3, 20, passim.

[38] Yaneer Bar-Yam, "Unifying Principles in Complex Systems," in ibid., 335–56.

[39] Roco and Bainbridge, *Converging Technologies for Improving Human Performance* (cit. n. 31), 11; cf. Jan Schmidt, "Unbounded Technologies," in *Discovering the Nanoscale*, ed. D. Baird, A. Nordmann, and J. Schummer (Amsterdam, 2004), 35–50.

[40] Mike Rogers, "ToR for a STRATA ETAN Group on Convergent Technologies," draft memo, 2003, Brussels.

[41] Wolfgang Bibel, Daniel Andler, Olivier da Costa et al., *Converging Technologies and the Natural, Social and Cultural World: A Report to the European Commission from an Expert Group on*

The initiative of Rogers and the Foresight Unit led to the establishment in December of 2003 of the High Level Expert Group (HLEG) on Foresighting the New Technology Wave. From the start, its mission relates to the uniqueness of Europe:

> The objective of the Group would be to assess the potential impact on the EU competitiveness and societal fabric, and the potential response of the EU and MS [Member States] to that, while examining what possibilities exist for a uniquely European approach to exploiting the potential synergies across these technologies.[42]

The wording "response of the EU" here refers to a response by Europe to the United States, suggesting that the European approach can be formulated only in contradistinction to that of the United States. However, the final statement of the group's mandate took care to avoid this interpretation:

> *In the broad, we want to find out what convergence is, how it will impact the future, and what Europe could do to meet its own policy objectives.* The starting point of this reflection was the US NSF report, which was analysed and discussed but does not constitute the focus point of reflection. It is a question of reflecting and proposing a European approach of the convergence of the sciences/technologies in relation to European cultural, ethical, socio-economic approaches; and European strengths and weaknesses in these technological fields. Cognitive sciences were considered as the most innovative research area for a European approach. Questions—sometimes profound reservations—need to be specified, often they express legitimate concern on the use of these technologies for ideological or military purposes. It is a priority to clarify the civil and societal benefits of this research to give them a new legitimacy and to put them firmly in a context of positive social dynamics. The principle of precaution should be taken into account to fix the framework of the research.[43]

Indeed, the group's final report hardly refers to the U.S. document at all but cites it as a background document and as one example among others for framing converging technologies. This was facilitated by the fact that in the meantime there had also appeared a Canadian document that outlined an approach to structure the convergence.

Foresighting the New Technology Wave (Brussels, 2004), 6f: ftp://ftp.cordis.europa.eu/pub/foresight/docs/ntw_sig4_en.pdf. The cited text is C. Svanfeldt, M. W. Rogers et al., "From the Human Genome to the Human Cognome?" Analysis and review of the U.S. report on *Converging Technologies for Improving Human Performance*, EC Internal Note (Brussels, 2003). In the context of German science policy, Susanne Giesecke elaborated the latter concern: "American visions are strongly oriented towards capabilities for optimizing the human being, and there is a danger that these visions diffuse into a Germany that lacks a developed science policy position of its own. Such a conception of the human being will find little acceptance in Germany. This might lead to a loss of the opportunities that can potentially arise from the convergence of advanced technologies. As an alternative to this, there must therefore be a broadly conceived public debate on a science policy which is compatible with the German mode of innovation and system of values and which clearly sets itself off from discussions in the US." Susanne Giesecke, "Verschläft Deutschland die Konvergenz der Spitzentechnologien?" *ips: innovation positioning system—Innovationspolitische Standpunkte,* newsletter of VDI/VDE Innovation+ Technik, 1 Nov. 2004, http://www.vdivde-it.de/ips/november2004.

[42] Rogers, *"ToR for a STRATA ETAN Group on Convergent Technologies"* (cit. n. 40).

[43] Foresight Unit, "Group Mandate," in HLEG, *Converging Technologies* (cit. n. 26), 56. The Foresight Unit provided the group with terms of reference that were updated as the group's work progressed. The final version of this document was published in September 2004 as an appendix to the final report. This passage is taken from its very beginning. The work of the group proceeded on the assumption that a European vision would emerge of itself once a group of Europeans joined together in reflection. Supposedly, the U.S. contrast was not necessary to articulate one's values.

This meaning of "converging technologies" was established in a December 2001 workshop organised by the US National Science Foundation and Department of Commerce.

- The title of the published workshop report suggests that converging technologies enable each other in the pursuit of a common goal: "Converging Technologies for Improving Human Performance—Nanotechnology, Biotechnology, Information Technology, and Cognitive Science."[6]

- A science and technology foresight report for the Canadian National Research Council soon followed the same pattern: Converging technologies for bio-health, eco and food system integrity, and disease mitigation—nanotechnology, ecological science, biotechnology, information technology, and cognitive sciences.[7]

- A third example was suggested by a Norwegian researcher. It repeats the pattern: "Converging technologies for salmon-productive aquatic environments—bioinformatics, environmental science, systems theory, salmon genomics, production biology, economics."[8]

- More examples of CT research were considered by the expert group. These include "Converging technologies for natural language processing—information and nanotechnology, linguistics, cognitive and social science," "Converging technologies for the treatment of obesity." and "Converging technologies for intelligent dwelling."

All these CTs agendas provide a list of enabling technologies and technology-enabling sciences, stating that these are converging technologies for the achievement of a more or less general goal. This shared pattern suggests the working definition of "Converging Technologies" that was adopted by the expert group.

Figure 1. HLEG, *Converging Technologies: Shaping the Future of European Societies (Luxembourg, 2004), 13–14.*

This Canadian document allowed the CTEKS report to define the U.S. NBIC vision as one among others.

From here, the CTEKS report took only a small step to establish its "uniquely European approach." Where the U.S. and Canadian reports wedded the convergence to a single overarching goal, the CTEKS report places the emphasis on the procedural aspect of agenda setting itself—and it thereby suggests that these agenda-setting processes would bring into being a politically transparent and participative Europe.

The report argues that the scheme "convergence for" can be completed in various ways. On this account, the mistake of the U.S. account is that it appears to be technology driven. It pretends to take NBIC convergence as a given starting point and apparently goes on to only articulate its meaning. And on the assumption that all technological development has always served to expand human powers, the converging powers of nano, bio, info, and cogno might clearly have a profound impact. In contrast, the CTEKS report draws on science and technology studies and posits a demand-based deliberative model: what might citizens want, and can this demand be matched by a convergence of basic capabilities from various fields of research?[44]

Converging technologies for salmon-productive aquatic environments, for natural

---

[44] With Daniel Andler, Kristine Bruland (the group's chair), Jean-Pierre Dupuy, Günter Küppers, Arie Rip, and myself, the Expert Group included numerous members from the field of science and technology studies, widely conceived. In total, there were twenty-three members, only seven of whom represented primarily the natural and engineering sciences. Several members (including Küppers and Rip) had "hybrid" careers.

DEFINING THE TERMS

"Enabling technologies" prepare the ground for a wide variety of technical solutions. Because they unlock vast potential and open the door to radically novel technological developments, they are also referred to as "key technologies." Nanotechnology is a prominent enabling technology. Biotechnology and information technology are also enabling, as is the knowledge base of cognitive, social, and other sciences.

"Converging Technologies (CTs)" refers to the convergence on a common goal by insights and techniques of basic science and technology: CTs are enabling technologies and scientific knowledge systems that enable each other for the achievement of a shared aim. Singly or together, NBIC-technologies (nano, bio, info, cogno) are likely to contribute to such convergence.

"NBIC-convergence for Improving Human Performance" is the name of a prominent agenda for CT research in the US. *"Bio-Systemics Synthesis"* suggests another agenda for CT research that was developed in Canada.

"Converging Technologies for the European Knowledge Society (CTEKS)" designates the European approach to CTs. It prioritizes the setting of a particular goal for CT research. This presents challenges and opportunities for research and governance alike, allowing for an integration of technological potential, recognition of limits, European needs, economic opportunities, and scientific interests.

Figure 2. *HLEG, Converging Technologies: Shaping the Future of European Societies (Luxembourg, 2004), 19.*

language processing, for the treatment of obesity, and for intelligent dwelling might meet such criteria.[45] In all of these cases, it is clear that nano, bio, info, and cogno may each play a role, although perhaps not a necessary one, and that other disciplines need to join in, including the social and human sciences and the humanities. Thus, whereas NBIC convergence is a convergence *of* enabling technologies, CTEKS calls for a convergence *upon* a set goal.

Since the acronym CTEKS does not refer to any specific configuration of technologies, and since it does not single out any specific common goal upon which these technologies converge, it designates only the deliberative process through which the convergence is organized. According to the CTEKS designation, these deliberative processes have as their goal the European knowledge society—that is, a society of Europeans who are jointly embarked on the project of solving problems and reforming their world and who place knowledge production as well as technical innovation

---

[45] During the time frame of FP6, none of these suggested programs had been implemented, and only two small-scale explorative projects were launched. M. Van Lieshout, C. Enzing, A. Hoffknecht et al., "Converging Technologies for Enabling the Information Society," *Converging Applications for Enabling the Information Society and Prospects of the Convergence of ICT with Cognitive Science, Biotechnology, Nanotechnology, and Material Sciences,* ed. Roman Compañó (Seville, Spain, 2006); and Roman Compañó, A.-K. Bock, J. C. Burgelman et al., "Converging Applications for Active Ageing Policy," *Foresight* 8, no. 2 (2006): 30–42. There appears to be greater emphasis in FP7: for a survey see Christopher Coenen, TAB (Office of Technology Assessment at the German Parliament), *Konvergierende Technologien und Wissenschaften: Der Stand der Debatte und politischen Aktivitäten zu "Converging Technologies,"* (Berlin, 2008), http://www.tab.fzk.de/de/projekt/zusammenfassung/hp16.pdf. One FP7 call for proposals concerns "converging technologies for clean water."

in the service of this project.[46] A substantial part of the report and of its recommendations is therefore devoted to integrative procedures and mechanisms for this deliberative process. As such, the European document reads to members of the scientific community as a blueprint for democratizing technological development rather than as a catalog of technical challenges and visions.

A CLASH OF CULTURES

The CTEKS report constructs a series of contrasts between the U.S. approach and its own that, so far, sound innocent and stereotypical enough. Here is the substantive vision of a final frontier, and there are procedural norms. Here is technological determinism, and there the coconstruction of technology and society. Here is American individualism, and there societal welfare. Here is the subservience of social science and the humanities, and there is their leading role. The two reports thus provide fertile ground for a perpetual production of presumed differences between the United States and Europe—both at a rather more concrete or descriptive level and at a rather more philosophically abstract level.[47]

The contrasts continued. Aside from the overview, the U.S. report is structured like a conference proceeding. The European document has a single author with endorsement by the entire group; in addition, subgroups and individual group members

[46] Since the term "knowledge society" has the reputation of being trite, the report sought to provide a definition that was based on the contributions to the group by economist Emilio Fontela: "Increasing emphasis on nontradable goods is a hallmark of the Lisbon Agenda's so-called 'European knowledge society' and one reason for the label CTEKS (Converging Technologies for the European Knowledge Society). Pharmaceutical companies, for example, are shifting from the manufacture of drugs to the development of diagnostic tools. For steel manufacturers, too, the production of bulk material is becoming subsidiary to the creation of targeted solutions. Such knowledge-based solutions consider the entire life-cycle of technology-based responses to consumer-specific needs." HLEG, *Converging Technologies* (cit. n. 26), 24; cf. Lisbon European Council, "Presidency Conclusions" (cit. n. 11).

[47] If only for this reason, the differences between the NBIC and the CTEKS reports have been a popular subject of analysis. See, e.g., Davis Baird, "Converging Technologies, Diverging Values? European and American Perspectives on NBIC," presentation at the AAAS Forum on Science and Technology Policy, Washington, D.C., 22 April 2004; G. Berthoud, "The Techno-Utopia of Human Performance Enhancement," in *Utopie Heute: Zur aktuellen Bedeutung, Funktion und Kritik des utopischen Denkens und Vorstellens*, ed. Beat Sitter-Liver (Fribourg, Germany, 2007), 1:279ff; Nigel Cameron, "Convergence and Divergence: European Union Responses to US Converging Technology Policy," presentation to the Converging Technologies Conference (NBIC 2005), 24–25 Feb. 2005, Hawaii. Coenen, Fleischer, and Rader, "Of Visions, Dreams, and Nightmares" (cit. n. 28); Emilio Fontela, *Convergencia NBIC 2005: El Desafío de la Convergencia de las Nuevas Tecnologías* (Madrid, 2006); Steve Fuller, "The Converging Technologies Agenda: The Stakes and the Prospects," *Knowledge Politics Converging Technologies Newsletter* no. 3 (2008): 1–3; Fuller, "Research Trajectories and Institutional Settings of New Converging Technologies," Deliverable 1 of *KNOWLEDGE NBIC Knowledge Politics and New Converging Technologies: A Social Science Perspective* (EC-funded project CIT6 no. 028334, 2008), http://www.converging-technologies.org/docs/Knowledge%20NBIC%20D1 .pdf; Liana Giorgi and Jacquelyne Luce, eds., *Converging Science and Technologies: Research Trajectories and Institutional Settings,* special issue of *Innovation: The European Journal of Social Science Research* 20, no. 4 (2007); Armin Grunwald, "Nanotechnologie als Chiffre der Zukunft," in *Nanotechnologien im Kontext,* ed. A. Nordmann, J. Schummer, and A. Schwarz (Berlin, 2006), 49–80; Grunwald, "Converging Technologies: Visions, Increased Contingencies of the Conditio Humana, and Search for Orientation," *Futures* 39 (2007): 380–92; George Khushf; L. Laurent, and J.-C. Petit, "Nanosciences and Its Convergence with Other Technologies: New Golden Age or Apocalypse?" *HYLE—International Journal for Philosophy of Chemistry* 11 (2006): 45–76; Richard Saage, "Konvergenztechnologische Zukunftsvisionen und der klassische Utopiediskurs," in Nordmann, Schummer, and Schwarz, *Nanotechnologien im Kontext,* 179–94; World Council of Churches, *Convergent Technologies,* vol. 1 of *Science, Faith, and Human Life—Transforming Life,* (n.p., 2005). See below on ongoing research projects.

produced a variety of supporting documents. NBIC convergence had science policy managers recruit a wide diversity of scientists, engineers, and humanists. The European Expert Group included twenty-three members, it was chaired by a historian of technology, and its final report was drafted by a philosopher of science. The group met six times over a nine-month period, and during these meetings it received testimony from a variety of perspectives. Where the U.S. report employs a constative future tense ("this will happen"), the European report considers a conditional future ("this might happen").[48] Acknowledging that all technological development affects the mental and physical organization of individuals and collective bodies, the European report still urges that a distinction be maintained between the NBIC program of engineering *of* body and mind and the CTEKS endorsed engineering *for* body and mind. This would finally allow for a juxtaposition of major philosophical commitments that inform each of the reports. They, too, appear stereotypical in the context of a contrastive enumeration: The NBIC program looks to technological innovation as a mean of realizing human potential for better communication, teamwork, and decision making. In contrast, the CTEKS report encourages social innovation as a way to promote technical development and thus to realize technological potential.[49] Accordingly, NBIC visions take technology to be a means of overcoming limits to produce a Second Creation, a New Eden, or human salvation. While this view of technology is expressed elsewhere,[50] historians of technology have found it to have special cultural resonance in the United States: It marries the ideal of liberated, emancipated individuals with a conception of transcendence, if not manifest destiny.[51] In contrast, again, CTEKS holds to the notion that technology ingeniously adapts nature to human limits and adapts human desires to the limits of nature. This ingenuity consists in achieving ever more with always the same limited means.

This stark juxtaposition of the NBIC and CTEKS reports clearly owes as much to the work of their readers as it does to the brief of the CTEKS report that it develop an alternative vision of converging technologies, one more compatible with European values than the overtly American NBIC-vision. Significantly, however, the further development of NBIC convergence in Europe broke out of this simple dichotomy.

APPROPRIATIONS

When Mike Rogers and Susanne Giesecke advocated the formulation of a European vision for converging technologies, they were motivated by a sense of excitement about the convergence of enabling technologies and the prospect of a nano- and bio-

---

[48] Grunwald, "Nanotechnologie"; and Grunwald, "Converging Technologies." (Both cit. n. 47.)

[49] The latter slogan ("social innovation to realize technological potential") was suggested after completion of the CTEKS report at a follow-up Brussels conference on the report of a Key Technologies Expert Group. It was formulated by Josephine Green of Philips Design, explaining why Royal Philips Electronics believed in Europe as a place for creative innovation of consumer products. Josephine Green, "Sense Making and Making Sense of the Future," presentation at the Key Technologies for Europe conference, DG Research, Science, and Technology Foresight Unit, Brussels, 19–20 Sept. 2005, slide 21, ftp://ftp.cordis.europa.eu/pub/foresight/docs/conf_kte_j_green.pdf.

[50] Arnold Gehlen, "Anthropologische Ansicht der Technik," in *Technik im technischen Zeitalter: Stellungnahmen zur geschichtlichen Situation*, ed. H. Freyer, J. C. Papalekas, and G. Weippert (Düsseldorf, Germany,1965), 101–18; Gerd Binnig, preface to Nils Boeing, *Alles Nano? Eine neue Epoche für Wissenschaft und Technik* (Berlin, 2006).

[51] David Noble, *The Religion of Technology* (New York, 1999); Thomas Hughes, *Human-Built World* (Chicago, 2004).

technological reach all the way into cognitive processes. Accordingly, the mandate of the expert group included the emphatic formulation, "Convergence is the driver, Europe the context!"[52] In other words, the expert group was charged not to redefine NBIC convergence but to appropriate or Europeanize it.

Although the expert group sidestepped the emphatic formulation and broadened the notion of convergence,[53] it did manage to dissociate NBIC research from the particular American agenda of improving human performance. The notion that convergence is the driver was sidestepped by the CTEKS report in various ways. First, it rejected its inherent technological determinism by insisting that the convergence is instituted only through an agenda-setting process. Second, it suggested that each convergence requires its own constellation of research fields and that the nano-, bio-, info-, cogno-constellation is just one among many. It turned out to be quite sufficient, however, for the CTEKS report to demonstrate that NBIC research can be framed in a more benign, possibly European manner. To the extent that the reach from nano to cogno appeals not only to the popular imagination but also to that of visionary policy makers and engineers, all that was required was proof of the concept that NBIC convergence can be compatible with European values. The CTEKS report tried to do much more, but it appears to have succeeded at least in that.

Since the reception of NBIC and CTEKS programs continues beyond the time frame of FP6, only a sampling of evidence for the Europeanization of NBIC research can be provided here.[54] In this time frame, no major funding initiatives were designated to converging technologies research in the United States or in the European Union. In the United States, three further conferences continued the work of the first one, but it is not clear whether and how the NSF- and DOC-sponsored NBIC initiative has continued beyond the NBIC 2005 meeting in Hawaii. Although there is no major dedicated funding line for converging technologies at the European Commission, the administrative unit responsible for nanotechnology has been renamed Nano- and Converging Sciences and Technologies. While there are some EC-funded technological projects (converging technologies for active aging, for enabling the information society, for clean water, for environmental protection), these are minor in the larger picture of industrial research.[55] Significant for its persistence as a Europeanized research activity is the rather large presence of NBIC convergence in the areas of technology foresight, science and society, and ELSA (ethical, legal, societal aspects) research.[56] These projects and some academically based activities

[52] Foresight Unit, "Group Mandate" (cit. n. 43), 57

[53] The title page of the CTEKS report features as a running head an expansion of NBIC (nano-bio-info-cogno) convergence: "Nano-Bio-Info-Cogno-Socio-Anthro-Philo-Geo-Eco-Urbo-Orbo-Macro-Micro-Nano-."

[54] For more complete documentation, see Coenen, TAB, *Konvergierende Technologien und Wissenschaften* (cit. n. 45).

[55] But see ibid. for evidence of increased emphasis in FP7 calls for proposals on "converging sciences" and "converging technologies."

[56] The European Commission sponsored the Key Technologies for Europe Expert Group to elaborate and concretize the CTEKS program. Funded projects with a focus on converging technologies include "Knowledge NBIC," "Contecs," and "Ethics School"; see also R. Berloznik, R. Casert, C. Enzing et al., STOA (European Parliament Scientific and Technological Options Assessment), *Technology Assessment on Converging Technologies* (Brussels, 2006), http://www.europarl.europa.eu/stoa/publications/studies/stoa183_en.pdf; Coenen, TAB, *Konvergierende Technologien und Wissenschaften* (cit. n. 45); Erdyn Consultants, SKEP ERA-net: Scientific Knowledge for Environmental Protection, *Converging Technologies and Environmental Regulations: Literature Review* (Brussels, 2008), http://www.skep-era.net/site/files/WP6.2_final%20report.pdf.

recapitulate the dialogue between the NBIC and CTEKS reports and thus use the discussion of "converging technologies" to rehearse the question of European (and U.S.) identity.[57]

Arguably, then, the highly politicized and publicly visible discussion of NBIC research in Europe served to legitimize, even stabilize, the concept and may have extended its relevance for research policy. The successful Europeanization of NBIC research relied not only on overt discussions of the difference between U.S. and European approaches, it also found two other avenues motivated by the CTEKS report. It rejects engineering *of* the mind and the body and encourages instead engineering *for* the mind and body. Rather than reinforce the contrast of NBIC and CTEKS convergence, these engineering ideals cut across the distinction: Europeanized NBIC research engages in engineering *for* the mind. Closely related to this is another way of embracing NBIC research from the European perspective. In scientific and technical terms, the weakness of the U.S. report was seen to lie in its very impoverished conception of cognition and mind. The NBIC report took cognitive science to mean little more than a neuroscientific investigation of the physical basis of thought. The technological convergence was thus to interface straightforwardly with neurons or nerve cells and thus to facilitate mind-machine or mind-mind communication.[58] Here, a specific opportunity was seen for European research. If there is to be convergent NBIC research at all, one needs first to build on European traditions to compensate for the deficits of an all too narrowly conceived cognitive science. This would include studies on cognition and cognitive functioning that draw on neuropharmacology, sociolinguistics, and group psychology, among other areas. Once the proper knowledge base is acquired, so the story went, the nano-, bio-, or information-technologically informed engineering approaches will work to support and expand cognitive functioning and will thus turn out to be engineering *for* rather than *of* the mind.

Finally, the story of a failure of implementation provides poignant testimony to the Europeanization of NBIC research. One of the first European activities immediately inspired by the CTEKS report was an exploratory investigation of prospects for converging technologies for active aging.[59] It set out to clearly dissociate converging technologies from the program of improving the performance of individual humans. Charged with addressing a broadly accepted and socially relevant challenge, the project was thought to be sufficiently Europeanized to fully appropriate the original NBIC program. This attempt to have it both ways proved to be a dead end, however. The technical imagination remained fixated on nano, info, and cogno and thus on bringing cognitive processes into the realm of technical control. Content with the

---

[57] Among academic investigations devoted to this study of contrasts include the Practis group in Madrid (Javier Echeverría) and a project at the University of Bergen (Roger Strand). For other comparative projects see, e.g., Khushf, "The Ethics of NBIC-Convergence" (cit. n. 29).

[58] See Sarewitz, "Will Enhancement Make Us Better?" (cit. n. 30); or Andy Clark, "Re-Inventing Ourselves: The Plasticity of Embodiment, Sensing, and Mind," *J. Med. & Phil.* 32 (2007): 263–82. In the context of the CTEKS report and the Key Technologies for Europe Expert Group, it was especially Daniel Andler who pursued this line of argumentation: Daniel Andler, *Cognitive Science*, report contributed to the Key Technologies for Europe Expert Group (2005), ftp://ftp.cordis.europa .eu/pub/foresight/docs/kte_cognitive.pdf.

[59] See the statements of Jean-Claude Burgelman in *Summary Report of the Conference and Roundtable of EPTA on Converging Technologies,* ed. Raf Casert, Robby Berloznik, and Robby Deboelpaep (Brussels, 2005); cf. also *Information and Communication Technologies for Active Ageing: Opportunities and Challenges for the European Union*, ed. Marcelino Cabrera and Norbert Malanowski (Amsterdam, 2009).

social relevance of this narrowly technical vision, one of the exploratory workshops was dedicated to brain-machine technologies and yielded the negative result that these technologies may have some utility for seriously impaired patients but nothing to offer to an aging European population at large.

EXPERIMENTAL REIFICATION

Even if CTEKS did not displace NBIC,[60] its introduction placed NBIC research into a European context. The "European knowledge society" is implicitly at issue when the two reports are contrasted, when "engineering for the mind" is invoked, when a deeply embedded cognitive science is pursued, and when public interests are brought into the process of setting an agenda for some convergence.

This brief history of the CTEKS report has indicated that the European knowledge society served as its telos and that it sought to realize this goal by making the organization of research subservient to the dynamics of European decision making. On the one hand, there is no convergence without a European goal to converge upon. And inversely, there is no European knowledge society until there can be a public agenda-setting process that integrates research into strategies for the solution of recognized societal problems. This construction is based on an unproven hypothesis, however, namely the hypothesis of the social shaping of technology or of the coevolution of technology and society. According to the CTEKS proposal, to become European is to bet on this hypothesis and enter into the large-scale experiment that could render it true.

This section will show that the CTEKS report did not invent but merely reified this rather peculiar way of fostering identification with Europe. It originated in the science studies literature and, in particular, in accounts of European science policy.

According to the CTEKS proposal, the European knowledge society can prove itself in a collective experiment. This experiment assumes optimistically that technological development is open to social shaping and that societies can actually assume and exercise the power to shape technological trajectories. It is an experiment precisely in that there is no certainty or guarantee that the underlying assumptions are true. In particular, the experiment cannot rely on simple technical means by which to systematically exercise the in-principle power to socially shape technological trajectories. Instead, the optimistic assumptions about social shaping must be humbly submitted to the vagaries of politics. It is this uncertain ground on which the European knowledge society must prove itself by way of an experiment.

The interpretation so far suggests that the CTEKS report aims for "Europeanness" in a peculiar way: It does not refer to or mobilize a European identity that merely needs to be discovered, remembered, and affirmed. Instead, it offers a procedural experiment that involves opportunities for identification with Europe. In this experiment, Europe does not appear as a bureaucratic center of political and technical control, as it might be exerted to protect its citizens from harm.[61] Instead, "Europe"

---

[60] Cf. Fuller, "Converging Technologies Agenda"; and Fuller, "Research Trajectories." (Both cit. n. 47.)

[61] This is not to say that Europe is not perhaps just such a bureaucratic center of power and control. Indeed, the procedural and experimental image of Europe is meant to counter just this perception. And thus, even the regulatory scheme of REACh (Registration, Evaluation, Authorisation and Restriction of Chemicals) departs from classical regulatory measures in that it was created through stakeholder

emerges from submission to the vagaries of politics, stakeholder dialogues, and the like. And to this submission corresponds an attitude of humility and uncertainty, as opposed to the hubris of technical control that might commandeer public acceptance of new technologies as required by the nation's destiny or mission.

The contrast between humility and hubris, between political experimentation and technical control, between European and U.S. approaches to science policy, figures prominently in the work of Sheila Jasanoff, especially in her 2002 paper "Citizens at Risk: Cultures of Modernity in the US and the EU," which predated her book *Designs on Nature* (2005). Another science studies scholar, Hans Glimell, summarized Jasanoff's argument at a fall 2003 conference on nanotechnology in Darmstadt, Germany. By way of his paraphrase, Jasanoff's analysis insinuated itself into the CTEKS report:

> Sheila Jasanoff has recently discussed the dedication of producing consent in relation to risks (Jasanoff 2002). She notices that even in the adversarial US environment, there has been an eagerness for processes such as consensus conferences to foster cooperation among disparate parties—"Getting to yes" has become a paramount goal. But as uncertainties mount and as science impinges upon the most intimate, even sacred, aspects of human life, it is no longer wise to assume that societies will or should always agree upon the instruments of governance. Jasanoff argues that, instead, a diversity of approaches can acknowledge that within modernity's complex socio-technical formations, safety comes from the heterogeneity of our accommodations with risk. Rather than seeking consensus, it may be more fruitful for authorities to learn how to foster "informed dissent" about risk among knowledgeable publics.
>
> According to Jasanoff, much of the analytical ingenuity of science policy has been directed toward devising predictive methods like risk assessment, cost-benefit analysis or climate modeling. For her, these represent *'technologies of hubris'*, achieving their power through claims of objectivity and by systematically overstating what is known about risks while downplaying uncertainty and conflict. There is instead a need for *'technologies of humility'*, capable of incorporating unforeseen consequences, plural viewpoints and mutual learning.[62]

By associating technologies of hubris with the United States and technologies of humility with European approaches in her analysis, Jasanoff offers her European readers an opportunity for identification.[63] At least in Glimell's hands, her distinction offers a template for the construction of the list we encountered above of more and less stereotypical contrasts between CTEKS and NBIC convergence. By subscribing to the hitherto unarticulated program of technologies of humility, European science policy, risk governance, and public engagement exercises can reinscribe into Europe the analysis provided by Jasanoff. Among those listening to Glimell's presentation

---

debates and is largely a collaborative reporting scheme that relies on quasi-voluntarily commitments. For a brief introduction, see http://www.chemicalspolicy.org/reachhistory.shtml.

[62] Hans Glimell, "Grand Visions and Lilliput Politics: Staging the Exploration of 'the Endless Frontier,'" in Baird, Nordmann, and Schummer, *Discovering the Nanoscale* (cit. n. 39), 242. Glimell refers to Jasanoff, "Citizens at Risk" (cit. n. 16), 3.

[63] Mariachiara Tallacchini, "Epistemology of the European Identity," *Journal of Biolaw and Business, Supplement Series Bioethix*, 2002, 60–66, describes the American model as "science-based regulation" (hubris) and the European model as "policy-related science." The latter term was proposed by Silvio Funtowicz, Iain Shepherd, David Wilkinson, and Jerry Ravetz , "Science and Governance in the European Union: A Contribution to the Debate," *Science and Public Policy* 27 (2000): 327–36; and Iain Shepherd, ed., *Science and Governance in the European Union: A Contribution to the Debate*, EC Joint Research Centre (EUR 19554 EN), Brussels, 9 March 2000, http://governance.jrc.it/scandg-eur .pdf, to improve European governance by humbly taking uncertainties as a point of departure.

were not only Mike Rogers and Elie Faroult, the EC program officers who shepherded the work of the converging technologies expert group, but also myself as the person who would be chosen to draft the CTEKS report. I subsequently offered as my personal contribution to the work of the expert group a reflection on "Technologies for Dealing with Technological Advance." In it I dealt with the possibility of appropriating technological developments within locally cultural, regional or national contexts, including the transnational context of Europe.

> Attention to technologies for dealing with technology foregrounds the dialectic between global and local effects. The ways of appropriating and contextualizing technologies localize the global forces that drive technological development. These local effects may serve as slight or profound resistance to the global drivers; they may serve to drastically alter or, more likely, cosmetically color future developments. (In some ways, the work of this HLEG negotiates precisely this dialectic as it seeks to identify European constraints and opportunities for the convergence of technologies.) . . . Sheila Jasanoff contrasts in ongoing work and a forthcoming book "technologies of hubris" and "technologies of humility" as different technologies for dealing with the risks posed by an uncertain technical future (Jasanoff 2002). Technologies of hubris bring an engineering attitude to the questions of forecasting, risk-assessment, and policy making. Their goal is consensus formation on the basis of a fundamental trust in our ability to always find yet another technological fix where things do not work according to plan. In contrast, technologies of humility modestly defer to our limits of knowledge and planning, they aim for an "informed dissent" regarding a variety of possible technological futures.[64]

Toward the beginning of the work of the expert group, the point was made that "Europeanness" must be sought in the mode of conceiving and appropriating technological developments. Jasanoff unwittingly provided a blueprint for this that would be reinforced by many of the themes of the CTEKS report—among them, collaborative agenda-setting processes, resistance as a social selection factor rather than an obstacle to development, CTEKS as a "tool for the development of local solutions that foster natural and cultural diversity," and the need to balance technological problem-solving approaches against low-tech or no-tech policy alternatives.[65]

This is but one example of how science studies helped identify an occasion for identification with Europe, creating in effect a self-fulfilling feedback loop.[66] A more or less distinctive feature of "Europe" becomes reinforced and reified. As with the

[64] Alfred Nordmann, "Technologies for Dealing with Technological Advance," in *Foresighting the New Technology Wave—Expert Group: State of the Art Reviews and Related Papers* (supporting material published with the HLEG's *Converging Technologies* report [cit. n. 26]), 14 June 2004, 223, http://ec.europa.eu/research/conferences/2004/ntw/pdf/soa_en.pdf.

[65] Cf. HLEG, *Converging Technologies* (cit. n. 26), 8.

[66] And it is the simplest, most straightforward example at that. Another example informs the next section: The notion of the society as a laboratory has been articulated by various European theorists and implicitly reflects conditions that, for geopolitical reasons, are especially pronounced in Europe. (Its paradigm example is the Chernobyl disaster that united all European scientists, citizens, and policy makers who lived downwind from the Chernobyl site.) See Matthias Groß, Holger Hoffmann-Riem, and Wolfgang Krohn, *Realexperimente: Ökologische Gestaltungsprozesse in der Wissensgesellschaft* (Bielefeld, Germany, 2005); Wolfgang Krohn and J. Weyer, "Society as a Laboratory: The Social Risks of Experimental Research," *Sci. Pub. Pol.* 21 (1994): 173–83; or Ulrich Beck, *Ecological Enlightenment: Essays on the Politics of the Risk Society* (Atlantic Highlands, N.J., 1995). Again, a weak notion of "Europeanness" ("we are all part of the experiments conducted in the European knowledge society") emerges as an occasion for identification. A third example was suggested to me by Kristine Bruland, the historian of technology who chaired the HLEG: the strong reliance on participative mechanisms reflects specific knowledge regarding the role of institutions in the development of science and technology and thus on a shared and understood history of European institutions.

European paradox, this identification of features follows the pattern "in weakness lies strength" and offers an argument that humility may just be good enough not only to usher new technologies into competitive knowledge societies but also to promote identification with Europe. The humble approach may appear risky at first, but it appears to be full of opportunity in that it promises a more sustainable integration of technology and society. In the CTEKS report, this uncertain promise was couched in the language of challenges. For example, CTEKS is said to present "challenges and opportunities for research and governance alike, allowing for an integration of technological potential, recognition of limits, European needs, economic opportunities, and scientific interests."[67]

COLLECTIVE EXPERIMENTATION

Three major dimensions have been identified so far that show how the CTEKS report works in support of Europe. First, the report was considered as an element of FP6 and thus of the Lisbon Agenda and the larger program of strengthening the European knowledge society in the research policy arena. Second, the report was considered in its more immediate context of providing a European vision for converging technologies. Third, the report was interpreted to suggest a particular experimental mechanism for the identification with Europe, and this was seen as the reinforcement of certain preexisting analyses of "Europeanness" in regard to the development of science and technology. This final section will show how the further development of the CTEKS ideas leads back to and generalizes again the notion that "Europeanness" is to be constructed upon the precarious and, indeed, dangerous terrain of collective experimentation.

TAKING EUROPEAN KNOWLEDGE SOCIETY SERIOUSLY

In many ways, the CTEKS report fulfilled its job to provide a European vision and thus to offer occasions for identification with Europe. In one respect, however, it rejected its brief. Rather than consider converging technologies as the driver and Europe as the context, it put Europeans in the driver's seat and tied the very fact of a convergence to an agenda-setting process. Following the science studies doctrine of a "co-production of science and society,"[68] the report viewed Europe as a product of this agenda-setting process, along with the convergence of technological trajectories. However, according to the original charge, Europe is a preexisting context, if only by way of the European institutions that will promote and regulate the convergence. And though the CTEKS report resonated with various strategies to produce identifications with Europe, it contested more robust and perhaps less humble ways to conceptualize Europe. By their very existence, the European Commission and other European institutions are set to articulate political and cultural values, produce economic benefits, celebrate diverse European traditions, and offer protections to consumers, thereby creating a European Union that citizens can identify with. In contrast, the CTEKS report promotes an experiment by which the European Knowledge Society is yet to invent itself—if the experiment succeeds. For the experiment to succeed, the chal-

[67] HLEG, *Converging Technologies* (cit. n. 26), 19.
[68] Sheila Jasanoff, ed., *States of Knowledge: The Co-Production of Science and Social Order* (London, 2002).

lenge posed by the idea of a technological convergence upon a common goal must be met.[69]

By recommending this experiment, the European expert group provided opportunities for identification but refrained from providing knowledge that anyone could act on. It withheld from European decision makers the illusion of control that—armed with proper knowledge of what is coming their way—they might then usher it into the European context.

> How credible are certain predictions about the state of technology in 2020? Will nanotechnology prove to be essential to CT research? Can the social and natural sciences come together in the formulation and evaluation of research programs? . . . only time can provide the answers. For the time being, the expert group adopted a proactive stance that does not foreclose future debate. The report's aim to outline the opportunities and challenges of CTs has to be distinguished from a study of their impacts. This report is not focused on existing or imminent products and processes that will impact European societies in one way or another. Instead, it considers CTs in terms of their specific potential to generate in the medium and long term new kinds of technological applications. Though it is too early to speak of their likely impacts, it is not too early to consider how the creative development of CTs might address and solve societal problems, how they can build on existing strengths in Europe, orient themselves to social and environmental needs and prompt ethical debate. It is also not too early to assess the promises that are made on behalf of CTs and to address concerns regarding their risks.[70]

This approach was taken a step further and expressed in a far more explicit and general manner by another European expert group on science and governance.[71] Asked to "provide insights which might improve the treatment" of governance challenges posed by "public unease with science," the group decided "to step back" and expand its mandate "beyond the range of immediate instrumental analysis."[72] Instead of providing decision makers with tools for a technology of hubris, the group tried to impress upon the European Commission the need for humility in light of contingencies and complexities.

> In the end, there are no simple answers to the pressing and apparently contradictory demands placed on European science and governance. Global economic imperatives to pursue science-led innovation as quickly and efficiently as possible conflict with the inevitable frictions and temporal demands of democratic governance. In response, we suggest that the main guide lies in trusting Europe's rich democratic and scientific traditions. It is in the realisation of diversity and multiplicity, and in the robust and distributed character of publics and their imaginations, that we may justly conceive different pathways of technoscientific development, and so achieve more mature and robust outcomes.
>
> In the perceived pressing need to encourage innovation, democratic governance has become dislocated in ways that cannot be remedied by technical methods and tools alone. Policy making should not stop at simple or mechanical solutions; it should address the complex issues of science and governance honestly, thoroughly, patiently and

[69] Cf., e.g., HLEG, *Converging Technologies* (cit. n. 26), 7f.

[70] Ibid., 12.

[71] The group included prominent European science studies scholars and one who is known for her sympathies for European science policy: Michel Callon, Maria Eduarda Gonçalves, Sheila Jasanoff, Maria Jepsen, Pierre-Benoît Joly, Zdenek Konopasek, Stefan May, Claudia Neubauer, Arie Rip, Karen Siune, Andy Stirling, and Mariachiara Tallacchini. The group was chaired by Brian Wynne; Ulrike Felt acted as rapporteur and drafted the report. Another member of the group was Isabelle Stengers, who did not sign on to the final document.

[72] Felt, *Taking European Knowledge Society Seriously* (cit. n. 26), 9, 14.

with humility. Only then will European policy take 'knowledge society' seriously—and fulfil its abundant promise.[73]

This call for honesty and humility culminates in the discussion of "collective experimentation" as the basic condition of European societies and as a hitherto unacknowledged starting point for public debate and governance. Two chapters, in particular, stress this notion. One chapter describes experimentation as one of two regimes of innovation. While the "regime of technoscientific promises" refers to future solutions of current problems to the vast potential of, say, nanotechnology, the "regime of collective experimentation" is modeled on the open-source movement and develops technical trajectories from a multitude of local interactions. While both regimes are said to be complementary rather than mutually exclusive, the expert group thinks "a vibrant European knowledge society must in the long-term be built on collective experimentation." To support this point and to provide an example of collective experimentation, the group refers to the CTEKS report.[74]

Significantly, however, this celebratory description of collective experimentation is based on a rather sobering view of a predicament for the governance of new technologies.

> Two generally accepted insights shape our view of the importance of experimental idioms of thought and practice for social learning for European science and governance. These are:
> - first, the *contingency* of scientific knowledge as considered for potential use in public arenas of all sorts, whether innovation and technologies, or regulatory policies, or combinations; and
> - second, the recognized [obsolescence][75] of the traditional framework which supposed that all technological innovations introduced into society were first tested under the controlled and isolated conditions of a laboratory, which left society protected from premature release of uncertain entities. Thanks to the incessant intensification and growing scale of technologies and technosciences, as Krohn and Weyer first put it in 1988, nowadays "society [and the larger environment] is the laboratory."[76]

The report points out that neither of these conditions is new but that they have become pervasive, especially in regard to the disappearance of the laboratory as a protected space in which the safety of products and processes can be tested before they enter society. With that disappearance, society has become the laboratory in which experiments with new technologies are conducted and observed:

> [I]f society is indeed now the experimental laboratory without walls, and by implication therefore, social subjects are also the subjects (guinea-pigs) of such open-ended techno-social-environmental experiments, it is necessary to begin discussion of the implications

---

[73] Ibid., 12.

[74] Ibid., 27f. The following passage from the CTEKS report is quoted as a call to collective experimentation: "Since enabling technologies are not dedicated to a specific goal or limited to a particular set of applications, they tend to be judged by the visions that go into them rather than the results they produce. Since these visions reach far beyond disciplinary perspectives, scientists and engineers, policy makers and philosophers, business and citizens are called upon to develop social imagination for CTEKS applications." HLEG, *Converging Technologies* (cit. n. 26), 42.

[75] In the place of "obsolescence," the report speaks somewhat misleadingly of "redundancy."

[76] Felt, *Taking European Knowledge Society Seriously* (cit. n. 26), 68.

for governance, science, publics and technology. What is meant by experiment here? And if everyone is in principle a guinea-pig, then who is participant in the experimental design, and interpretation—and who has right to its veto?[77]

To take "European Knowledge Society seriously" would be to openly confront what is anxiously described and to turn it into an exhilarating opportunity for genuine collectivity, that is, for the collective exercise of responsible experimental design.[78] Where human test subjects enter into traditional (laboratory-) controlled experiments by giving their informed consent, European knowledge societies should foster "informed dissent" for a vigilant pursuit of their collective experiments.

> If society is now the laboratory, then everyone is an experimental guinea-pig, but also a potential experimental designer and practitioner. Whose experiments we are involved in, and what is being tested, are mostly confused, blind and inadvertent, and open-ended. We have not yet even acknowledged that this is the state we are in, as a prelude to defining what kinds of experiment, to what ends, under what conditions, are acceptable. Basic democratic principles require that this new realization be acknowledged, and acted-upon. We suggest that in early 21st century conditions this societally distributed capacity is in need of deliberate development, in the face of intensifying techno-scientific demands on our trust and credulity.[79]

LIMITS OF ARTICULATION

Taking European Knowledge Society Seriously heeds its own lessons when it introduces its recommendation to acknowledge the condition of society as a laboratory under the heading Risking Collective Experimentation.[80] After all, the concession that Europeans ought to openly deal with the uncertainties of collective experimentation carries with it its own risks. Who wants to tell the European public that they are guinea pigs in experiments, even when these are, to some extent, experiments of their own design? Is collective experimentation a foundation upon which to build a sense of identity—even if the European Union is a political experiment, and even if according to thinkers like Dewey and Popper, democratically open societies are always engaged in collective experimentation?[81] And in particular, even if collective experimentation holds the promise of an integration of science, society, and technology and thus of robust technological development, will this solve or aggravate the European paradox? Here, the report of the Science and Governance expert group offers a display of true daring. After rejecting the dubious rhetoric regarding a "competitive race for economic advantage," it changes gear entirely:

---

[77] Ibid.

[78] One criticism of the report notes: "The human being, the population and its life disappear as 'objects' of governance—and a world of participating citizens appears." Petra Gehring, "Biopolitik: Eine 'Regierungskunst'" (unpublished typescript, Technische Universität Darmstadt, 2007), 28.

[79] Felt, *Taking European Knowledge Society Seriously* (cit. n. 26), 71.

[80] Ibid., 67.

[81] The Science and Governance Expert Group refers to Dewey but not to Popper (ibid., 26). To be sure, various authors refer to the fears associated with modernization processes as a tenuous basis for European identity: Robert Picht, "Disturbed Identities: Social and Cultural Mutations in Contemporary Europe," in Garcia, *European Identity* (cit. n. 17), 82–94; Ralph Grillo, "European Identity in a Transnational Era," in *The European Puzzle: The Political Structuring of Cultural Identities at a Time of Transition*, ed. Marion Demossier (New York, 2007), 79.

> The regime of collective experimentation faces challenges because such embedded in-
> novation is laborious, typically loosely-coordinated and slow; as it should be, because
> users and other stakeholders have their own contexts and logics to consider. Inspired by
> the "slow food" movement, one can now proclaim a "slow innovation" program.[82]

It is here, at the latest, that science policy as a testing ground for European identity becomes a dangerous terrain. It would appear that Europe can endure this notion of collective experimentation only just as long as it can endure the European paradox: for the sake of economic competitiveness, the paradox needs to be overcome, and Europe must move from its position of producer of ideas to that of efficient industrial implementation. For purposes of identity formation, however, Europe might as well remain old-fashioned and slow, if that means that ideas, technologies, traditions, and cultural values are well integrated.

In respect to a particular European institution (the European Environment Agency), Claire Waterton and Brian Wynne have described how this tension plays out. They showed that it corresponds to "different notions of society and the European polity." Where ignorance and uncertainty are acknowledged, where the deliberation of tech-nologies moves from downstream considerations of calculable and manageable risks to upstream negotiations of societal needs, and where integrative precautionary ap-proaches displace expert rulings, "civil society is called upon to play a much larger role in articulating public values, supplementing the formal representative (and ad-ministrative) institutions of parliamentary democracy."[83]

> This more upstream focus was the preferred idiom of the EEA [of the CTEKS and Sci-
> ence and Governance reports], against Commission disapproval because it more directly
> identified and implied possible policy initiatives and needs. It also happened to be at this
> more upstream level that a non-universalistic, non-standardized and non-unified Europe
> became more visible.[84]

Since the tension described by Waterton and Wynne was deeply inscribed into the European Commission's own Sixth Framework Programme for research fund-ing, there was no longer a question simply of Commission approval and disapproval. Pushing the boundaries meant simply to run up against certain limits of articulation. In the terms of FP6, the CTEKS report succeeded by Europeanizing NBIC conver-gence and by opening avenues for further research, allowing at the same time a lib-eral disregard of some of its suggestions. Although the CTEKS report and that of the Science and Governance expert group adopt very similar views on the European Knowledge Society, the latter articulates the European experiment with an all too painful clarity, forcing Europeans to acknowledge and embrace their manifold but rather weak ties to Europe, ties that also originate in uncertainty and ignorance re-garding the outcome of the experiment that holds them together.

[82] Felt, *Taking European Knowledge Society Seriously* (cit. n. 26), 27.
[83] Claire Waterton and Brian Wynne, "Knowledge and Political Order in the European Environment Agency," in Jasanoff, *States of Knowledge* (cit. n. 68), 104, 100.
[84] Ibid., 97.

Notes on Contributors

Ross Bassett is an associate professor of history at North Carolina State University. He is the author of *To the Digital Age: Research Labs, Start-Up Companies and the Rise of MOS Technology* (Washington, D.C., 2002) and is currently working on a book on the history of Indian graduates of MIT.

Pratik Chakrabarti is the Wellcome Lecturer in the History of Modern Medicine at the University of Kent at Canterbury, UK. His research interests are in the history of imperial medicine and science. He is the author of *Western Science in Modern India: Metropolitan Methods, Colonial Practices* (Delhi, 2004). He has just completed his second book, *Materials and Medicine; Trade, Conquest and Therapeutics in the Eighteenth Century*, and is writing his third book, *Laboratories in the Empire: Medical Research in British India, 1890–1940*.

Michael D. Gordin is Associate Professor of History at Princeton University. He specializes in the history of modern science with an emphasis on science in Russia and the Soviet Union. He is the author of *A Well-Ordered Thing: Dmitrii Mendeleev and the Shadow of the Periodic Table* (New York, 2004), *Five Days in August: How World War II Became a Nuclear War* (Princeton, N.J., 2007), and *Red Cloud at Dawn: Stalin, Truman, and the End of the Atomic Monopoly* (Farrar, Straus and Giroux, forthcoming 2009). He is also a coeditor (with Karl Hall and Alexei Kojevnikov) of *Osiris* 23: *Intelligentsia Science: The Russian Century, 1860–1960* (2008).

Carol E. Harrison is Associate Professor of History at the University of South Carolina, specializing in the history of women and gender. She is the author of *The Bourgeois Citizen in Nineteenth-Century France: Gender, Sociability, and the Uses of Emulation* (Oxford, 1999), and she is currently writing a book on gender and Catholicism in postrevolutionary France.

Ann Johnson is an Assistant Professor of History at the University of South Carolina. In 2009, Duke University Press will publish her *Hitting the Brakes: Engineering Design and the Production of Knowledge*, an examination of the development of antilock braking systems for automobiles. Her current research focuses on the role of engineers in the construction of national identity in the early American republic. Her next project will look at recent (post-1980) computer simulations and the exploratory and predictive modes of research they facilitate. The thread that connects these projects is an interest in the way engineers construct communities of knowledge.

Edward Jones-Imhotep is Assistant Professor of Science and Technology Studies at York University in Toronto. He specializes in the history and philosophy of science and technology and in technoscientific identity. He is currently working on two book manuscripts, one focusing on the place of northern radio communications in postwar Canadian national identity, and the other on the cultural history of electronic reliability during the cold war.

Suzanne Moon is an Assistant Professor in the Department of the History of Science at the University of Oklahoma. She specializes in the history of technology in colonial and postcolonial Southeast Asia. Her 2007 book, *Technology and Ethical Idealism: A History of Development in Netherlands East Indies* (Leiden) explores the history and politics of technological development in the late colonial period in Indonesia.

Chandra Mukerji is Professor of Communication and Science Studies at the University of California, San Diego. She is author of *A Fragile Power: Science and the State* (Princeton, N.J., 1989), which won the Robert K. Merton Award in 1991, and *Territorial Ambitions and the Gardens of Versailles* (Cambridge, UK, 1997), which won the Mary Douglas prize in 1998. She is also author of *From Graven Images: Patterns of Modern Materialism* (New York, 1983) and with Michael Schudson coeditor of *Rethinking Popular Culture* (Berkeley, Calif., 1991). She has just completed a book on the Canal du Midi, examining the relationship between infrastructural engineering and political territoriality in seventeenth-century France. Titled *Impossible Engineering: The Canal du Midi and the Politics of Impersonal Rule*, the book will appear in spring or summer 2009 from Princeton University Press.

Alfred Nordmann is Professor of Philosophy at Technische Universität Darmstadt and Centenary Visiting Professor at the University of South Carolina. He studies the writings of Immanuel Kant, Charles Sanders Peirce, and Ludwig Wittgenstein, and also treats as philosophers scientists such as Georg Christoph Lichtenberg, Antoine Lavoisier, William Bateson, and Heinrich Hertz. In recent years, he has begun articulating a comprehensive philosophy of technoscience. His

303

studies of nanotechnological research, in particular, have involved him in various European research projects.

Katherine Pandora is an Associate Professor of History and Philosophy of Science at the University of Oklahoma and the author of *Rebels within the Ranks: Psychologists' Critique of Scientific Authority and Democratic Realities in New Deal America*. She is currently completing a work titled *Science in the American Vernacular: Improvisations in Natural History across the Twentieth Century* and pursuing research on popular science in the nineteenth-century United States by looking at children's books on science by American authors.

Grace Yen Shen is Assistant Professor of Humanities at York University in Toronto. Her work explores the intersection of science and nationalism in modern China. She recently organized a Focus Section on Chinese Science for *Isis*, and she is preparing a manuscript on the development of modern geology in Republican China.

Asif A. Siddiqi is Assistant Professor of History at Fordham University in New York. He specializes in the history of science and technology and modern Russian history. His forthcoming book, *The Red Rockets' Glare: Soviet Imaginations and the Birth of Sputnik* (Cambridge University Press), focuses on the social, cultural, and technological roots of space exploration in early twentieth-century Russia.

Bruno J. Strasser is Assistant Professor in the Program for the History of Science and Medicine at Yale University. He is the author of a book on the history of molecular biology in Switzerland, *La fabrique d'une nouvelle science: La biologie moléculaire à l'âge atomique, 1945–1964* (Florence, 2006).

Index

Abbott, Jacob: female characters, 90; *The Little Philosopher*, 89; place in children's literature, 77–78; *Rollo* series, 80, 84–85, 91–92; teaching children to question, 88–89

Albring, Werner, 122, 125

Amaldi, Edoardo, 176

American engineering practice and material strength research: and national character, 53–55; publications on, 57

Amir, Sulfikar, 257

Anderson, Benedict O'G, 5, 15, 254, 255, 256, 271

Anderson, William, 39

Andreossy, François, 29

Andrews, Thomas: background of, 110; career, 105, 110–111; and Queen's College Belfast, 112–113; struggles for a scientific Ireland, 111–116

Andrews, Thomas, 99, 102; critical point research of, 106–107

Anglo-Irish Protestant Ascendancy, 99; careers of scientists in, 100

antebellum children's literature: background, 75–79, 97–98; critics of genre, 86–87; influence of, 81–82; literary techniques in, 82–83; rhetorical style, 84

Appleton, Edward, 152

Auger, Pierre, 170, 172, 176

Bache, Alexander Dallas, 70–71

Bakker, Cornelis Jan, 177

Ball, George W., 169

Banks, Joseph, 39–40; and d'Entrecasteaux voyage collections, 48; *Florilegium*, 41

Barnard, J. G., 61, 64

Barnes, Albert, 57

Bartlett, William, 65–66

Bassett, Ross, 212–230, 303

Baudin, Nicolas, 35; on glories of scientific exploration, 38; journal observations of, 44; personnel mix on ship of, 40; qualifications as naturalist, 44–45, 50

Bell, David, 36

Beriia, Lavrentii, 126, 133

Bernal, J.D., 205

Bernard, Simon, 61, 62

Bhatt, Mahendra, 208–209

Blasing, Manfred, 125

Blass, Josef, 132

Bodin, Jean, 16

Bonaparte, Napoleon: scientific expeditions and, 35; Egyptian campaign of, 38

Brain, Robert, 146

Brèvedant, Léon, parodic journal of, 43

Britain and France in the Pacific, 38–41

Brown, T.S., 64, 66, 74

Brückner, Martin, 55

Bunsen, Robert Wilhelm, 108

Byron, John, 46

Cai, Yuanpei, 244, 245, 249

Campmas, Pierre, 21

Canada: arctic ionosphere of, 161–162; communications problems in North, 149–151; Defence Research Board, 162; Department of External Affairs, 144–145, 148–149; ionosphere, 145; meteorological projects in North, 149; nordicity index and, 161–162; North, defined, 162–163; OIC/6, 151, 163; Radio Physics Laboratory, 154; role of nature in, 147; Sector Principle, 149; Yukon report on radio broadcasts, 149–151

Canadian Broadcasting Company, 151, 162

Canal du Midi: maps of, 29–30; objections to building of, 18–19; opening ceremony for, 25–26; social collaboration process of, 21; technical design of, 17–18

Carnot, Lazare, 36

Carteret, Philip, 46

Cattell, James McKeen, 75

Central Medical Research Institute, 195–199; background, 195–199

CERN: Communist objections to, 174–175; creation of, 170–176

Chakrabarti, Pratik, 12, 188–211, 303

Chappey, Jean-Luc, 35

Chen, Grace Yen, 231–252

Chénier, André, 36

Chertok, Boris, 125, 129, 131

Chevalier de Clerville, 21

China and geological fieldwork, 237–238; challenges due to Chinese nature, 231–234, 237–238; effect of Opium Wars, 248; Geological Section, Bureau of Mines, 232; inadequacies of, 236–237; *Student-Teacher Studies*, 243–244

Coen, Deborah, 146–147

Cohen, I. Bernard, 54

Colbert, Jean-Baptiste, 16; and canal's slow pace, 27–28

Cook, James: as model explorer, 38; publications of, 46; voyages of, 39–40

Crandall, John, 82–83

d'Aguesseau, 27–28

Dalal, Ardeshir, 226–227

Dalal plan, 226–227

d'Antic, Louis-Augustin Bosc, 33

Darmaputera, Eka, 255, 275

Das Varshnei, Ishwar, 216–217

305

SUGGESTIONS FOR CONTRIBUTORS TO OSIRIS

OSIRIS is devoted to thematic issues, conceived and compiled by guest editors who submit volume proposals for review by the OSIRIS Editorial Board in advance of the annual meeting of the History of Science Society in November. For information on proposal submission, please write to the Editor at osiris@etal.uri.edu.

1. Manuscripts should be submitted electronically in Rich Text Format using Times New Roman font, 12 point, and double-spaced throughout, including quotations and notes. Notes should be in the form of footnotes, also in 12 point and double-spaced. The manuscript style should follow *The Chicago Manual of Style*, 15th ed.

2. Bibliographic information should be given in the footnotes (not parenthetically in the text), numbered using Arabic numerals. The footnote number should appear as superscript. "Pp." and "p." are not used for page references.

　　a. References to books should include the author's full name; complete title of book in *italics*; place of publication; date of publication, including the original date when a reprint is being cited; and, if required, number of the particular page cited (if a direct quote is used, the word "on" should precede the page number). *Example*:

　　　[1] Mary Lindemann, *Medicine and Society in Early Modern Europe* (Cambridge, 1999), 119.

　　b. References to articles in periodicals or edited volumes should include the author's name; title of article in quotes; title of periodical or volume in *italics*; volume number in Arabic numerals; year in parentheses; page numbers of article; and, if required, number of the particular page cited. Journal titles are spelled out in full on the first citation and abbreviated subsequently according to the journal abbreviations listed in *Isis Current Bibliography*. *Example*:

　　　[2] Lynn K. Nyhart, "Civic and Economic Zoology in Nineteenth-Century Germany: The 'Living Communities' of Karl Möbius," *Isis* 89 (1999): 605–30, on 611.

　　c. Journal articles are given in full in the first reference. For succeeding citations, use an abbreviated version of the title with the author's last name. *Example*:

　　　[3] Nyhart, "Civic and Economic Zoology" (cit. n. 2), 612.

3. Special characters and mathematical and scientific symbols should be entered electronically.

4. A small number of illustrations, including graphs and tables, may be used in each volume. Hard copies should accompany electronic images. Images must meet the specifications of The University of Chicago Press "Artwork General Guidelines" available from the Editor.

5. Manuscripts are submitted to OSIRIS with the understanding that upon publication copyright will be transferred to the History of Science Society. That understanding precludes consideration of material that has been previously published or submitted or accepted for publication elsewhere, in whole or in part. OSIRIS is a journal of first publication.

OSIRIS (ISSN 0369-7827) is published once a year.

Single copies are $33.00.

Address subscriptions, single issue orders, claims for missing issues, and advertising inquiries to *Osiris*, The University of Chicago Press, Journals Division, PO Box 37005, Chicago, IL 60637.

Postmaster: Send address changes to *Osiris*, The University of Chicago Press, Journals Division, PO Box 37005, Chicago, IL 60637.

OSIRIS is indexed in major scientific and historical indexing services, including *Biological Abstracts, Current Contexts, Historical Abstracts*, and *America: History and Life*.

Copyright © 2009 by the History of Science Society, Inc. All rights reserved. The paper in this publication meets the requirements of ANSI standard Z39.48-1984 (Permanence of Paper). ⊗

Paperback edition, ISBN 978-0-226-31778-6

 **A RESEARCH JOURNAL DEVOTED TO THE HISTORY OF SCIENCE AND ITS CULTURAL INFLUENCES**

A PUBLICATION OF THE HISTORY OF SCIENCE SOCIETY

EDITOR
ANDREA RUSNOCK
University of Rhode Island

MANUSCRIPT EDITOR
JARELLE S. STEIN

PAST EDITOR
KATHRYN OLESKO
Georgetown University

PROOFREADER
JENNIFER PAXTON

OSIRIS EDITORIAL BOARD

SONJA BRENJTES
Aga Khan University

ANN JOHNSON
University of South Carolina

TARA NUMMENDAL
Brown University

MICHAEL GORDIN
Princeton University

MORRIS LOW
Johns Hopkins University

BERNARD LIGHTMAN
York University
EX OFFICIO

HSS COMMITTEE ON PUBLICATIONS

KAREN PARSHALL
University of Virginia

KEN ADLER
Northwestern University

PAULA FINDLEN
Stanford University

PAUL FARBER
Oregon State University

ELIZABETH G. MUSSELMAN
Southwestern University

EDITORIAL OFFICE
DEPARTMENT OF HISTORY
80 UPPER COLLEGE ROAD, SUITE 3
UNIVERSITY OF RHODE ISLAND
KINGSTON, RI 02881 USA
osiris@etal.uri.edu